Genetics and Breeding in Aquaculture

Genetics and Breeding in Aquaculture

Guest Editor

Xidong Mu

MDPI

Basel • Beijing • Wuhan • Barcelona • Belgrade • Novi Sad • Cluj • Manchester

Guest Editor
Xidong Mu
Pearl River Fisheries Research
Institute
Chinese Academy of Fishery
Sciences
Guangzhou
China

Editorial Office
MDPI AG
Grosspeteranlage 5
4052 Basel, Switzerland

This is a reprint of the Special Issue, published open access by the journal *Fishes* (ISSN 2410-3888), freely accessible at: https://www.mdpi.com/journal/fishes/special_issues/C4K283I09D.

For citation purposes, cite each article independently as indicated on the article page online and as indicated below:

Lastname, A.A.; Lastname, B.B. Article Title. *Journal Name* **Year**, *Volume Number*, Page Range.

ISBN 978-3-7258-6019-7 (Hbk)
ISBN 978-3-7258-6020-3 (PDF)
https://doi.org/10.3390/books978-3-7258-6020-3

Contents

About the Editor

Xidong Mu

Xidong Mu is a professor at the Pearl River Fisheries Research Institute, Chinese Academy of Fishery Sciences. He specializes in fish genetics and breeding, with a focus on the molecular mechanisms of sex determination, germplasm resource conservation, and the development of novel breeding technologies in aquaculture. He has led numerous national and provincial research projects aimed at improving the genetic quality and reproductive efficiency of economically important freshwater species. His research integrates multi-omics approaches and field-based breeding practices, contributing significantly to the advancement of precision aquaculture and sustainable resource management. He also serves as an Editorial Board Member and scientific advisor for several journals and institutions in the field.

Preface

This Reprint presents a collection of ten peer-reviewed research articles that highlight recent advances in aquaculture genetics and breeding. The scope covers both fundamental and applied studies aimed at improving key traits such as sex differentiation, growth performance, and disease resistance. The featured works employ a broad range of molecular and quantitative approaches, including RNA interference, gene expression profiling, mutagenesis, and quantitative trait analysis, reflecting the growing integration of omics technologies and functional validation into modern breeding programs. The studies span a variety of aquaculture species, from widely farmed tilapia and grass carp to non-model species such as bullfrog and zig-zag eel, illustrating the integration of species-targeted and generalizable strategies in aquaculture breeding innovation.

Xidong Mu
Guest Editor

fishes

Editorial

Genetic and Molecular Approaches for Breeding Improvement in Aquaculture

Xidong Mu [1,2]

1 Key Laboratory of Prevention and Control for Aquatic Invasive Alien Species, Ministry of Agriculture and Rural Affairs, Guangdong Modern Recreational Fisheries Engineering Technology Center, Pearl River Fisheries Research Institute, Chinese Academy of Fishery Sciences, Guangzhou 510380, China; muxd@prfri.ac.cn

2 Guangdong Provincial Key Laboratory of Aquatic Animal Immunology and Sustainable Aquaculture, Guangzhou 510380, China

1. Introduction

Aquaculture has become an increasingly vital sector for global food security, contributing significantly to the supply of high-quality, sustainable animal protein. As the industry intensifies and diversifies, there is growing demand for breeding strategies that are not only efficient and precise, but also adaptable to environmental and production challenges [1,2]. In response, genetic, transcriptomic, and molecular approaches have emerged as transformative tools, enabling the dissection of complex traits and accelerating stock improvement beyond the limits of traditional selective breeding alone [3–5].

This Special Issue, "Genetics and Breeding in Aquaculture," brings together ten original research articles that highlight recent advances in the application of molecular biology, functional genomics, and omics-based technologies in aquaculture breeding. These studies span a wide range of cultured species and address traits such as growth, sex differentiation, pigmentation, disease resistance, and environmental stress response. The contributions can be thematically grouped into four major categories: (1) sex determination and endocrine regulation, (2) growth-trait genetics and breeding technologies, (3) phenotypic regulation and transcriptomic networks, and (4) immune response and environmental adaptation.

2. Synopsis of Special Issue

2.1. Sex Determination and Endocrine Regulation

Sex determination and differentiation are critical biological processes with substantial implications for aquaculture productivity, especially for species exhibiting sexual dimorphism in growth or reproduction [6,7]. Understanding the genetic and hormonal underpinnings of sex control can enable the development of all-male or all-female stocks with improved yield and management characteristics [8,9]. In the zig-zag eel (*Mastacembelus armatus*), Cui et al. [10] conducted a comparative gonadal transcriptomic analysis of males and females across both undifferentiated and differentiated developmental stages. They identified sex-biased differentially expressed genes (DEGs), including male-biased genes such as *sox9*, *gsdf*, and *dmrt2b*, as well as female-biased genes such as *foxl2*, *rspo1*, *gdf9*, *bmp15*, and *wnt4*. Several key signaling pathways were enriched, including the MAPK, Wnt, and TGF-β pathways. These findings provide valuable insights into the molecular basis of sex differentiation in this economically important species and highlight candidate targets for sex-control breeding strategies. Notably, further investigation into the potential crosstalk between these signaling pathways, particularly the interaction between the Wnt and TGF-β pathways during ovarian development, may yield novel insights into the evolutionary diversity of sex determination

 https://doi.org/10.3390/fishes10090434

mechanisms in teleosts [11,12]. Zhang et al. [13] identified and analyzed three insulin-like growth factor (IGF) genes, *CmIGF1-1*, *CmIGF1-2*, and *CmIGF2*, in blotched snakehead (*Channa maculata*), showing that their expression was primarily liver-specific, higher in males, developmentally regulated, and significantly influenced by sex steroid hormones (17α-ethynylestradiol (EE_2) and 17α-methyltestosterone (MT)), suggesting a key role in sex-specific growth patterns. These findings not only clarify the molecular basis of sex-differentiated growth in blotched snakehead but also offer valuable genetic targets for hormone-regulated selective breeding strategies aimed at enhancing aquaculture productivity. Liu et al. [14] conducted a comparative analysis of liver enzyme activities and gene expression in yellowfin tuna (*Thunnus albacares*), revealing that females exhibited higher digestive and lipid metabolism capacities, while males showed enhanced sugar metabolism, antioxidant defenses, and elevated expression of growth hormone-related genes. Understanding the distinct metabolic and transcriptional profiles between male and female fish offers valuable insights for precision aquaculture, potentially enabling targeted nutritional and breeding strategies tailored to sex-specific physiological needs [15].

2.2. Growth-Trait Genetics and Breeding Tools

Growth performance constitutes a primary target trait in the genetic improvement of virtually all aquaculture species. Contemporary breeding programs increasingly integrate genomic tools, targeted mutagenesis, and quantitative genetic approaches to enhance growth efficiency while preserving genetic diversity and adaptive potential [16]. Wang et al. [17] developed an integrative growth model combining phenotypic measurements and genome-wide SNP genotyping in a hybrid pufferfish (*Takifugu obscurus* ♀× *Takifugu rubripes* ♂). They identified 13 candidate genes associated with growth traits and confirmed the hybrid's superior growth performance, underscoring its genetic potential for genome-assisted selection. By elucidating the genetic basis underlying heterosis, this study demonstrates the utility of interspecific hybridization as a strategy to combine favorable traits from both parental lines for improved aquaculture productivity. Jiang et al. [18] demonstrated that atmospheric-pressure room-temperature plasma (ARTP) mutagenesis effectively induced genetic variation in Songpu mirror carp, enhanced their growth potential and antioxidant capacity, and showed promise as a tool for selective fish breeding. The implementation of ARTP mutagenesis in Songpu mirror carp represents an innovative application of plasma-based techniques in aquaculture, providing a controllable and efficient method to enhance genetic diversity for selective breeding programs [19]. Xu et al. [20] investigated the genetic parameters of growth and developmental traits in American bullfrog (*Rana catesbeiana*) tadpoles and found that body weight at later developmental stages exhibited moderate heritability and a positive genetic correlation with metamorphosis rate. These findings suggest that late-stage body weight could serve as a reliable selection criterion for breeding high-performance bullfrog lines. By linking growth traits to metamorphic success, this study offers a practical framework for trait-based selection, supporting more targeted and efficient genetic improvement in bullfrog aquaculture.

2.3. Phenotypic Regulation and Transcriptomic Networks

Understanding how gene regulatory networks influence visible traits such as color and size is essential for both breeding and consumer preference-driven markets [21,22]. Wu et al. [23] identified and characterized microRNAs involved in the morphological color change in Midas cichlids (*Amphilophus citrinellus*), suggesting that miR-133-x and miR-183-x may regulate key pigmentation genes during the transition from black to gold coloration. This finding highlights the potential of miRNA-mediated regulatory networks as crucial molecular mechanisms driving pigment cell differentiation and color pattern evolution in vertebrates. Tönißen et al. [24] found that early growth variation in pikeperch (*Sander* lucioperca) larvae is not directly linked to size-dependent gene expression, but rather to developmental stage-specific differences in genes related to muscle formation and energy metabolism, highlighting complex regulatory factors that may inform strategies to reduce size-related mortality in aquaculture.

2.4. Immune Response and Environmental Adaptation

With intensifying aquaculture systems and increasing environmental stressors, understanding host–pathogen interactions and pollutant responses at the molecular level has become a research priority [25,26]. Zhang et al. [27] cloned and characterized the *CiSRB1* gene in grass carp (*Ctenopharyngodon idellus*) and demonstrated that its age-dependent expression may modulate susceptibility to grass carp reovirus (GCRV) infection. Their findings offer valuable insights into the molecular basis of age-related antiviral resistance in grass carp and identify *CiSRB1* as a promising target for genetic improvement and disease management in aquaculture. Li et al. [28] assessed the effects of nonylphenol (NP), a pervasive endocrine-disrupting compound (EDC), on BMP2 and BMP4 expression in red crucian carp (*Carassius auratus* red var.) and found that NP exposure downregulated these genes via the BMP-Smad pathway, impairing bone morphogenesis and highlighting the importance of incorporating environmental toxicology into breeding and risk assessment.

3. Conclusions

This Special Issue highlights the diverse and advancing landscape of genetic and breeding research in aquaculture, spanning fundamental investigations into sex differentiation mechanisms to applied approaches for growth-trait enhancement through mutagenesis and genomic selection. Collectively, the ten featured studies provide not only robust empirical evidence and methodological advancements, but also practical insights with direct implications for selective breeding strategies and aquaculture production. Looking forward, the integration of multi-omics technologies, computational biology, and functional validation is expected to significantly accelerate genetic improvement and adaptive capacity across cultured species. As the aquaculture sector navigates the dual imperatives of enhancing productivity and ensuring sustainability, the research presented herein will be instrumental in informing and shaping the development of next-generation, science-driven breeding programs.

Funding: This research was funded by the National Freshwater Genetic Resource Center, FGRC18537, and the China-ASEAN Maritime Cooperation Fund, CAMC-2018F.

Institutional Review Board Statement: Not applicable.

Informed Consent Statement: Not applicable.

Acknowledgments: We sincerely thank all the authors for their valuable contributions and for their patience in addressing the rigorous revisions suggested by peer reviewers and guest editors. We are also grateful to the editorial team of *Fishes* for their support in finalizing the format of this Special Issue.

Conflicts of Interest: The author declares no conflicts of interest.

References

1. Sae-Lim, P.; Kause, A.; Mulder, H.A.; Olesen, I. Breeding and genetics symposium: Climate change and selective breeding in aquaculture. *J. Anim. Sci.* **2017**, *95*, 1801–1812. [CrossRef] [PubMed]
2. Yanez, J.M.; Barria, A.; Lopez, M.E.; Moen, T.; Garcia, B.F.; Yoshida, G.M.; Xu, P. Genome-wide association and genomic selection in aquaculture. *Rev. Aquac.* **2023**, *15*, 645–675. [CrossRef]
3. Yáñez, J.M.; Newman, S.; Houston, R.D. Genomics in aquaculture to better understand species biology and accelerate genetic progress. *Front. Genet.* **2015**, *6*, 128. [CrossRef] [PubMed]
4. Chandhini, S.; Rejish Kumar, V.J. Transcriptomics in aquaculture: Current status and applications. *Rev. Aquacult.* **2019**, *11*, 1379–1397. [CrossRef]

5. Wang, J.; Cheng, Y.; Su, B.; Dunham, R.A. Genome manipulation advances in selected aquaculture organisms. *Rev. Aquac.* **2025**, *17*, e12988. [CrossRef]
6. Toyota, K.; Miyakawa, H.; Hiruta, C.; Sato, T.; Katayama, H.; Ohira, T.; Iguchi, T. Sex determination and differentiation in decapod and cladoceran crustaceans: An overview of endocrine regulation. *Genes* **2021**, *12*, 305. [CrossRef]
7. Wang, T.; Yu, Y.; Li, S.; Li, F. Molecular mechanisms of sex determination and differentiation in decapod crustaceans for potential aquaculture applications: An overview. *Rev. Aquac.* **2024**, *16*, 1819–1839. [CrossRef]
8. Martínez, P.; Viñas, A.M.; Sánchez, L.; Díaz, N.; Ribas, L.; Piferrer, F. Genetic architecture of sex determination in fish: Applications to sex ratio control in aquaculture. *Front. Genet.* **2014**, *5*, 340. [CrossRef]
9. Chen, J.; Zhu, Z.; Hu, W. Progress in research on fish sex determining genes. *Water Biol. Secur.* **2022**, *1*, 100008. [CrossRef]
10. Cui, F.; Wang, Y.; Liang, H.; Yang, Y.; Jiang, Z.; Song, J.; Liu, C.; Wu, Y.; Mu, X.; Liu, Y. Comparative transcriptomic analysis of male and female gonads in the zig-zag eel (*Mastacembelus armatus*). *Fishes* **2025**, *10*, 117. [CrossRef]
11. Wu, L.; Wu, F.; Xie, L.; Wang, D.; Zhou, L. Synergistic role of β-catenin1 and 2 in ovarian differentiation and maintenance of female pathway in Nile tilapia. *Mol. Cell. Endocrinol.* **2016**, *427*, 33–44. [CrossRef]
12. Yu, H.; Du, X.; Chen, X.; Liu, L.; Wang, X. Transforming growth factor-β (TGF-β): A master signal pathway in teleost sex determination. *Gen. Comp. Endocrinol.* **2024**, *355*, 114561. [CrossRef] [PubMed]
13. Zhang, X.; Wu, Y.; Zhang, Y.; Zhang, J.; Chen, K.; Liu, H.; Luo, Q.; Fei, S.; Zhao, J.; Ou, M. Molecular characteristics, expression patterns, and response of insulin-like growth factors gene induced by sex steroid hormones in blotched snakehead (*Channa maculata*). *Fishes* **2024**, *9*, 120. [CrossRef]
14. Liu, D.; Yang, H.; Li, S.; Huang, H.; Li, G.; Chen, H. Comparative analysis of enzymatic activities and transcriptional profiles of various hepatic enzymes between male and female yellowfin tuna (*Thunnus albacares*). *Fishes* **2024**, *9*, 184. [CrossRef]
15. Cai, C.; Yang, P.; Shi, Y.; Wang, X.; Chen, G.; Zhang, Q.; Cheng, G.; Kong, W.; Xu, Z. Transcriptomic and metabolomic analysis revealed potential mechanisms of growth and disease resistance dimorphism in male and female common carp (*Cyprinus carpio*). *Fish Shellfish Immunol.* **2025**, *158*, 110150. [CrossRef]
16. D'Agaro, E.; Favaro, A.; Matiussi, S.; Gibertoni, P.P.; Esposito, S. Genomic selection in salmonids: New discoveries and future perspectives. *Aquac. Int.* **2021**, *29*, 2259–2289. [CrossRef]
17. Wang, C.; Shi, Y.; Gao, Y.; Shi, S.; Wang, M.; Yao, Y.; Sun, Z.; Wang, Y.; Zhao, Z. Construction of a growth model and screening of growth-related genes for a hybrid puffer (*Takifugu obscurus*♀ × *Takifugu rubripes*♂). *Fishes* **2024**, *9*, 404. [CrossRef]
18. Jiang, X.; Li, C.; Shang, M.; Hu, X.; Ge, Y.; Jia, Z. A new mutagenesis tool for Songpu mirror carp (*Cyprinus carpio* L.) for selective breeding: Atmospheric-pressure room-temperature plasma mutagenesis technology. *Fishes* **2024**, *9*, 448. [CrossRef]
19. Su, X.L.; Zhao, S.S.; Xu, W.J.; Shuang, L.; Zheng, G.D.; Zou, S.M. Efficiently whole-genomic mutagenesis approach by ARTP in blunt snout bream (*Megalobrama amblycephala*). *Aquaculture* **2022**, *555*, 738241. [CrossRef]
20. Xu, W.; Wang, Y.; Wang, G.; Zhang, L.; Zhang, G.; Huo, Z.; Ge, H. Heritability estimates for growth traits and correlation analysis between weight and metamorphosis rate in the bullfrog *Rana* (*Aquarana*) *catesbeiana*. *Fishes* **2024**, *9*, 105. [CrossRef]
21. Wu, L.; Yang, Y.; Wang, X.; Weng, Z.; Hua, S.; Li, D.; Xia, J.; Liu, X.; Meng, Z. Genome-wide QTL mapping and RNA-seq reveal the genetic variation influencing growth traits in giant grouper (*Epinephelus lanceolatus*). *Aquaculture* **2023**, *563*, 738944. [CrossRef]
22. Liu, J.; Yin, M.; Ye, Z.; Hu, J.; Bao, Z. Harnessing hue: Advances and applications of fish skin pigmentation genetics in aquaculture. *Fishes* **2024**, *9*, 220. [CrossRef]
23. Wu, G.; Mu, X.; Liu, Y.; Liu, C.; Wang, X.; Yang, Y.; Song, H. Identification and characterization of microRNAs in morphological color change of polychromatic Midas cichlids (*Amphilophus citrinellus*). *Fishes* **2024**, *9*, 194. [CrossRef]
24. Tönißen, K.; Franz, G.P.; Rebl, A.; Lutze, P.; Grunow, B. Does size matter? Small and large larvae of pikeperch (*Sander lucioperca*) in a comparative gene expression analysis. *Fishes* **2024**, *9*, 33. [CrossRef]
25. Molbert, N.; Agostini, S.; Alliot, F.; Angelier, F.; Biard, C.; Decencière, B.; Leroux-Coyau, M.; Millot, A.; Ribout, C.; Goutte, A. Parasitism reduces oxidative stress of fish host experimentally exposed to PAHs. *Ecotoxicol. Environ. Saf.* **2021**, *219*, 112322. [CrossRef]
26. Zhang, M.; Shan, C.; Tan, F.; Limbu, S.M.; Chen, L.; Du, Z.Y. Gnotobiotic models: Powerful tools for deeply understanding intestinal microbiota-host interactions in aquaculture. *Aquaculture* **2020**, *517*, 734800. [CrossRef]
27. Zhang, Y.; Shi, J.; Lu, Y.; Luo, Q.; Chu, P.; Huang, R.; Chen, K.; Zhao, J.; Wang, Y.; Ou, M. Molecular cloning and characterization of scavenger receptor Class B Type 1 in grass carp (*Ctenopharyngodon idellus*) and its expression profile following grass carp reovirus challenge. *Fishes* **2024**, *9*, 276. [CrossRef]
28. Li, D.; Cui, X.; Chen, S.; Xu, J.; Li, Y.; Zhang, Q.; Sun, Y. Mechanistic insights into nonylphenol stress on BMP2 and BMP4 gene expression in red crucian carp (*Carassius auratus* red var.). *Fishes* **2024**, *9*, 159. [CrossRef]

fishes

Article

Comparative Transcriptomic Analysis of Male and Female Gonads in the Zig-Zag Eel (*Mastacembelus armatus*)

Fangyu Cui [1,2], Yuanyuan Wang [1], Haiyan Liang [3], Yexin Yang [1], Zhiyong Jiang [4], Jiahuan Song [1], Chao Liu [1], Yuli Wu [4], Xidong Mu [1,*] and Yi Liu [1,*]

1 Key Laboratory of Prevention and Control for Aquatic Invasive Alien Species, Ministry of Agriculture and Rural Affairs, Guangdong Engineering and Technology Research Center of Modern Recreational Fishery, Pearl River Fisheries Research Institute, Chinese Academy of Fishery Sciences, Guangzhou 510380, China; cuifangyu803@163.com (F.C.)
2 College of Fisheries and Life Science, Shanghai Ocean University, Shanghai 201306, China
3 Agricultural and Rural Bureau of Zengcheng District, Guangzhou 511300, China
4 Agro-Tech Extension Center of Guangdong Province, Guangzhou 510520, China
* Correspondence: muxd@prfri.ac.cn (X.M.); liuyi@prfri.ac.cn (Y.L.)

Abstract: The zig-zag eel (*Mastacembelus armatus*) is a unique economic fish species in China and exhibits significant dimorphism of male and female phenotypes. Cultivating all-male seedlings can significantly improve production efficiency. To investigate sex differentiation and gonadal development in *M. armatus*, high-throughput sequencing technology was used to analyze the transcriptomes of male and female gonads at different developmental stages, both before and after sex differentiation. We identified key genes involved in sex differentiation, male-specific differentially expressed genes (DEGs), including *dmrt1*, *amh*, *sox9*, *gsdf*, and *dmrt2b*, and female-biased DEGs, including *foxl2*, *rspo1*, *gdf9*, *bmp15*, and *wnt4*. GO and KEGG enrichment analyses revealed that signaling pathways such as MAPK, Wnt, and TGF-β play significant roles in sex differentiation in *M. armatus*. The expression levels of 13 sex-related genes, including *dmrt1*, *sox9*, *amh*, *foxl2*, *rspo1*, and *wnt4*, were determined by RT–qPCR in addition to RNA sequencing. RT-qPCR validation results were consistent with the transcriptomic data, confirming the reliability of our findings. This research provides valuable insights into the mechanisms of sex differentiation in *M. armatus* and lays a foundation for developing all-male populations in aquaculture.

Keywords: *Mastacembelus armatus*; transcriptome; sex differentiation; gonadal development; differentially expressed genes

Key Contribution: This study identifies key genes involved in sex differentiation in *M. armatus* and highlights the critical signaling pathways regulating this process. The findings deepen our understanding of the molecular mechanisms underlying sex differentiation and provide a basis for the development of all-male populations in aquaculture.

1. Introduction

Sexual dimorphism refers to the differences between males and females of the same species, including traits such as body size, morphology, coloration, physiology, and behavior [1]. Such differences are common across numerous species and are particularly pronounced in fishes, where sexual size dimorphism (SSD) is frequently notable [2]. For instance, in species such as Nile tilapia (*Oreochromis niloticus*) [3], yellow catfish (*Tachysurus fulvidraco*) [4], and northern snakehead (*Channa argus*) [5], males generally grow faster than females. In contrast, females exhibit faster growth in species like common carp (*Cyprinus carpio*) [6], turbot (*Scophthalmus maximus*) [7], and Chinese tongue sole (*Cynoglossus semilaevis*) [8]. Modifying the sex ratio or producing monosex populations in aquaculture can significantly enhance growth rates and improve profitability, making these approaches valuable for optimizing production.

Transcriptome sequencing is a powerful technique that generates comprehensive sequences of mRNA transcripts from specific tissues or organs under controlled conditions, facilitating the identification of functional genes. This approach is particularly effective for investigating the molecular mechanisms underlying sex differentiation and gonadal development in fishes [9]. In fishes, gonadal development is regulated by various sex-related genes and signaling pathways. The transcriptomic analysis of gonads offers a comprehensive understanding of sex-specific gene expression profiles, enabling the identification of differentially expressed genes (DEGs) closely associated with sex differentiation. This analysis provides foundational insights into the regulatory networks involved in sex differentiation in gonadal development. For example, Tao et al. [10] conducted transcriptome sequencing on the gonads of Nile tilapia (*Oreochromis niloticus*) at various developmental stages from 5 to 180 days post-hatching (dph), revealing the relationship between gene expression patterns and the processes of sex differentiation and gonadal development. Their study found minimal differences in the transcriptomes from 5 to 20 dph, but significant differential gene expression emerged from 90 to 180 dph. Based on these results, key genes closely related to sex differentiation, such as *cyp19a1a*, *foxl2*, and *amh*, were identified. Similarly, Ribas et al. [11] carried out transcriptome analysis on 3-month-old turbot (*Scophthalmus maximus*), identifying 56 DEGs related to sex differentiation, including 44 genes associated with ovarian differentiation (e.g., *cd98*, *gpd1*, *cry2*) and 12 genes related to testicular differentiation (e.g., *ace*, *capn8*, *nxph1*). Fan et al. [12] performed transcriptome sequencing of the ovaries and testes of Chinese longsnout catfish (*Leiocassis longirostris*), screening 71 sex-related candidate genes. Among these genes, 50 were highly expressed in the testes (e.g., *dmrt1*, *cyp17a1*, *samd7*, *wnt6*, *wt1*) and 21 in the ovaries (e.g., *foxl2*, *gdf9*, *zp3*, *zp1*, *figla*, *bmp15*). Additionally, they identified 16 signaling pathways involved in gonadal development, including ovarian steroidogenesis, the TGF-β signaling pathway, and the GnRH signaling pathway.

Mastacembelus armatus, belonging to the order Synbranchiformes, family Mastacembelidae, and genus *Mastacembelus* [13], is a bottom-dwelling freshwater fish adapted to warm waters. It is mainly distributed in major river systems across Southeast Asia, particularly in tropical and subtropical regions [14]. The species has an elongated, eel-like body that is laterally compressed with a flattened tail. Its pointed snout features a tubular rostrum at the tip; although the oral cavity is small, the pharynx is wide, and the sharp teeth allow it to prey on small benthic aquatic animals such as shrimp and worms [15]. Owing to its tender meat and delicious taste, *M. armatus* is highly favored by consumers. However, in recent years, environmental pollution and overfishing have significantly reduced the population size of wild *M. armatus*. Consequently, in China, provinces such as Guangdong and Fujian have designated *M. armatus* as a protected wild aquatic species [16]. Moreover, large-scale aquaculture of *M. armatus* has not yet been successfully established. *M. armatus* has a male heterogametic sex determination system of XX/XY [17]. *M. armatus* exhibits notable sexual dimorphism, with males growing significantly faster than females [18]. This advantage makes males particularly valuable for aquaculture production, suggesting that cultivating all-male populations could further enhance production yields. However, in current aquaculture practices, the high female-to-male ratio hampers the development of the *M. armatus* industry. Additionally, the mechanisms of sex determination and differentiation are still not fully understood.

Therefore, this study utilized high-throughput sequencing technology to analyze the transcriptomic expression profiles of the ovaries and testes of *M. armatus* before and after gonadal differentiation, identifying differentially expressed genes related to sex differentiation. Additionally, several signaling pathways potentially involved in the sex differentiation of *M. armatus* were uncovered. In the future, gene editing (CRISPR/Cas9) technology can be further used to verify the function of key genes and explore methods of sex control. These findings provide valuable data for advancing our understanding of sex differentiation mechanisms and offer a basis for the targeted cultivation of all-male *M. armatus* populations in the future.

2. Materials and Methods

2.1. Sample Collection and Handling

The gonadal samples of *M. armatus* used in this experiment were sourced from Guangzhou Heshenghui Agricultural Technology Co., Ltd. (Guangzhou, China). All samples were collected from caudal fin tissue for DNA extraction and sex identification. PCR amplification was performed using sex-linked markers developed by Qin et al. to identify the genetic sex of *M. armatus* [19]. Pre-differentiation samples were derived from whole fish. At the pre-differentiation stage, due to the extremely small size of the gonads and the difficulty of avoiding contamination from surrounding tissues, we collected whole fish samples at 10 days post-hatch (dph) to ensure sufficient RNA for transcriptome sequencing and to capture transcriptomic information related to sex differentiation as comprehensively as possible. At the post-sex differentiation stage, three male and three female individuals, each three months old, were selected, and their gonadal tissues (testes and ovaries) were carefully dissected and collected. The fish were anesthetized with MS-222 (Sigma-Aldrich, St. Louis, MO, USA) before rapid dissection to remove the gonads. Each gonad was divided into two portions: one was immediately frozen in liquid nitrogen and stored at −80 °C for subsequent RNA extraction, and the other was fixed in 4% paraformaldehyde (Servicebio, Wuhan, China) for paraffin embedding and histological analysis.

2.2. Paraffin Sectioning of *M. armatus* Gonads

The gonadal tissues were kept in 4% paraformaldehyde for 24 h before being processed into paraffin sections for further examination. The paraffin sections underwent ethanol gradient dehydration, xylene transparency treatment, wax infiltration, and embedding. Continuous 5–6 μm sections were prepared and stained with hematoxylin–eosin (HE) (Servicebio, Wuhan, China). Finally, the sections were mounted with neutral resin, and the gonadal morphology was observed and photographed under a microscope.

2.3. Total RNA Extraction and Library Construction for Sequencing

Total RNA was extracted using the TRIzol reagent (Life Technologies, Foster City, CA, USA) according to the manufacturer's protocol. RNA concentration and purity were assessed with a NanoDrop 2000 spectrophotometer (Thermo Fisher Scientific, Wilmington, DE, USA). This study employed Illumina RNA-seq technology for transcriptome sequencing, and the library construction was performed following the method described by Mortazavi et al. [20]. mRNA was isolated using poly-T oligo-conjugated magnetic beads, followed by cDNA synthesis using the Hieff NGS® Ultima Dual-mode mRNA Library Prep Kit for Illumina® (Yeasen Biotechnology, Shanghai, China) according to the manufacturer's instructions. The products were then amplified by PCR and evaluated with an Agilent Bioanalyzer 2100 system (Agilent Technologies, Santa Clara, CA, USA) to assess their quality and integrity.

2.4. Transcriptome Sequencing and Analysis

The prepared libraries were sequenced on the Illumina NovaSeq platform (Yeasen Biotechnology, Shanghai, China), generating 150 bp paired-end reads. Raw reads were filtered with SOAPnuke (version 2.1.0) to remove low-quality sequences and adapters [21]. Data quality was evaluated based on Q20, Q30, and GC content. Hisat2 software (version 2.2.1) (Johns Hopkins University, Baltimore, MD, USA) was used to map with the reference genome. Differential gene expression between male and female gonads was analyzed using DESeq2 (version 1.36.0) [22], with gene expression levels quantified as FPKM (fragments per kilobase of exon model per million mapped reads) and fold change calculated as FPKMtestes/FPKMovaries. Genes with $|\log 2 \text{ fold change}| \geq 1$ and FDR (false discovery rate) < 0.05 were considered differentially expressed.

Gene Ontology (GO) enrichment analysis and KEGG pathway enrichment analysis of DEGs (differentially expressed genes) were performed using the clusterProfiler package in R (version 4.4.4) [23].

2.5. Real-Time Quantitative PCR (RT-qPCR) Validation

RT-qPCR was used to validate sex-related DEGs. Specific primers were designed using Primer Premier 5 (Premier, Vancouver, BC, Canada). All primers were synthesized by TianyiHuayu Gene Technology Co., Ltd. (Wuhan, China) (Table 1). Total RNA (1 μg) from male and female gonads was reverse-transcribed into cDNA using the PrimeScript RT Master Mix Kit (Takara, Shiga, Japan). The qPCR reactions were conducted on a 7500 Real-Time PCR system (ABI StepOnePlus, San Diego, CA, USA) using TB Green Fast qPCR Mix (Takara, Shiga, Japan). Each 20 μL reaction contained 10 μL of 2× TB Green Fast qPCR Mix, 0.4 μL of ROX Reference Dye II (50×), 0.8 μL each of the forward and reverse primers (10 μmol/L), 1 μL of cDNA template, and 7 μL of sterile water. The qPCR cycling conditions were as follows: 95 °C for 2 min, followed by 40 cycles of 95 °C for 15 s and 60 °C for 60 s. Each sample was tested in triplicate. Gene expression levels were calculated using the $2^{-\Delta\Delta Ct}$ method for relative quantification. Statistical analyses were conducted using GraphPad Prism 9.5 (GraphPad Software, San Diego, CA, USA), with results presented as mean ± standard deviation (x ± SD).

Table 1. Primer sequences used for real-time quantitative PCR.

Gene	Sequences From (5′-3′)	Gene	Sequences From (5′-3′)
dmrt1	F:CGGCCCAGGTTGCCTTGAG R:CCAGCTTCATTCTTCACCATCA	*foxl2*	F:TCCGTCCCAGAAACCACCGTAT R:CCTGATGCTGTTCTGCCAACCT
sox9	F:GAAGGACGAGGACGATAAGTT R:TGGCATAGGCACGAGGGT	*rspo1*	F:AAAGGCTCACAATCTCGG R:CCTCCTCTACTGCCATCC
amh	F:TCTGGCACTCAGCTTATCC R:CCATCTCCTCCTCCCTTA	*wnt4*	F:TGGGCAACATCATCAAGG R:TGGATCATAGTCGCAGAAA
wt1	F:TGCGTTCACCGTCCACTT R:CGACCGTGCTGTAACCTG	*gdf9*	F:ATCTACACCTGCTCATCAA R:ACTGACTACTGAACCCTGAT
bmp4	F:GTATCGGCTACAGTCAGGG R:ATCTTCGGGAATGGTGCT	*bmp15*	F:GCAGAAAGCGGACCAGAA R:CGAGGGAAGAGTGTCAAGC
sox4	F:GGGACTTGGATTTGAACTTTG R:TCGCTCACTTCGGGCGTA	*dmrt2a*	F:CGGGAATACAAAGAACGAGA R:CGCTGACATTGGAGGAGAT
dmrt2b	F:AACCAGGGAGGATAAGGA R:GCTGACGTGCTATTTGAGT	*β-actin*	F:TCATGAGGTAGTCTGTGAGGTCCC R:GCCTCTGGTCGTACCACTGGTATT
gsdf	F:TCCAAGGAAGAACCTGCAACCT R:CAGGCATCCATGGCTCAGACTC		

3. Results and Analysis

3.1. Histological Observation

Histological observations indicated that the gonads of the fish 10 days post-hatch (dph) were undifferentiated, presenting as primordial gonads containing primordial germ cells. Histological observations indicated that the gonads of three-month-old *M. armatus* contained differentiated cells (e.g., spermatogonia in males or primary oocyte in females), suggesting that sex differentiation had been completed (Figure 1). At 90 dph, male gonads were in developmental stage II, with elongated and branched seminiferous tubules containing spermatogonia. The spermatogonia nuclei were darkly stained, while the cytoplasm appeared more transparent. Similarly, female gonads were also in developmental stage II at 90 dph, primarily containing primary oocytes with large, centrally, or near centrally positioned nuclei that were darkly stained with prominent nucleoli. Follicular cells closely surrounded the irregularly shaped oocytes.

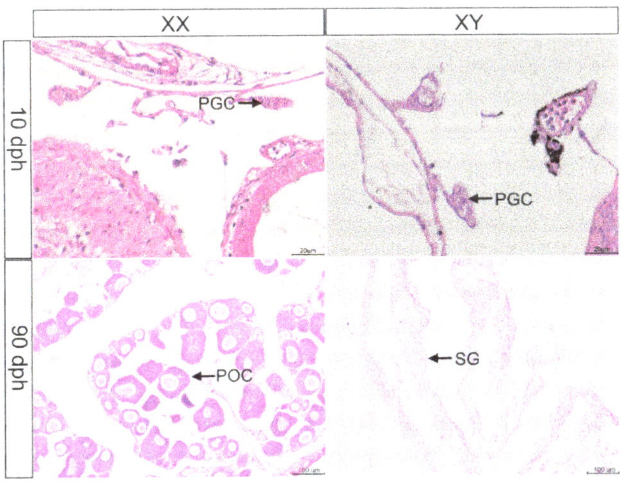

Figure 1. HE staining of paraffin sections from gonads of *Mastacembelus armatus*. PGC: primordial germ cell; POC: primary oocyte; SG: spermatogonia.

3.2. Transcriptome Sequencing Results

High-throughput sequencing was performed on 12 samples (3 male and 3 female samples before and after sex differentiation), yielding a total of 80.81 Gb of clean data. The clean data for each sample exceeded 6.14 Gb, with a Q20 base percentage above 98.60%, a Q30 base percentage above 96.00%, and GC content ranging from 44.79% to 48.59%, indicating a high level of data integrity (Table S1).

Clean reads from 12 samples were mapped to the reference genome (accession number: GCA_019455525.1) [17]. The overall mapping rates for all samples exceeded 84.40%, and the unique mapping rates were all above 70.54% (Table S2). These high mapping rates support accurate gene annotation for *M. armatus* gonadal samples.

3.3. Differential Gene Expression Analysis

In the gonadal transcriptome of *M. armatus*, differential expression analysis was based on a threshold of $|\log 2 \text{ fold change}| \geq 1$ and FDR ≤ 0.05. M1/F1 represents males/females before sex differentiation, while M2/F2 represents males/females after sex differentiation. Compared to the F1 group, 10 genes were upregulated and 49 genes were downregulated in the M1 group (Figure 2A). Compared to the F1 group, 5939 genes were upregulated and 9003 genes were downregulated in the F2 group (Figure 2B). Compared to the M1 group, 1368 genes were upregulated and 586 genes were downregulated in the M2 group (Figure 2C). Lastly, the comparison of F2 vs. M2 showed that 5051 genes were upregulated and 3320 genes were downregulated in the M2 group (Figure 2D).

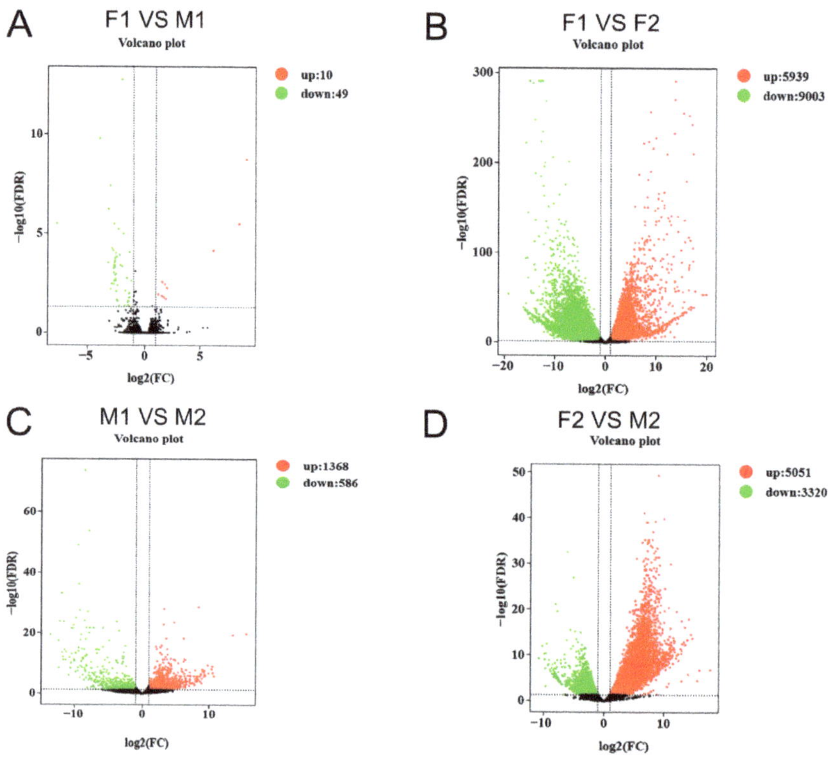

Figure 2. Volcano plot of differentially expressed genes. (**A**) F1 vs. M1; (**B**) F1 vs. F2; (**C**) M1 vs. M2; (**D**) F2 vs. M2. F represents female; M represents male; 1/2 denotes before or after sex differentiation, respectively. The horizontal dashed line represents log10 (FDR = 0.05), and the vertical dashed line represents |log2 fold change| = 1. Black dots represent unchanged genes.

3.4. GO and KEGG Enrichment Analysis

The DEGs were mapped to the GO and KEGG databases for enrichment analyses. The results of the GO enrichment analyses are shown in Figure 3. Comparison of male and female samples prior to differentiation showed significant enrichment of the expression of genes involved in biological processes, such as cellular processes, biological regulation, and metabolic processes. Molecular functions such as binding and catalytic activity were prominently represented, highlighting key differences in transcriptional regulation and molecular interactions during the pre-differentiation phase (Figure 3A). A comparison of samples before and after female differentiation showed a notable upregulation of genes associated with developmental processes, multicellular organismal processes, and reproductive processes. Molecular functions related to protein binding, receptor activity, and transcription factor activity were particularly enriched, emphasizing molecular changes during ovarian differentiation (Figure 3B). A comparison of samples before and after male differentiation revealed significant changes in biological processes like reproductive processes, immune system processes, and cellular communication. Molecular functions such as catalytic activity, structural molecule activity, and transporter activity were enriched, reflecting the transcriptional shifts involved in testicular differentiation (Figure 3C). A comparison of male and female samples after differentiation showed prominent enrichment of the expression of genes involved in processes such as growth, developmental processes, and cellular processes. Molecular functions like signal transduction, transcription regulator activity, and antioxidant activity were differentially expressed, underlining the functional specialization of male and female gonads after differentiation (Figure 3D).

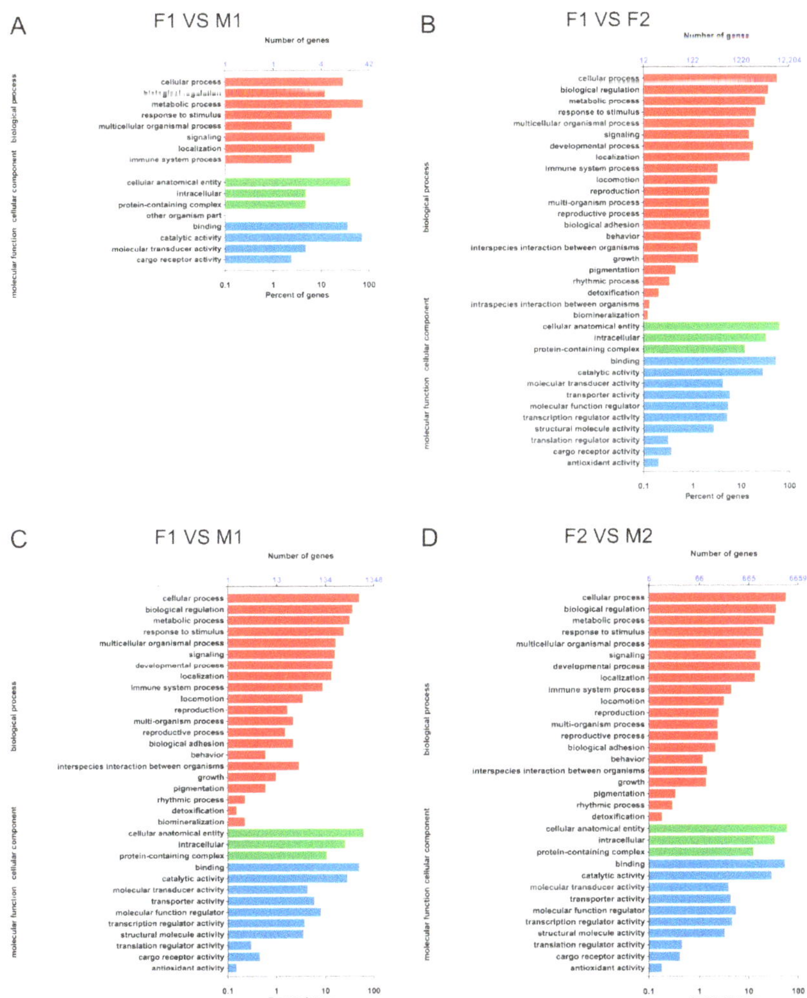

Figure 3. GO enrichment analyses of differentially expressed genes in the gonads of *M. armatus*. (**A**) F1 vs. M1; (**B**) F1 vs. F2; (**C**) M1 vs. M2; (**D**) F2 vs. M2. F represents female; M represents male; 1/2 denotes before or after sex differentiation, respectively.

The results of the KEGG enrichment analyses are shown in Figure 4. A comparison of samples before and after male differentiation showed that the main pathways were categorized under metabolism, such as glycolysis/gluconeogenesis, fatty acid metabolism, and protein processing in the endoplasmic reticulum, along with signaling pathways such as the NOD-like receptor signaling pathway (Figure 4A). A comparison of male and female samples prior to differentiation showed that the main pathways were categorized under cellular processes, such as endocytosis, the regulation of the actin cytoskeleton, and the cell cycle, as well as apoptosis-related pathways like the autophagy–animal and p53 signaling pathway (Figure 4B). A comparison of samples before and after female differentiation showed that the main pathways were categorized under environmental information processing, including the MAPK signaling pathway, neuroactive ligand-receptor interaction, and calcium signaling pathway, as well as metabolic pathways like purine metabolism and pathways related to viral infection such as Herpes Simplex Virus 1 Infection (Figure 4C). A comparison of male and female samples after differentiation showed that the main pathways spanned multiple categories, including cellular processes such as focal adhesion and tight junction, environmental information processing like the MAPK signaling pathway, metabolic pathways such as amino acid metabolism, and disease-related pathways such as Herpes Simplex Virus 1 Infection (Figure 4D).

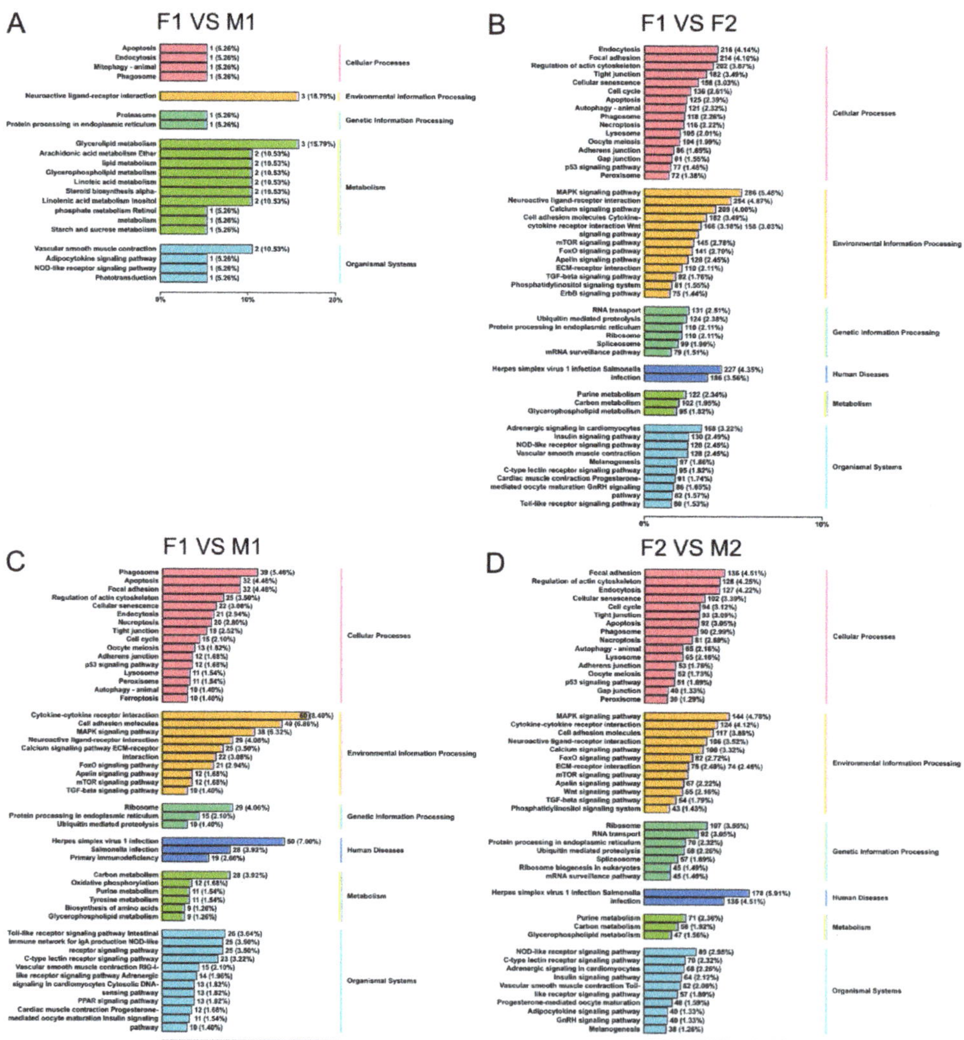

Figure 4. KEGG enrichment analyses of differentially expressed genes in the gonads of *M. armatus*. **(A)** F1 vs. M1; **(B)** F1 vs. F2; **(C)** M1 vs. M2; **(D)** F2 vs. M2. F represents female; M represents male; 1/2 denotes before or after sex differentiation, respectively.

Among all of the identified pathways, 10 were found to be potentially related to sex differentiation and gonadal development, with 519 DEGs mapped to these pathways (Figure 5). The MAPK signaling pathway showed the highest enrichment, both in terms of DEG count and the enrichment ratio. Other sex-related pathways included the cell cycle, FoxO signaling, mTOR signaling, TGF-β signaling, Wnt signaling, oocyte meiosis, progesterone-mediated oocyte maturation, calcium signaling, and the GnRH signaling pathways.

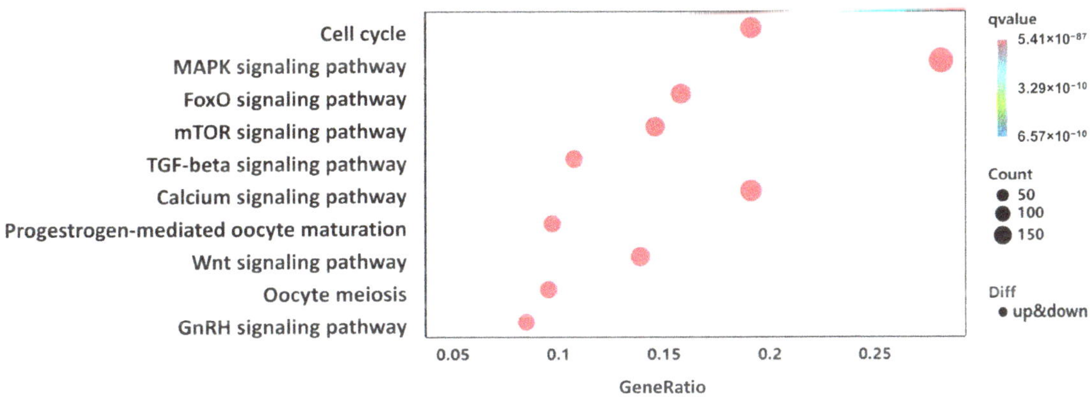

Figure 5. KEGG pathways related to sex differentiation and gonadal development in *M. armatus*.

3.5. Screening and Analysis of Key Sex-Related DEGs

Through GO and KEGG enrichment analyses, we identified 41 key sex-related genes. These included classic sex-regulating gene families such as the DMRT gene family (*dmrt1*, *dmrt2a*, *dmrt2b*), the SOX gene family (*sox4*, *sox11b*, *sox7*, *sox9a*, *sox10*, *sox17*, *sox4b*), and the TGF-β superfamily (*amh*, *gsdf*, *gdf9*, *bmp4*, *bmp15*), as well as the 17β-hydroxysteroid dehydrogenase gene family (*hsd3b1*, *hsd17b4*, *hsd17b12a*). Additionally, genes related to reproduction and regulation were identified, including sex hormone receptors (*fshr*, *pgrmc1*, *gnrhr4*, *paqr6*, *paqr8*, *paqr7a*, *esrrb*), hormone regulatory factors (*figla*, *fem1b*, *wt1a*, *wt1b*), the Rspo1/Wnt/β-catenin signaling pathway (*rspo1*, *wnt4*, *wnt9a*, *wnt10b*, *wnt11*), the DEAD-box gene family (*ddx4-vasa*), genes involved in spermatogenesis (*spata2l*, *spata5*, *spata6*, *spata22*), and transcription factors such as *foxl2*.

The gene expression analysis of the aforementioned sex-related genes identified 24 genes highly expressed in the testes and 17 genes highly expressed in the ovaries; genes such as *dmrt1*, *sox9*, *amh*, and wt1 were predominantly expressed in the testes, whereas *foxl2*, *rspo1*, *wnt4*, *gdf9*, and *bmp15* were significantly expressed in the ovaries (Figure 6).

3.6. RT-qPCR Validation

To validate the transcriptome sequencing results, 14 key genes related to sex differentiation were selected from the previously mentioned gene families and pathways for RT-qPCR analysis. The results showed that *dmrt1*, *sox9*, *amh*, *wt1*, *bmp4*, *sox4*, *dmrt2b*, and *gsdf* were highly expressed in the testes, while *foxl2*, *rspo1*, *wnt4*, *gdf9*, *bmp15*, and *dmrt2a* were highly expressed in the ovaries. These findings were consistent with the RNA-Seq data (Figure 7), confirming the reliability of the sequencing data and the accuracy of the DEG screening results.

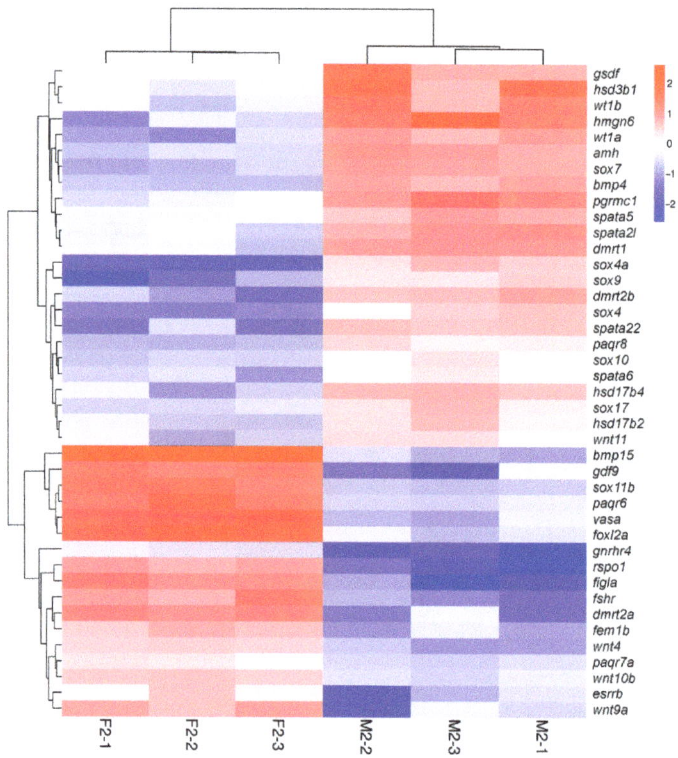

Figure 6. Heatmap of results of clustering analysis of sex-related genes in the gonads of *M. armatus*. F and M represent female and male, respectively. Numbers (1-1/2/3) refer to individuals before differentiation, and (2-1/2/3) refer to individuals after differentiation.

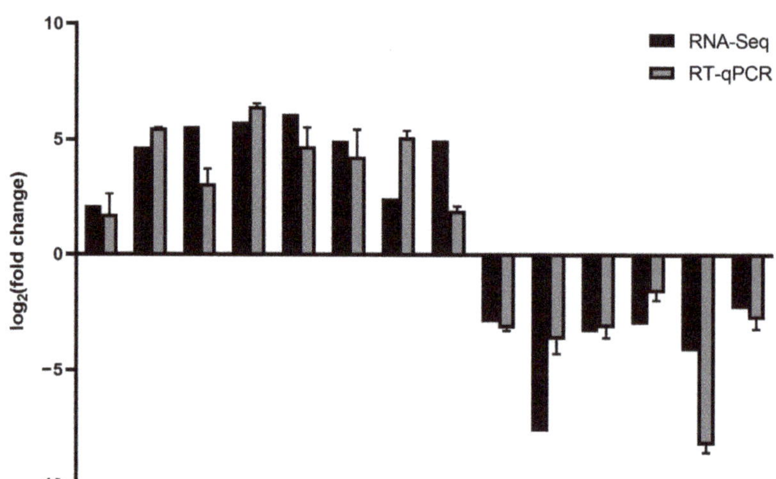

Figure 7. RT-qPCR validation of 14 key sex-related genes in *M. armatus*. Fold change = relative expression of male/relative expression of female.

4. Discussion

In this study, we conducted an in-depth analysis of the transcriptomes of male and female gonads in *M. armatus* before and after sex differentiation using high-throughput sequencing technology. Since the pre-differentiation samples were derived from whole fish tissues, the DEG analysis inevitably reflected overall gene expression differences. Therefore, we first manually filtered the DEGs to identify candidate sex-related genes and further compared them with the literature to select key genes with strong evidence supporting their roles in sex determination and gonadal development. Finally, we identified key sex-related genes and pathways involved in the sex differentiation of *M. armatus*.

4.1. Sex-Related DEGs in M. armatus

In fishes, gonadal development is regulated by complex, interconnected gene regulatory networks [24]. It is well known that *dmrt1*, *sox9* and *amh* play pivotal roles in male gonadal development [25]. *Dmrt1*, a member of the highly conserved DMRT family of sex-regulating factors, is essential for initiating testicular differentiation [26], partly by promoting the expression of *sox9* [27,28]. During this process, *dmrt1* is highly expressed, activating *sox9* and triggering downstream testicular development [29]. In females, *sox9* induces *amh* expression, which suppresses the development of reproductive structures while activating genes involved in testicular differentiation and inhibiting ovarian differentiation pathways involving β-catenin [30]. *Amh*, a TGF-β superfamily gene, prevents Müllerian duct development and indirectly promotes male differentiation [31]. Mustapha's study indicates that genes related to male gonadal development in spotted scat, *Scatophagus argus*, such as *dmrt1*, *gsdf*, and *amh*, are significantly more highly expressed in the testes [32], which is consistent with the findings of our study. In this study, prior to sex differentiation, although the male gonads had not fully differentiated morphologically, transcriptomic analysis revealed weak expression of male marker genes such as *dmrt1* and *sox9*. As differentiation progressed, *dmrt1*, *sox9*, *amh*, and *wt1* were predominantly expressed in the testes of *M. armatus*, playing critical roles in promoting testicular development. Additionally, in common carp (*Cyprinus carpio*), *dmrt1* and *sox9* not only promote male gonadal development but also inhibit the expression of female-specific genes such as *foxl2*, thereby facilitating male differentiation [33].However, further functional studies are needed to confirm the specific roles of these sex-related genes in the gonadal development and sex differentiation of *M. armatus*.

During female gonadal development, *foxl2* is a key regulatory factor, functioning to suppress the expression of male-related genes while promoting ovary-specific gene expression [34,35]. Research on marble goby, *Oxyeleotris marmorata*, indicates that *foxl2a* was highly expressed during the female gonadal differentiation stage. It is suggested that these genes may play a role in oocyte polarity establishment, early oogenesis, and ovarian differentiation [36], which is consistent with the findings of this study. *rspo1* and *wnt4* are essential for ovarian differentiation, and the absence of the expression of either gene can result in sex reversal in Japanese rice fish (*Oryzias latipes*) [37]. In male medaka embryos, the overexpression of *rspo1* can induce a transformation to female characteristics [38]. *foxl2* upregulates *rspo1* and *wnt4*, thereby activating the Wnt-β-catenin signaling pathway, which plays a key role in ovarian differentiation [39]. *Wnt4* is vital for normal ovarian development. Once β-catenin is activated, it translocates to the nucleus, where it regulates genes involved in folliculogenesis and granulosa cell development [40–42]. Through the activation of the Rspo1-Wnt4 regulatory axis, *foxl2* effectively prevents masculinization and supports normal ovarian development. In addition, *foxl2* supports follicle development by regulating *gdf9* and *bmp15*, which are important factors secreted by oocytes, to maintain ovarian function by regulating granulosa cell differentiation and follicle maturation [43,44]. In this study, prior to the onset of sex differentiation, although the morphological differentiation of female gonads was incomplete, transcriptomic analysis detected low-level expression of key female-specific genes, including *foxl2*, *rspo1*, and *gdf9*. As differentiation advanced, the expression levels of *foxl2*, *rspo1*, and *wnt4* were markedly elevated in the ovaries, with *bmp15* and *gdf9* playing pivotal roles in regulating and promoting ovarian development. Nevertheless, further functional investigations are required to elucidate the precise roles of these sex-related genes in gonadal development and sex differentiation in *M. armatus*.

4.2. Role and Interaction of Signaling Pathways in Gonadal Development

KEGG enrichment analysis of DEGs related to sex differentiation and gonadal development in male and female gonads of *M. armatus* revealed several pathways involved in gonadal development. The MAPK signaling pathway is widely involved in cell proliferation and differentiation [45]. A previous study suggests that inhibiting the activity of p38 MAPKs in *Brachymystax lenok* leads to abnormal levels of gonadotropin-releasing hormone (GnRH), follicle-stimulating hormone (FSH), and estradiol, thereby affecting follicular development [46]. Further studies have also suggested that the MAPK pathway may be involved in regulating the development of female nuclei and the meiotic process of diploid crucian carp, regulating nuclear replication and chromosome doubling, thus playing a role in the process of gametogenesis [47]. The TGF-β signaling pathway regulates germ cell differentiation by activating Smad proteins and plays a role in sex differentiation during early gonadal development by promoting extracellular matrix remodeling and germ cell migration [48]. This pathway also interacts with other signaling pathways, such as the Wnt and MAPK pathways, to coordinate gonadal development. For example, cross-regulation between TGF-β and Wnt signaling is critical for ovarian development in mice, where these pathways collaborate to inhibit masculinization and support female gonadal differentiation [49].The GnRH (gonadotropin-releasing hormone) signaling pathway indirectly affects gonadal development by regulating hormone secretion within the hypothalamic–pituitary–gonadal axis [50]. GnRH stimulates the pituitary to release luteinizing hormone (LH) and follicle-stimulating hormone (FSH), which subsequently regulate sex hormone secretion, influencing processes such as follicle development, ovulation, and spermatogenesis [51]. The GnRH pathway also interacts with the progesterone-mediated oocyte maturation pathway to promote oocyte maturation and the ovulation process [52].

These signaling pathways involved in gonadal development interact closely by regulating cell proliferation, differentiation, and metabolism. Gaining insight into how these pathways interact is crucial for understanding gonadal development and sex differentiation mechanisms in *M. armatus*.

This study offers important resources for investigating sex differentiation mechanisms in *M. armatus* and lays the groundwork for developing all-male breeding populations. The findings also provide valuable insights for the conservation of *M. armatus* populations and have important implications for aquaculture practices aimed at improving breeding efficiency.

5. Conclusions

In this study, transcriptomic analysis was performed on male and female gonads before and after *M. armatus* differentiation. Key genes such as *dmrt1*, *sox9*, and *amh* were found to be essential for testicular development, while *foxl2*, *rspo1*, and *wnt4* were identified as crucial for ovarian development. Furthermore, this study also identified that the MAPK, Wnt, and TGF-β signaling pathways are associated with sex differentiation. These results suggest that during sex differentiation, the activation of sex-specific genes and signaling pathways plays a crucial role in gonadal development. After sex differentiation, the aforementioned genes were significantly expressed in the different sexes, suggesting their crucial roles in the sex differentiation of *M. armatus*. RT-qPCR validation supported the RNA-seq findings, deepening our understanding of how these sex-related genes contribute to gonadal development and sex differentiation in *M. armatus*.

Supplementary Materials: The following supporting information can be downloaded at https://www.mdpi.com/article/10.3390/fishes10030117/s1: Table S1: Transcriptome sequencing data information; Table S2: Alignment results of male and female gonadal transcriptome data of *M. armatus* with the reference genome.

Author Contributions: Conceptualization, X.M. and Y.L.; methodology, F.C. and Y.W. (Yuanyuan Wang); software, Y.W. (Yuanyuan Wang); validation, F.C., C.L. and Y.W. (Yuanyuan Wang); formal analysis, F.C. and Y.W. (Yuanyuan Wang); investigation, F.C., Y.Y., Y.W. (Yuli Wu) and Z.J.; resources, Y.W. (Yuanyuan Wang); data curation, F.C. and J.S.; writing—original draft preparation, F.C.; writing—review and editing, F.C., Y.W. (Yuanyuan Wang), H.L., Y.Y., Z.J., J.S., C.L., Y.W. (Yuli Wu), X.M. and Y.L.; visualization, F.C. and Y.W. (Yuanyuan Wang); supervision, Y.L; project administration, X.M. and Y.W. (Yuanyuan Wang); funding acquisition, X.M. All authors have read and agreed to the published version of the manuscript.

Funding: This research was funded by the Central Public-Interest Scientific Institution Basal Research Fund, CAFS (NO.2023SJHX6), the Guangdong Agricultural Technology Service—Major Agricultural Technology Rural Promotion Project (2130106), China-ASEAN Maritime Cooperation Fund (CAMC-2018F), the Rural Revitalization Strategy Special Provincial Organization and Imple-mentation Project Funds (2023SBH00001), and the National Freshwater Genetic Resource Center (FGRC18537).

Institutional Review Board Statement: This study was conducted in accordance with ethical guidelines and received approval from the Laboratory Animal Ethics Committee of the Pearl River Fisheries Research Institute (code: LAEC-PRFRI-2024-02-01; date: 26 February 2024).

Informed Consent Statement: Not applicable.

Data Availability Statement: The data presented in this study are openly available in NCBI at UTL, reference number BioProject PRJNA1177591.

Conflicts of Interest: The authors declare no conflicts of interest.

References

1. Mei, J.; Gui, J.F. Genetic basis and biotechnological manipulation of sexual dimorphism and sex determination in fish. *Sci. China Life Sci.* **2015**, *58*, 124–136. [CrossRef]
2. Wang, H.P.; Gao, Z.X.; Rapp, D.; O'Bryant, P.; Yao, H.; Cao, X.J. Effects of temperature and genotype on sex determination and sexual size dimorphism of bluegill sunfish *Lepomis macrochirus*. *Aquaculture* **2014**, *420*, S64–S71. [CrossRef]
3. Tao, W.J.; Zhu, X.; Cao, J.M.; Xiao, H.S.; Dong, J.J.; Kocher, T.D.; Lu, M.X.; Wang, D.S. Screening and characterization of sex-linked DNA markers in Mozambique tilapia (*Oreochromis mossambicus*). *Aquaculture* **2022**, *557*, 738331. [CrossRef]
4. Wang, L.Y.; Qi, P.P.; Chen, M.; Yuan, Y.C.; Shen, Z.G.; Fan, Q.X. Effects of sex steroid hormones on sexual size dimorphism in yellow catfish (*Tachysurus fulvidraco*). *Acta Hydrobiol. Sin.* **2020**, *44*, 379–388. [CrossRef]
5. Dai, S.; Chen, M.; Zheng, S.; Su, J.; Wu, J.; Han, L.; Zhou, C.; Zou, Y.; Wang, D.; Li, M. Sex-linked DNA marker screening and characterization in albino northern snakehead (*Channa argus* var.) via third-generation sequencing and pool resequencing. *Aquaculture* **2025**, *594*, 741449. [CrossRef]
6. Zhai, G.; Shu, T.; Chen, K.; Lou, Q.; Jia, J.; Huang, J.; Shi, C.; Jin, X.; He, J.; Jiang, D. Successful production of an all-female common carp (*Cyprinus carpio* L.) population using cyp17a1-deficient neomale carp. *Engineering* **2021**, *8*, 181–189. [CrossRef]
7. Li, X.Y.; Mei, J.; Ge, C.T.; Liu, X.L.; Gui, J.F. Sex determination mechanisms and sex control approaches in aquaculture animals. *Sci. China Life Sci.* **2022**, *65*, 1091–1122. [CrossRef]
8. Wang, N.; Wang, R.; Wang, R.; Chen, S. Transcriptomics analysis revealing candidate networks and genes for the body size sexual dimorphism of Chinese tongue sole (*Cynoglossus semilaevis*). *Funct. Integr. Genom.* **2018**, *18*, 327–339. [CrossRef]
9. Hrdlickova, R.; Toloue, M.; Tian, B. RNA-Seq methods for transcriptome analysis. *Wiley Interdiscip. Rev. RNA* **2017**, *8*, e1364. [CrossRef]

10. Tao, W.; Chen, J.; Tan, D.; Yang, J.; Sun, L.; Wei, J.; Conte, M.A.; Kocher, T.D.; Wang, D. Transcriptome display during tilapia sex determination and differentiation as revealed by RNA-Seq analysis. *BMC Genom.* **2018**, *19*, 363. [CrossRef]
11. Ribas, L.; Robledo, D.; Gómez-Tato, A.; Viñas, A.; Martínez, P.; Piferrer, F. Comprehensive transcriptomic analysis of the process of gonadal sex differentiation in the turbot (*Scophthalmus maximus*). *Mol. Cell. Endocrinol.* **2016**, *422*, 132–149. [CrossRef] [PubMed]
12. Fan, J.H.; Ye, H.; Song, X.H.; Yue, H.M.; Huang, L.; Ruan, R.; Gao, W.H.; Li, C.J. Comparative transcriptomic analysis of male and female gonads of the Chinese longsnout catfish (*Leiocassis longirostris*). *J. Fish. Sci. China* **2024**, *31*, 129–143. [CrossRef]
13. Nelson, J.S.; Grande, T.C.; Wilson, M.V. *Fishes of the World*; John Wiley & Sons: Hoboken, NJ, USA, 2016.
14. Talwar, P.K.; Jhingran, A.G. *Inland Fishes of India and Adjacent Countries*; CRC Press: Boca Raton, FL, USA, 1991.
15. Anand, S.; Gedam, A.; Sonawane, S. Structures associated with feeding in *Mastacembelus armatus* (Lacepede, 1800) from Kaigaon Toka (MS). *J. Exp. Sci* **2012**, *3*, 17–20. [CrossRef]
16. Xue, L. Observation on the embryonic development of *Mastacembelue armatus*. *Freshw. Fish.* **2016**, *44*, 101–104. [CrossRef]
17. Xue, L.; Gao, Y.; Wu, M.; Tian, T.; Fan, H.; Huang, Y.; Huang, Z.; Li, D.; Xu, L. Telomere-to-telomere assembly of a fish Y chromosome reveals the origin of a young sex chromosome pair. *Genome Biol.* **2021**, *22*, 203. [CrossRef]
18. Zhou, H.Q.; Li, F.; Shu, H.; Zhong, D.M.; He, P.Y.; Huang, X.Q.; Chen, Z.K. Analysis on morphological indexes and discrimination of male and female *Mastacembelus armatus*. *J. Guangdong Ocean Univ.* **2019**, *39*, 1–6. [CrossRef]
19. Qin, W.; Han, C.; Yang, J.; Yu, Z.; Feng, Y.; Wu, Y.; Lu, B.; Cui, M.; Shu, H. Development of genetic sex markers of zig-zag eel (*Mastacembelus armatus*) by a NGS method. *Aquaculture* **2023**, *571*, 739498. [CrossRef]
20. Mortazavi, A.; Williams, B.A.; McCue, K.; Schaeffer, L.; Wold, B. Mapping and quantifying mammalian transcriptomes by RNA-Seq. *Nat. Methods* **2008**, *5*, 621–628. [CrossRef]
21. Chen, Y.; Chen, Y.; Shi, C.; Huang, Z.; Zhang, Y.; Li, S.; Li, Y.; Ye, J.; Yu, C.; Li, Z. SOAPnuke: A MapReduce acceleration-supported software for integrated quality control and preprocessing of high-throughput sequencing data. *Gigascience* **2018**, *7*, gix120. [CrossRef]
22. Niedziela, G.; Szabelska-Beręsewicz, A.; Zyprych-Walczak, J.; Graczyk, M. Application of edgeR and DESeq2 methods in plant experiments based on RNA-seq technology. *Biom. Lett.* **2022**, *59*, 127–139. [CrossRef]
23. Young, M.D.; Wakefield, M.J.; Smyth, G.K.; Oshlack, A. Gene ontology analysis for RNA-seq: Accounting for selection bias. *Genome Biol.* **2010**, *11*, R14. [CrossRef] [PubMed]
24. Guiguen, Y.; Fostier, A.; Herpin, A. *Sex Determination and Differentiation in Fish: Genetic, Genomic, and Endocrine Aspects*; John Wiley & Sons: Hoboken, NJ, USA, 2018.
25. Matson, C.K.; Zarkower, D. Sex and the singular DM domain: Insights into sexual regulation, evolution and plasticity. *Nat. Rev. Genet.* **2012**, *13*, 163–174. [CrossRef] [PubMed]
26. Lavery, R.; Chassot, A.-A.; Pauper, E.; Gregoire, E.P.; Klopfenstein, M.; de Rooij, D.G.; Mark, M.; Schedl, A.; Ghyselinck, N.B.; Chaboissier, M.-C. Testicular differentiation occurs in absence of R-spondin1 and Sox9 in mouse sex reversals. *PLoS Genet.* **2012**, *8*, e1003170. [CrossRef] [PubMed]
27. Zarkower, D.; Murphy, M.W. DMRT1: An ancient sexual regulator required for human gonadogenesis. *Sex. Dev.* **2022**, *16*, 112–125. [CrossRef]
28. Minkina, A.; Matson, C.K.; Lindeman, R.E.; Ghyselinck, N.B.; Bardwell, V.J.; Zarkower, D. DMRT1 protects male gonadal cells from retinoid-dependent sexual transdifferentiation. *Dev. Cell* **2014**, *29*, 511–520. [CrossRef]
29. Wagner, S.; Whiteley, S.L.; Castelli, M.; Patel, H.R.; Deveson, I.W.; Blackburn, J.; Holleley, C.E.; Marshall Graves, J.A.; Georges, A. Gene expression of male pathway genes *sox9* and *amh* during early sex differentiation in a reptile departs from the classical amniote model. *BMC Genom.* **2023**, *24*, 243. [CrossRef]
30. Vining, B.; Ming, Z.; Bagheri-Fam, S.; Harley, V. Diverse regulation but conserved function: SOX9 in vertebrate sex determination. *Genes* **2021**, *12*, 486. [CrossRef]
31. Zhou, Y.; Sun, W.; Cai, H.; Bao, H.; Zhang, Y.; Qian, G.; Ge, C. The role of anti-Müllerian hormone in testis differentiation reveals the significance of the TGF-β pathway in reptilian sex determination. *Genetics* **2019**, *213*, 1317–1327. [CrossRef]
32. Mustapha, U.F.; Peng, Y.-X.; Huang, Y.-Q.; Assan, D.; Zhi, F.; Shi, G.; Huang, Y.; Li, G.-L.; Jiang, D.-N. Comparative transcriptome analysis of the differentiating gonads in *Scatophagus argus*. *Front. Mar. Sci.* **2022**, *9*, 962534. [CrossRef]
33. Wang, M.; Chen, L.; Zhou, Z.; Xiao, J.; Chen, B.; Huang, P.; Li, C.; Xue, Y.; Liu, R.; Bai, Y. Comparative transcriptome analysis of early sexual differentiation in the male and female gonads of common carp (*Cyprinus carpio*). *Aquaculture* **2023**, *563*, 738984. [CrossRef]
34. Uhlenhaut, N.H.; Jakob, S.; Anlag, K.; Eisenberger, T.; Sekido, R.; Kress, J.; Treier, A.-C.; Klugmann, C.; Klasen, C.; Holter, N.I. Somatic sex reprogramming of adult ovaries to testes by FOXL2 ablation. *Cell* **2009**, *139*, 1130–1142. [CrossRef]
35. Georges, A.; Auguste, A.; Bessiere, L.; Vanet, A.; Todeschini, A.-L.; Veitia, R.A. FOXL2: A central transcription factor of the ovary. *J. Mol. Endocrinol.* **2014**, *52*, R17–R33. [CrossRef] [PubMed]
36. Liu, W.; Zhang, H.; Xiang, Y.; Jia, K.; Luo, M.; Yi, M. A novel germline and somatic cell expression of two sexual differentiation genes, Dmrt1 and Foxl2 in marbled goby (*Oxyeleotris marmorata*). *Aquaculture* **2020**, *516*, 734619. [CrossRef]

37. Zhou, L.; Charkraborty, T.; Yu, X.; Wu, L.; Liu, G.; Mohapatra, S.; Wang, D.; Nagahama, Y. R-spondins are involved in the ovarian differentiation in a teleost, medaka (*Oryzias latipes*). *BMC Dev. Biol.* **2012**, *12*, 36. [CrossRef] [PubMed]
38. Tomizuka, K.; Horikoshi, K.; Kitada, R.; Sugawara, Y.; Iba, Y.; Kojima, A.; Yoshitome, A.; Yamawaki, K.; Amagai, M.; Inoue, A. R-spondin1 plays an essential role in ovarian development through positively regulating Wnt-4 signaling. *Hum. Mol. Genet.* **2008**, *17*, 1278–1291. [CrossRef]
39. Nicol, B.; Estermann, M.A.; Yao, H.H.; Mellouk, N. Becoming female: Ovarian differentiation from an evolutionary perspective. *Front. Cell Dev. Biol.* **2022**, *10*, 944776. [CrossRef]
40. Kim, Y.; Kobayashi, A.; Sekido, R.; DiNapoli, L.; Brennan, J.; Chaboissier, M.-C.; Poulat, F.; Behringer, R.R.; Lovell Badge, R., Capel, B. Fgf9 and Wnt4 act as antagonistic signals to regulate mammalian sex determination. *PLoS Biol.* **2006**, *4*, e187. [CrossRef]
41. Kocer, A.; Pinheiro, I.; Pannetier, M.; Renault, L.; Parma, P.; Radi, O.; Kim, K.-A.; Camerino, G.; Pailhoux, E. R-spondin1 and FOXL2 act into two distinct cellular types during goat ovarian differentiation. *BMC Dev. Biol.* **2008**, *8*, 36. [CrossRef]
42. Cederroth, C.R.; Pitetti, J.L.; Papaioannou, M.D.; Nef, S. Genetic programs that regulate testicular and ovarian development. *Mol. Cell. Endocrinol.* **2007**, *265*, 3–9. [CrossRef]
43. Chen, W.; Liu, L.; Ge, W. Expression analysis of growth differentiation factor 9 (Gdf9/*gdf9*), anti-Müllerian hormone (Amh/*amh*) and aromatase (Cyp19a1a/*cyp19a1a*) during gonadal differentiation of the zebrafish, *Danio rerio*. *Biol. Reprod.* **2017**, *96*, 401–413. [CrossRef]
44. Halm, S.; Ibáñez, A.J.; Tyler, C.R.; Prat, F. Molecular characterisation of growth differentiation factor 9 (*gdf9*) and bone morphogenetic protein 15 (*bmp15*) and their patterns of gene expression during the ovarian reproductive cycle in the European sea bass. *Mol. Cell. Endocrinol.* **2008**, *291*, 95–103. [CrossRef] [PubMed]
45. Lu, C. *Molecular Regulation Mechanism of MKPs Toward MAPKs in the MAPK Pathway*; Tsinghua University: Beijing, China, 2018.
46. Huang, T.; Gu, W.; Liu, E.; Zhang, L.; Dong, F.; He, X.; Jiao, W.; Li, C.; Wang, B.; Xu, G. Screening and validation of p38 MAPK involved in ovarian development of *Brachymystax lenok*. *Front. Vet. Sci.* **2022**, *9*, 752521. [CrossRef] [PubMed]
47. Wang, Z.L.; Luo, Y.R.; Luo, Z.W.; Wang, J.; Xiao, Y.M. Expression of MAPK in the Gonadal Tissue of Different Reproductive Characteristics of the Hybrid Crucian Carp. *Life Sci. Res.* **2016**, *20*, 95–101.
48. Ten Dijke, P.; Goumans, M.J.; Itoh, F.; Itoh, S. Regulation of cell proliferation by Smad proteins. *J. Cell. Physiol.* **2002**, *191*, 1–16. [CrossRef]
49. Shi, Y.; Massagué, J. Mechanisms of TGF-β signaling from cell membrane to the nucleus. *Cell* **2003**, *113*, 685–700. [CrossRef]
50. Sower, S.A. Landmark discoveries in elucidating the origins of the hypothalamic-pituitary system from the perspective of a basal vertebrate, sea lamprey. *Gen. Comp. Endocrinol.* **2018**, *264*, 3–15. [CrossRef]
51. Stamatiades, G.A.; Carroll, R.S.; Kaiser, U.B. GnRH—A key regulator of FSH. *Endocrinology* **2019**, *160*, 57–67. [CrossRef]
52. Hu, K.L.; Chen, Z.M.; Li, X.X.; Cai, E.; Yang, H.Y.; Chen, Y.; Wang, C.Y.; Ju, L.P.; Deng, W.H.; Mu, L.S. Advances in clinical applications of kisspeptin-GnRH pathway in female reproduction. *Reprod. Biol. Endocrinol.* **2022**, *20*, 81. [CrossRef]

fishes

MDPI

Article

Molecular Characteristics, Expression Patterns, and Response of Insulin-like Growth Factors Gene Induced by Sex Steroid Hormones in Blotched Snakehead (*Channa maculata*)

Xiaotian Zhang [1,2], Yuxia Wu [1,2], Yang Zhang [1,3], Jin Zhang [1,2], Kunci Chen [1,2], Haiyang Liu [1], Qing Luo [1], Shuzhan Fei [1], Jian Zhao [1,2,*] and Mi Ou [1,3,*]

1 Key Laboratory of Tropical and Subtropical Fishery Resources Application and Cultivation, Ministry of Agriculture and Rural Affairs, Pearl River Fisheries Research Institute, Chinese Academy of Fishery Sciences, Guangzhou 510380, China; zxt13733320610@163.com (X.Z.); wyx07260726@163.com (Y.W.); jinspethap@icloud.com (J.Z.); hyliu@prfri.ac.cn (H.L.); luoqing@prfri.ac.cn (Q.L.); feisz@prfri.ac.cn (S.F.)
2 College of Fisheries and Life Science, Shanghai Ocean University, Shanghai 201306, China
3 School of Life Sciences, Hunan University of Science and Technology, Xiangtan 411201, China
* Correspondence: zhaojian@prfri.ac.cn (J.Z.); om1990@prfri.ac.cn (M.O.); Tel.: +86-020-816165099 (J.Z.); +86-020-81174522 (M.O.)

Abstract: Insulin-like growth factors (IGFs) play central roles in the growth and development of vertebrates. Blotched snakehead (*Channa maculata*), an economically significant fish, exhibits obvious sexual dimorphism and achieves sexual maturity in one year. However, the role of IGFs in *C. maculata* remains unknown. Three *IGF* genes were identified in *C. maculata*, designated as *CmIGF1-1*, *CmIGF1-2*, and *CmIGF2*. The cDNA sequences of these genes are 1184, 655, and 695 bp, encoding putative proteins of 168, 131, and 215 amino acids, respectively, and all three proteins contain a conserved IGF domain. Quantitative real-time PCR (qPCR) revealed the predominant expression of *CmIGFs* in the liver of adult fish, with higher expression levels observed in males. Notably, *CmIGF1-1*, *CmIGF1-2*, and *CmIGF2* displayed analogous expression profiles in the liver across various developmental stages, peaking at 365 days after hatching (dah). Subsequently, 600 individuals at 75 dah, at an early developmental stage, were randomly divided equally into six groups and reared in aerated 2 m × 2 m × 2 m cement ponds at 26.0 ± 1.0 °C. Following a one-week acclimatization period, fish without observed abnormalities were intraperitoneally injected with either 17α-ethynylestradiol (EE_2) or 17α-methyltestosterone (MT) at a dose of 10 μg/g body weight. Three groups underwent short-term hormone treatment, and the remaining three groups underwent long-term hormone treatment, which included five injections at two-week intervals over ten weeks. The analysis of *CmIGFs* expression levels in the liver under different hormone treatments revealed that EE_2 suppressed the expression of *CmIGF1-1* and *CmIGF1-2* while promoting *CmIGF2* expression. In females, MT up-regulated the expression of *CmIGF1-1* and *CmIGF2* in a time-dependent manner, but consistently inhibited *CmIGF2* expression. In males, MT promoted the expression of *CmIGFs* in a time-dependent manner, reaching peak levels for *CmIGF1-1*, *CmIGF1-2*, and *CmIGF2* after 8, 10, and 2 weeks of injection, respectively. Additionally, *CmIGF1* and *CmIGF2* might exhibit a complementary relationship, with a compensatory increase in *CmIGF2* expression in response to low *CmIGF1* concentration. These findings highlight the potential key role of IGFs upon growth and their regulation by sex steroid hormones in *C. maculata*, providing a crucial foundation for future research aimed at elucidating the molecular mechanisms underlying the growth dimorphism between female and male blotched snakeheads.

Keywords: IGFs; gene cloning; growth dimorphism; expression analysis; hormone induction

Key Contribution: This study shows, for the first time, the presence of three insulin-like growth factor genes (*CmIGF1-1*, *CmIGF1-2*, and *CmIGF2*) in the economically significant blotched snakehead (*Channa maculata*), and elucidates their predominant expression in liver, with higher levels observed in males. Furthermore, it demonstrates the regulatory impact of sex steroid hormones (EE_2 and MT) on the expression of *CmIGFs*. These findings highlight the potential key role of IGFs in growth and their regulation by sex steroid hormones in blotched snakehead (*C. maculata*), providing a foundation for future research aimed at elucidating the molecular mechanisms underlying the growth dimorphism between female and male blotched snakeheads.

1. Introduction

The growth hormone/insulin-like growth factor (GH/IGF) axis plays a vital role in vertebrate growth, encompassing key molecules such as growth hormone (GH), growth hormone receptor (GHR), insulin-like growth factor (IGF), and their associated binding proteins and receptors, including IGF-binding proteins (IGFBPs) and IGF receptors (IGF1R and IGF2R) [1]. GH secretion is regulated by the anterior pituitary gland, which responds to various hormones released from the hypothalamus, notably growth hormone-releasing hormone (GHRH) and growth hormone-releasing peptide (GHRP). GH travels to the liver through the bloodstream, where it rapidly binds to GHR, initiating a cascade of cellular signaling events that culminate in the secretion of IGFs [2]. Subsequently, IGFs travel through body fluids to various tissues, promoting cell growth and differentiation. GH, GHR, and IGFs orchestrate biological growth and development through close coordination during the different development stages of an organism's lifespan [2,3].

IGFs have garnered considerable attention across various species because of their vital role in growth regulation. IGFs are highly conserved and primarily include two types, IGF-1 and IGF-2 [4]. In mammals, IGF-1 primarily governs postnatal growth and development, while IGF-2 plays a crucial role during embryonic development, influencing processes such as cell proliferation, differentiation, migration, and apoptosis [5]. The *IGF-1* gene in fish was first cloned from Coho salmon (*Oncorhynchus kisutch*) [6] and subsequently identified in other fish species, such as gopher rockfish (*Sebastes carnatus*) [7], Nile tilapia (*Oreochromis niloticus*) [8], zebrafish (*Danio rerio*) [9], tongue sole (*Cynoglossus semilaevis*) [10], yellow catfish (*Pelteobagrus fulvidraco*) [11], and spotted scat (*Scatophagus argus*) [12]. *IGF-1* exhibits widespread expression in various tissues, particularly expressed in the liver of gopher rockfish (*S. carnatus*) [7], Nile tilapia (*O. niloticus*) [8], yellow catfish (*P. fulvidraco*) [11], and spotted scat (*S. argus*) [12]. Additionally, in Nile tilapia (*O. niloticus*) and yellow catfish (*P. fulvidraco*), males exhibit higher *IGF-1* expression levels than females. On the other hand, *IGF-2* in fish was initially cloned from rainbow trout (*Oncorhynchus mykiss*) [13] and subsequently identified in other fish species, such as zebrafish (*D. rerio*) [9], medaka (*Oryzias latipes*) [14], and spotted scat (*S. argus*) [12]. Studies have reported the expression patterns of *IGF-2* during embryonic development in Nile tilapia (*O. niloticus*) [8], zebrafish (*D. rerio*) [9], medaka (*O. mykiss*) [14], and spotted scat (*S. argus*) [12]. In zebrafish (*D. rerio*), the expression of *IGF-2* initiates at the gastrula stage and exhibits a gradual increase throughout embryonic development [9], and the disruption of *IGF-2* results in embryonic lethality [15]. The expression profile of *IGF-2* mRNA closely mirrors that of *IGF-1*, expressed in diverse tissues, and particularly pronounced in the liver. Furthermore, *IGF-1* or *IGF-2* exhibit different subtypes in a few fish species, including *IGF-1a* and *IGF-1b* in giant grouper (*Epinephelus lanceolatus*) [16] and *IGF-2a* and *IGF-2b* in grass carp (*Ctenopharyngodon idellus*) [17].

The synthesis and secretion of sex steroid hormones by the gonads are pivotal in regulating the hypothalamus–pituitary–liver axis in vertebrates [18]. Cumulative evidence indicates that sex steroid hormones can impact the growth and development of vertebrates by modulating the synthesis and secretion of IGFs [18–20]. In rainbow trout (*O. mykiss*) [21], testosterone (T) induction has been shown to stimulate the mRNA expression of *IGF1* and *IGF2*, while treatment with 17α-estradiol (E_2) decreases *IGF1* expression. In Nile tilapia (*O. niloticus*), the long-term injection of E_2 in females dramatically increases *IGF-1* expression levels in the liver [22], and MT strongly enhances *IGF-1* expression in males, albeit without significant effects on females [23]. These findings suggest that sex steroid hormones can regulate *IGFs* expression in fishes, with potential variation among species [24,25]. However, the underlying mechanisms remain unclear and warrant further investigation.

Blotched snakehead (*Channa maculata*), a member of the Channidae family, is widely farmed in China for its palatable taste, high protein content, and pharmacological properties [26]. Male snakeheads exhibit faster growth rates and bigger sizes than females; given the demand for large individuals in the market, it is imperative to understand the molecular mechanism of growth disparities between male and female individuals, which will provide a theoretical basis for selective breeding in production practices [27]. Previous studies have demonstrated the crucial role of the *GH* gene in the growth of blotched snakehead (*C. maculata*) [28]. This research aims to illustrate the function of the IGFs in the GH/IGF axis and explore their expression patterns during ontogeny and in adult tissues. Moreover, the effects of exogenous hormone induction via EE_2 and MT on the expression levels of *IGFs* in both male and female individuals were investigated. This work endeavors to provide insight into the basic function of IGFs and offers novel perspectives for exploring the molecular mechanisms underlying the growth disparities between male and female blotched snakehead.

2. Materials and Methods

2.1. Fish and Sampling

Blotched snakeheads in this study were reared in the fish laboratory of the Model Animal Research Center, Pearl River Fisheries Research Institute (Guangzhou City, Guandong Province, China). The fish were cultivated in an open system, with water quality parameters meeting aquaculture standards. Specifically, dissolved oxygen levels exceeded 6–8 mg/L, ammonia nitrogen remained below 0.5 mg/L, nitrite levels were less than 0.01 mg/L, and pH values were 7.0–7.5. Water exchange occurred every three days at a 20% exchange rate. To mimic natural environmental conditions, the lighting followed a 12 h day and night cycle. The fish were fed with a commercial diet provided by Nanhai Bairong Improved Aquatic Seed Co., Ltd. (FoShan City, Guangdong Province, China), comprising 45% protein, 5% fat, and 27% carbohydrates, ensuring the balanced nutritional intake that is suitable for snakehead growth requirements. The feeding ratio was adjusted according to the growth stage and size of the fish, following the manufacturer's recommendations. One-year-old male and female fish were randomly selected, respectively, and a small piece of fin tissue was sampled for the subsequent genomic DNA (gDNA) extraction and the cloning of gDNA sequences of *IGF* genes. Additionally, twelve tissue samples (gill (G), liver (L), spleen (S), intestines (I), middle kidney (MK), muscle (M), head kidney (HK), gonad [ovary (O)/testis (T)], heart (H), pituitary (P), hypothalamus (HY) and brain (B)) were collected from the blotched snakeheads (n = 3). Tissues were collected under sterile conditions, with each tissue individually removed using sterile instruments to prevent cross-contamination. The collected tissues were immediately snap-frozen in liquid nitrogen and stored at −80 °C to maintain RNA integrity for subsequent RNA extraction, cDNA cloning, and the assessment of *IGF* distribution patterns.

To analyze *IGF* expression levels in the liver, samples were collected from male and female blotched snakeheads (n = 5) at various developmental stages (45, 75, 105, 135, 165, 195, and 365 dah) after anesthetization. Genetic sex was identified by sex-specific molecular markers, as previously described [29]. All fish experiments were conducted following the regulations outlined in the National Institutes of Health guide for the care and use of laboratory animals (https://olaw.nih.gov/resources/publications/guide-care-2011.htm, accessed on 1 January 2024).

2.2. Hormone Treatment

Healthy individuals at 75 dah (n = 600; body length: 15.1 ± 0.4 cm; body weight: 67. 9 ± 4.3 g) were randomly divided into six groups, each comprising 100 individuals. These fish were reared in 2 m × 2 m × 2 m aerated cement ponds at 26.0 ± 1.0 °C. After a one-week acclimatization period, the fish with no abnormal clinical signs were subjected to further studies. Sex steroid hormones, EE_2 and MT (Aladdin, Shanghai, China) were dissolved in corn oil prior to intraperitoneal injection. Three groups underwent short-term hormone treatment: the first group received an intraperitoneal injection of MT (10 μg/g body weight); the second group was intraperitoneally injected with EE_2 (10 μg/g body weight); and the third group served as the control, receiving an intraperitoneal injection of corn oil [26]. Liver samples were collected from male and female individuals ($n = 5$) at 24, 48, 72, 96, 120, and 144 h post-injection. Genetic sex was determined using the same method as previously described [29]. The remaining three groups underwent a long-term hormone treatment, which consisted of five injections at two-week intervals over a ten-week period. At each injection, the fourth group received an intraperitoneal injection of MT (10 μg/g body weight), the fifth group received EE_2 (10 μg/g body weight) intraperitoneally, and the sixth group served as the control, receiving an intraperitoneal injection of corn oil. Liver samples were collected from male and female individuals ($n = 5$) at 2, 4, 6, 8, and 10 weeks post-injection, and genetic sex identification was performed as previously described [29]. Subsequently, all samples were immediately frozen in liquid nitrogen and stored at −80 °C before RNA extraction.

2.3. RNA Extraction and cDNA Synthesis

The total RNA was extracted from tissue samples utilizing TRIzol reagent (Invitrogen, Carlsbad, CA, USA) following the manufacturer's protocol. The quality and quantity of RNA were assessed using a spectrophotometer (ThermoFisher, Waltham, MA, USA), and RNA integrity was detected by electrophoresis using a 1.0% agarose gel. Subsequently, total RNA extracted from tissue samples of different individuals under the same conditions was pooled in equal amounts for cDNA synthesis. The synthesis of the first-strand cDNA was performed utilizing the pooled RNA, a random hexamer primer (Takara, Osaka, Japan), and M-MLV Reverse Transcriptase (Promega, Fitchburg, WI, USA). To obtain both 5′- and 3′-RACE Ready cDNA, the SMART™ RACE cDNA Amplification Kit (Takara, Japan) was employed. Finally, cDNA template for qPCR was synthesized using the ReverTra Ace qPCR RT Kit (Toyobo, Tokyo, Japan) with the pooled RNA as the template.

2.4. Full-Length cDNA Cloning and Sequence Analysis for IGFs in Blotched Snakehead (C. maculata)

Three *IGF* genes were identified in the blotched snakehead genome (SRA Accession No. PRJNA730430) [30]. Primers were specifically designed to verify the predicted cDNA sequences. To obtain the 5′ and 3′ untranslated regions (UTRs), specific and adapter primers were designed using 5′-RACE- and 3′-RACE-ready cDNA as templates, with primers within the UTRs designed accordingly (Table 1). PCR amplification was performed by 2 × Super Pfx MasterMix (Cwbio, Taizhou, China). The amplified products of the expected size were purified using a Gel Extraction Kit (Omega, Norwalk, CT, USA). The purified products were cloned into a pMD-19T vector (Takara, Japan) and transformed into competent *Escherichia coli* DH5α cells (Takara, Japan). Positive colonies including the target fragments were sequenced by a commercial company (Tianyihuiyuan, Guangzhou City, Guangdong Province, China).

Sequence analysis of *CmIGFs* was conducted using the Sequence Manipulation Suite (STS) (http://www.bio-soft.net/sms/, accessed on 10 January 2024). Domain features of the IGF proteins were predicted utilizing the Simple Modular Architecture Research Tool (SMART) (http://smart.embl-heidelberg.de, accessed on 15 January 2024). Physicochemical properties of the IGF proteins were predicted by ExPASy-Protparam Tool (https://web.expasy.org/protparam/, accessed on 15 January 2024). IGF protein sequences from other species were retrieved from the NCBI database (http://www.ncbi.nlm.nih.gov/, accessed on 15 January 2024). Multiple amino acid sequence alignments were performed using ClustalX 2.1 (http://www.ebi.ac.uk/tools/clustalx2.1, accessed on 15 January 2024), and a phylogenetic tree was constructed using Mega 5.0 software (http:

//www.megasoftware.net/index.html, accessed on 15 January 2024) with the neighbor-joining method and bootstrap resampling (1000 replicates) [31].

Table 1. Primers used for *IGF* genes cloning and qPCR in blotched snakehead (*C. maculata*).

Primer Name	Sequence (5′~3′)	Application
IGF1-1-F1	CTCCTGTAGCCACACCCTC	Partial sequence obtaining
IGF1-1-R1	GAATGACTGTGTCCAGGTAAAG	
IGF1-2-F1	ATGGGCTGTATCTCCTGTAGTC	
IGF1-2-R1	AGTATTCTCGGCAAGTCGGT	
IGF2-F1	AGCAAAGATACGGACAGCAC	
IGF2-R1	GTTGACATAGTTATCCGTGGC	
IGF1-1-5′R-out	AAGCCTCTTTCTCCACACACAAACTGC	5′-RACE PCR amplification
IGF1-1-5′R-in	GCAGTGAGAGGGTGTGGCTACAGGAG	
IGF1-2-5′R-out	AGCGTGTGGGTTTACT	
IGF1-2-5′R-in	TGTCGACCAGCTCCACCC	
IGF1-1-3′F-out	CCTGCCAAGACTAACAAGCCAACTCG	3′-RACE PCR amplification
IGF1-1-3′F-in	GAGAACAACTAAGAGACCTTTACCTGGACACA	
IGF1-1-F2	CCTGTTCGCTAAATCTCACTTCTC	ORF identification
IGF1-1-R2	CATTTGTCCATTCGCTCCATC	
IGF1-2-F2	GGACTACAAGAGAGACGG	
IGF1-2-R2	TTTGTCCCTTCGCTCCAT	
IGF2-F2	AGCCAAATAACCCCCAACA	
IGF2-R2	AGCGGGCTCATTTGTGG	
IGF1-1-DL-F	CGCTCTTTCCTCTCAGTGGC	qPCR amplification
IGF1-1-DL-R	CCATAGCCTGTGGGTTTACTGA	
IGF1-2-DL-F	GTTTGTGTGTGGAGACAGAGGC	
IGF1-2-DL-R	GCACGCACAGAGTGAGTTGG	
IGF2-DL-F	GTCTTCGTCCAGTCGTTCGC	
IGF2-DL-R	TGTTGCCCCTGCTGGTTG	
β-actin-F	GCAAGCAGGAGTATGATGAG	
β-actin-R	TTGGGATTGTTTCAGTCAGT	
EF1α-F	GGGAGACCCACAATAACATCG	
EF1α-R	CCAGGCATACTTGAAGGAGC	

2.5. Cloning the Genomic Sequence of IGFs Genes in Blotched Snakehead (C. maculata)

The gDNA was extracted from fin tissues using the General AllGen Kit (CWBio, Taizhou, China) following the manufacturer's protocol. Specific primers (Table 2) were designed according to the predicted genome sequences of *IGFs* obtained from the blotched snakehead genome [30]. Subsequently, the genomic sequences of *CmIGFs* were amplified using these primers and the fin gDNA as the template. The PCR products were sequenced to obtain the genomic sequences of *CmIGFs*, which were then compared with the cDNA sequences to identify exons and introns based on the GT/AG principle.

Table 2. Primers used for genomic sequence amplification of blotched snakehead (*C. maculata*) *IGFs* gene.

Primer Name	Sequence (5′~3′)	Length (bp)
IGF1-1-gDNA-F1	TTTATGATTGGGTCACAGCA	1674
IGF1-1-gDNA-R1	GGACTCAGCAGGAATTACTCT	
IGF1-1-gDNA-F2	GTTACTTACTGGCAGGTTTT	1827
IGF1-1-gDNA-R2	TGTTTGGGTTCTACTCAATT	
IGF1-1-gDNA-F3	GGCGGCAAATTAGAGTTGTG	1848
IGF1-1-gDNA-R3	ATGGACGAACTGAGGTTACAAG	

Table 2. *Cont.*

Primer Name	Sequence (5′~3′)	Length (bp)
IGF1-1-gDNA-F4	ACAAACGCTGTGAAGTGGTC	1764
IGF1-1-gDNA-R4	CAGGGAGCTACTTAATGCTTA	
IGF1-1-gDNA-F5	GGGTGATTTCACTGGGATGT	1962
IGF1-1-gDNA-R5	AACCTGTGGATTCTTGGAGC	
IGF1-1-gDNA-F6	TAGTCCCTGCCCAGCCGTAA	1997
IGF1-1-gDNA-R6	AGAAACAAAGCATAGGTGAA	
IGF1-1-gDNA-F7	TCGATTCCCCTGTCCCCTAA	2008
IGF1-1-gDNA-R7	GCCTGCGTTTCGACTTCACG	
IGF1-1-gDNA-F8	GATGTTTTAGGCAGCGTCTG	1588
IGF1-1-gDNA-R8	AACCGTGTTTTACTCTTTTAG	
IGF1-2-gDNA-F1	TGCCTTTGTAGTTTACCTTT	1829
IGF1-2-gDNA-R1	TAATTTGTCCCCTTTATTCG	
IGF1-2-gDNA-F2	ATGCTAGGACTGAAATGCTA	1727
IGF1-2-gDNA-R2	TAGATGATAAATAACGGGTA	
IGF1-2-gDNA-F3	GTCAGTGCTGTTCTTTCCAA	1155
IGF1-2-gDNA-R3	AAAAGGGGCTGTGCCTTGTT	
IGF1-2-gDNA-F4	AAGTGAAGCATTTCAAACTT	1546
IGF1-2-gDNA-R4	TAACTGGCAGAAGATGACTA	
IGF2-gDNA-F1	GGAGGAGCCGATGGGTGGTGG	1677
IGF2-gDNA-R1	AGCGGCCCATTGTCAGTCCG	
IGF2-gDNA-F2	GTCTCAAGACTTCGTCCAGG	1623
IGF2-gDNA-R2	GTCTCAAGACTTCGTCCAGG	
IGF2-gDNA-F3	CCACTATGGGAAACAATGCC	1742
IGF2-gDNA-R3	AGGACTGCCACAGAAATCAC	

2.6. Quantitative Real-Time PCR (qPCR)

Gene-specific primers were designed for quantifying the expression of *CmIGFs* using qPCR. *β-actin* and *EF1α* genes were selected as reference genes according to our previous study [26,32]. qPCR was performed utilizing the StepOnePlus™ Real-Time PCR System (ABI, Los Angeles, CA, USA) by SYBR® Green Realtime PCR Master Mix (Toyobo, Osaka, Japan), and each sample was subjected to three replications. For normalization, the geometric mean of the Ct values of both *β-actin* and *EF1α* was calculated for each sample. The expression levels of *CmIGFs* in adult tissues and developmental stages were calculated by the $2^{-\Delta\Delta Ct}$ method [33]. The expression levels of *CmIGFs* in female gills were used as the baseline (1.0) for tissue expression pattern analysis, and the expression levels of *CmIGFs* in the liver of 45 dah males were used as the baseline (1.0) for the developmental expression pattern analysis. The expression levels of *CmIGFs* after hormone treatment were calculated using the $2^{\Delta Ct}$ method [26].

2.7. Statistical Analysis

The experimental data were displayed as mean (n = 3) ± standard error of the mean (S.E.M). Two-way ANOVA was conducted in SPSS (version 22.0; SPSS, Chicago, IL, USA). When significant differences were observed among factors, one-way ANOVA was conducted, followed by Tukey's test. $p < 0.05$ was considered to indicate statistical significance.

3. Results

3.1. Characterization of CmIGFs

After PCR amplification and sequencing, a 1184-bp *CmIGF1-1* cDNA sequence (GenBank No. MW715800) was obtained. It included a 255-bp 5′-UTR, a 507-bp open reading frame (ORF) encoding 168 amino acids (aa), and a 452-bp 3′-UTR with RNA instability motifs (ATTTA), and a poly(A) tail. The analysis of the putative protein revealed the presence of a signal peptide (aa 1–43) and an IGF domain (aa 56–112) (Figure S1a). The genomic sequence of *CmIGF1-1*, obtained through amplification, sequencing, and alignment, spanned 13,504 bp. It consisted of five exons with lengths of 48, 187, 176, 36 and 60 bp, and four introns measuring 1301, 8800, 250 and 789 bp, respectively (Figure 1). It was in line with the GT/AG rule.

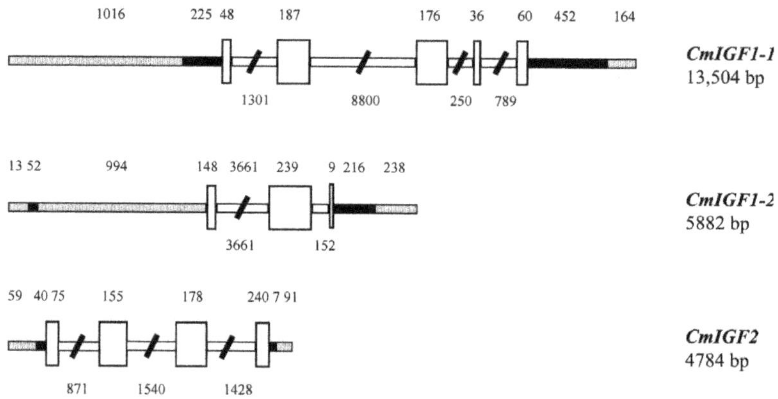

Figure 1. Genomic structure of *CmIGFs*. The shadow represents the flanking region, the black horizontal frame represents the untranslated region, the white vertical frame represents the exon, the white horizontal frame represents the intron, and the numbers above and below each schematic represent the length, respectively.

The homologous *CmIGF1-2* was also determined to be 675 bp in length (GenBank No. MW715801), comprising a 58-bp 5′-UTR, a 396-bp ORF encoding 131 aa, and a 211-bp 3′-UTR with a typical AATAA and a poly (A) tail. The bioinformatics analysis of the putative protein indicated the presence of a signal peptide (aa 1–26) and an IGF domain (aa 27–83) (Figure S1b). *Cm*IGF1-2 shared an amino acid sequence homology of 83.29% with *Cm*IGF1-1. After sequence identification, a genomic sequence of 5882 bp in length was obtained, comprising three exons (148, 239, and 9 bp) and two introns (3661 and 152 bp) following the GT/AG consensus rule (Figure 1).

The sequence of *CmIGF2* was determined to be 695 bp in length (GenBank No. MW715802) with an ORF of 648 bp encoding 215 aa, a 5′-UTR of 40 bp, and a 3′-UTR of 7 bp. The putative protein was determined to have an IGF domain (aa 53–110) and an IGF2_C domain (aa 147–202) (Figure S1c). When comparing the amino acid sequences of *Cm*IGF2 and *Cm*IGF1-1 with *Cm*IGF1-2, the similarities were 50.40% and 54.08%, respectively. The genomic sequence of *CmIGF2* was 4784 bp long, comprising four exons (75, 155, 178, and 240 bp) and three introns that were 871, 1540, and 1428 bp in size, respectively (Figure 1). All introns began with GT and ended with AG, which was consistent with the splicing rules between exons and introns in eukaryotes.

3.2. Multiple Alignments and Phylogenetic Analysis

*Cm*IGF1-1 exhibited the highest similarity to *Po*IGF1a (IGF1a in olive flounder (*Paralichthys olivaceus*)) (96.97%), followed by *Po*IGF1b (IGF1b in *P. olivaceus*) (91.82%) and *Sa*IGF1 (IGF1 in silthed sea bream (*Sparus aurata*)) (87.50%). Conversely, the lowest similarity was observed between *Cm*IGF1-1 and *Hs*IGF1b (IGF1b in human (*Homo sapiens*)) (50.05%) (Figure S2a). The protein sequence similarities between *Cm*IGF1-2 and other species varied, ranging from its highest homology with *Po*IGF1a (84.85%) to its lowest homology with *Hs*IGF1a (52.67%) (IGF1a in *H. sapiens*), as shown in Figure S2b. Further analysis indicated that the sequence similarity between *Cm*IGF1-1 and *Cm*IGF1-2 was only 73.28%. *Cm*IGF2 showed the highest identity with *Ca*IGF2 (IGF2 in *C. argus*) (98.68%) and *At*IGF2 (IGF2 in climbing bass (*Anabas testudineus*)) (90.73%), while it was the least similar to *Xl*IGF2a (IGF2a in African clawed frog (*Xenopus laevis*)) (44.98%) (Figure S2c). These results were corroborated by phylogenetic analysis (Figure 2). The phylogenetic tree suggested that these homologous proteins can be classified into two clades: IGF1 and IGF2. Each main clade consisted of four subclades, encompassing fishes, amphibians, birds, and mammals. Interestingly, the IGF1 and IGF2 proteins of fish appeared distinctly separated from those of other vertebrates. The phylogenetic tree reflected the consistent genetic relationships and evolution among species.

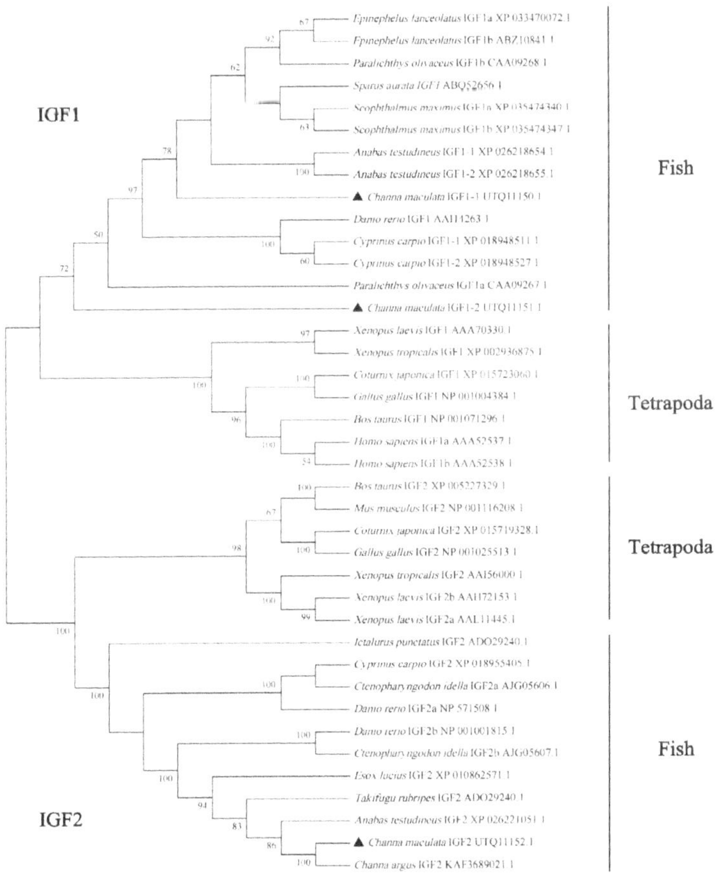

Figure 2. Phylogenetic relationship between the IGFs proteins in different species. A neighbor-joining phylogenetic tree was constructed using MEGA 5.0 software. The bootstrap values of the branches were obtained by testing the tree 1000 times and values were over 50% percent marked. The GenBank accession numbers of IGF proteins are given after the species names in the tree. The triangle represents IGF in blotched snakehead (*C. maculata*).

3.3. Tissue Distribution of CmIGFs

The expression levels of *CmIGFs* in male and female blotched snakeheads were assessed through qPCR. As depicted in Figure 3, similar expression patterns of *CmIGF1-1* (Figure 3a) and *CmIGF1-2* (Figure 3b) were observed, primarily expressed in liver (L), with low levels in spleen (S), and undetectable in other tissues. Interestingly, the expression levels of *CmIGF1-1* and *CmIGF1-2* in the liver (L) exhibited significant sexual dimorphism between males and females ($p < 0.01$). *CmIGF2* transcripts were strongly expressed in the liver (L), with higher expression in males compared to females ($p < 0.01$). Additionally, moderate expression levels were detected in spleen (S), intestines (I), and heart (H), with significant sex differences ($p < 0.01$), while low levels were observed in other tissues (Figure 3c).

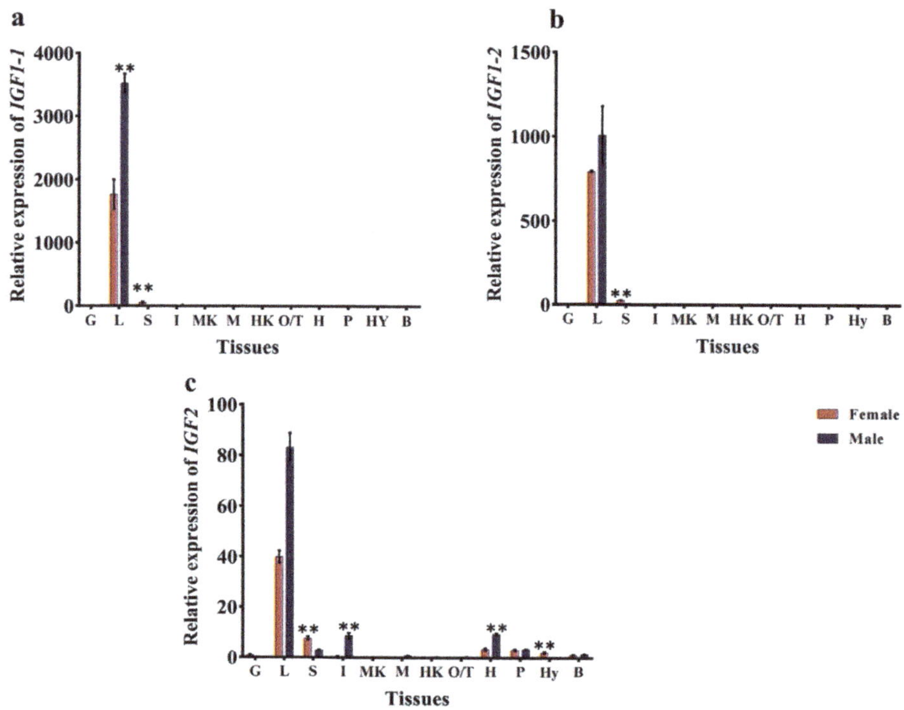

Figure 3. Gene expression patterns of *CmIGFs* in twelve tissues of males and females. (**a**) *CmIGF1-1*; (**b**) *CmIGF1-2*; (**c**) *CmIGF2*. The tissues are designated as follows: G (gills), L (liver), S (spleen), I (intestine), MK (middle kidney), M (muscle), HK (head kidney), O/T (ovary/testis), H (heart), P (pituitary), HY (hypothalamus), B (brain). *β-actin* and *EF1α* were the internal controls. Results represent the mean ± S.E.M ($n = 3$), and the relative expression levels in different tissues were the ratio of the expression in the female gills (G). Asterisks represent significant differences between males and females. ** $p < 0.01$.

3.4. *CmIGFs* Expression Patterns at Different Developmental Stages

In order to investigate the role of *CmIGFs* in growth, the expression levels of *CmIGFs* in liver (L) were analyzed across seven different developmental stages. Males reached body lengths of 9.6 ± 0.8 cm at 45 dah, 14.3 ± 0.8 cm at 75 dah, 21.3 ± 0.8 cm at 105 dah, 27.1 ± 1.2 cm at 135 dah, 29.1 ± 1.9 cm at 165 dah, 31.1 ± 1.3cm at 195 dah, and 44.7 ± 1.6 cm at 365 dah, and females reached body lengths of 9.7 ± 0.4 cm at 45 dah, 14.3 ± 0.7 cm at 75 dah, 20.5 ± 1.1 cm at 105 dah, 24.9 ± 1.5 cm at 135 dah, 28.0 ± 1.5 cm at 165 dah, 27.5 ± 2.8 at 195 dah, and 40.1 ± 1.3 cm at 365 dah. As shown in Figure 4, *CmIGFs* expression was higher in males than that in females across different developmental stages. In females, *CmIGF1-1* maintained low expression levels from 45 to 195 dah and peaked at 365 dah ($p < 0.01$). For male individuals, the transcription of *CmIGF1-1* started increasing from 75 to 105 dah and reached a tiny peak at 105 dah ($p < 0.01$). Subsequently, the expression levels decreased significantly from 105 dah to 195 dah before reaching the highest level at 365 dah ($p < 0.01$) (Figure 4a). The expression patterns of *CmIGF1-2* were similar to those of *CmIGF1-1* at developmental stages, peaking at 365 dah in both male and female individuals. From 45 to 195 dah, *CmIGF1-2* expression maintained a low level in females, while exhibiting a parabolic trend and peaking at 105 dah in males ($p < 0.01$) (Figure 4b). As for *CmIGF2*, the expression levels showed a considerable decline from the initial sample in both females and males, reaching the lowest levels at 135 dah. Then, they were gradually up-regulated and peaked at 365 dah ($p < 0.01$) (Figure 4c).

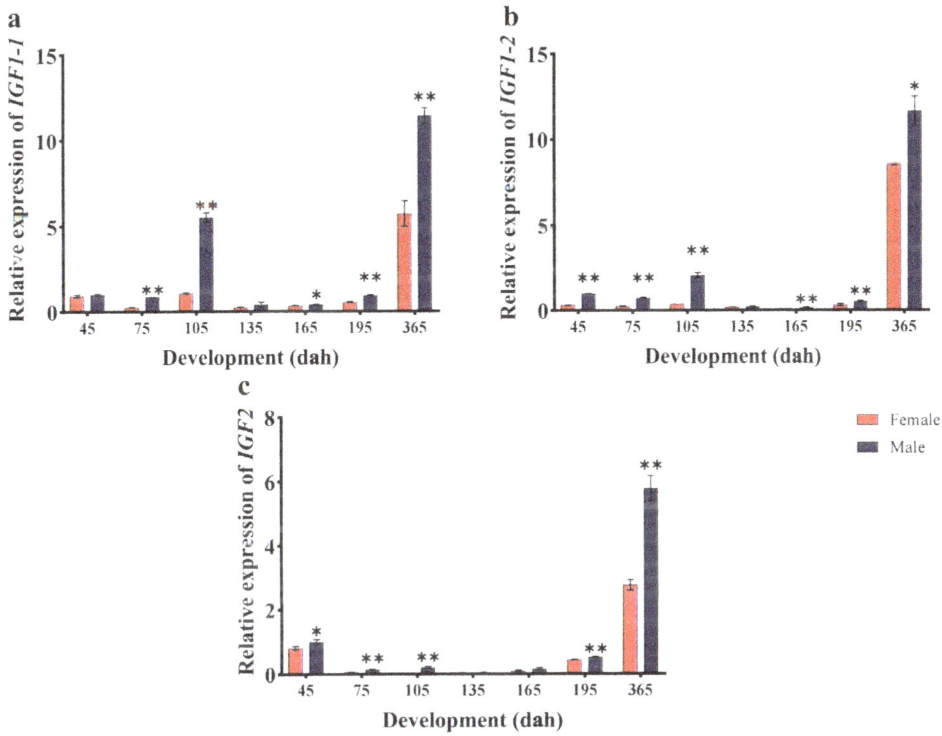

Figure 4. Gene expression profiles of *CmIGFs* in the liver of male and female individuals at different development stages. (**a**) *CmIGF1-1*; (**b**) *CmIGF1-2*; (**c**) *CmIGF2*. *β-actin* and *EF1α* were the internal controls. Results represent the mean ± S.E.M (n = 3), and *CmIGFs* transcription in the liver of 45 dah female served as the baseline (1.0). Asterisks represent significant differences between males and females. * $p < 0.05$, and ** $p < 0.01$.

3.5. Effects of Short-Term Hormone Treatment on CmIGFs in Females

In female blotched snakehead, the expression of *CmIGF1-1* in the control group fluctuated from 24 h to 144 h and peaked at 48 h and 120 h ($F_{5,10} = 84.254$, $p < 0.01$), respectively. Upon the administration of EE_2, *CmIGF1-1* transcription was inhibited, remaining lower than that in the control group ($F_{2,10} = 35.081$, $p < 0.01$) until 144 h, when the expression levels of *CmIGF1-1* were comparable between the control and the EE_2 treatment group (Figure 5a). After MT treatment, the highest peak of *CmIGF1-1* expression occurred at 48 h, while it remained significantly lower than that in the control group ($F_{2,10} = 35.081$, $p < 0.01$) (Figure 5a). Both EE_2 and MT administration down-regulated the expression of *CmIGF1-2* ($F_{2,10} = 48.841$, $p < 0.01$), which was lower than that in the control group from 48 h to 120 h (Figure 5b). A marked increase was observed in *CmIGF1-2* expression at 144 h when it was significantly higher ($F_{5,10} = 60.233$, $p < 0.01$) than that in the control group (Figure 5b). *CmIGF2* transcripts in EE_2-treated group were consistently lower ($F_{2,10} = 61.829$, $p < 0.01$) than those in the control group until 120 h, when *CmIGF2* displayed equivalent levels in the control and the EE_2 treatment group. After treatment with MT, *CmIGF2* showed higher levels than those in the control group from 24 h to 48 h ($F_{2,10} = 61.829$, $p < 0.01$), reaching the highest level at 120 h, after which it sharply dropped to the lowest level at 144 h ($F_{5,10} = 30.592$, $p < 0.01$) (Figure 5c).

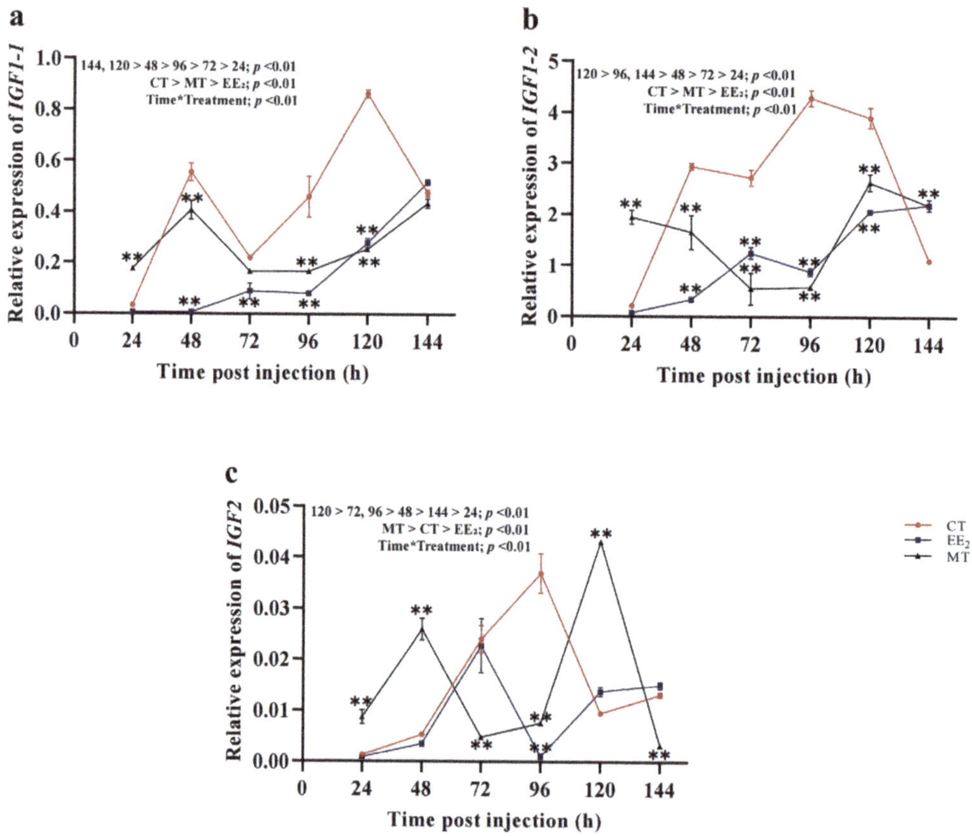

Figure 5. Effects of short-term hormone treatment on *CmIGFs* in female blotched snakehead. (**a**) *CmIGF1-1*; (**b**) *CmIGF1-2*; (**c**) *CmIGF2*. *β-actin* and *EF1α* were used as the internal controls for qPCR. Data are presented as mean ± S.E.M (n = 3). Asterisks represent significant differences between treatments and control at the corresponding time point for each dataset. ** $p < 0.01$.

3.6. Effects of Short-Term Hormone Treatment on CmIGFs in Males

In male individuals, *CmIGF1-1* transcripts showed a similar expression pattern, with the parabolic trends in both the control and EE_2-treated groups, gradually increasing and reaching their peak at 120 h, then sharply dropping to the initial level at 144 h ($F_{5,10}$ = 32.762, $p < 0.01$) (Figure 6a). However, it showed fluctuating variation in the MT-treated group ($F_{2,10}$ = 54.339, $p < 0.01$) (Figure 6a). In the control group, *CmIGF1-2* expression reached the highest value at 72 h and then gradually decreased and returned to the starting level at 144 h ($F_{5,10}$ = 54.207, $p < 0.01$). *CmIGF1-2* expression displayed fluctuating changes in the EE_2-treated and MT-treated groups ($F_{2,10}$ = 33.880, $p < 0.01$), with the expression level at 144 h being significantly higher than that in the control group (Figure 6b). *CmIGF2* expression exhibited a similar dynamic pattern of change in both the control and EE_2-treated groups, with *CmIGF2* being sharply down-regulated in the control group from 120 h, while showing a slight up-regulation ($F_{2,10}$ = 55.904, $p < 0.01$) in the EE_2-treated group. After MT treatment, *CmIGF2* transcription gradually decreased and reached the lowest level at 96 h and then quickly rose to the highest level at 144 h, which was significantly higher ($F_{5,10}$ =92.615, $p < 0.01$) than that in the control group ($p < 0.01$) (Figure 6c).

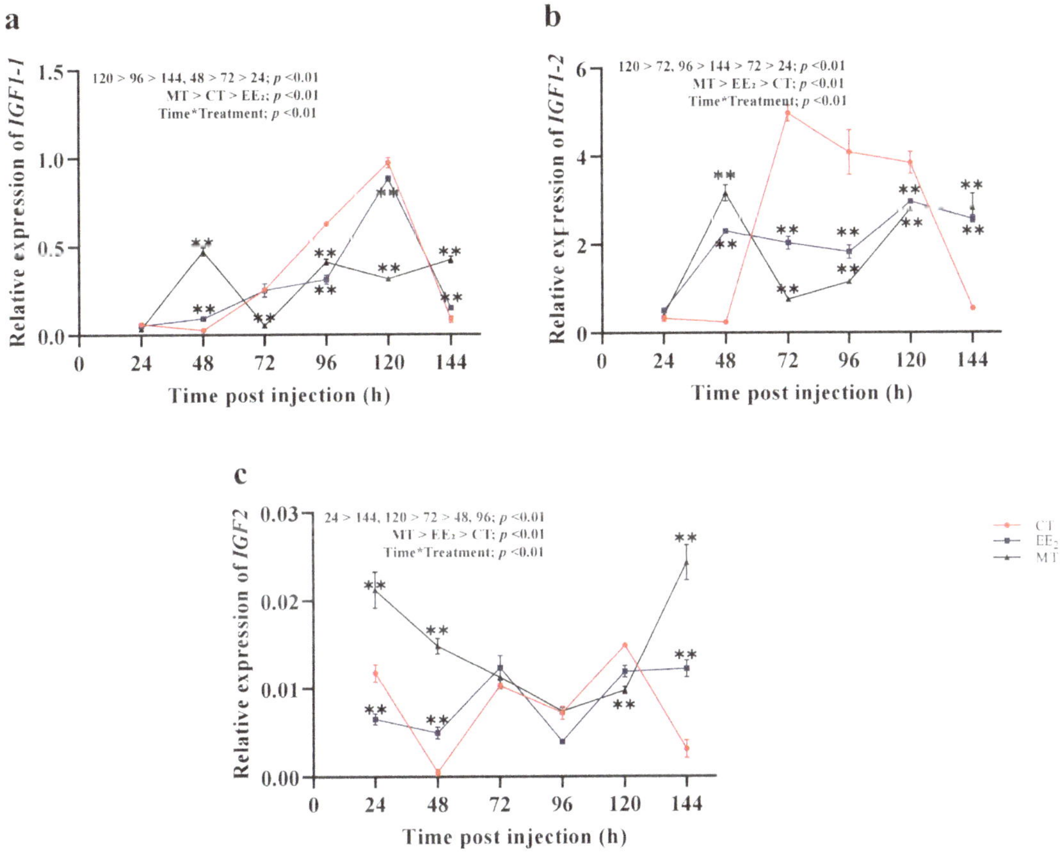

Figure 6. Effects of short-term hormone treatment on *CmIGFs* in male blotched snakehead. (**a**) *CmIGF1-1*; (**b**) *CmIGF1-2*; (**c**) *CmIGF2*. *β-actin* and *EF1α* were used as the internal controls for qPCR. Data are presented as mean ± S.E.M (n = 3). Asterisks represent significant differences between treatments and control at the corresponding time point for each dataset. ** $p < 0.01$.

3.7. Effects of Long-Term Hormone Treatment on CmIGFs in Females

In females, MT considerably boosted *CmIGF1-1* mRNA levels in comparison to the control group at 2 weeks ($F_{2,8} = 55.732$, $p < 0.01$), followed by a sudden drop to the lowest level at 4 weeks (Figure 7a). The highest level was detected at 8 weeks, which was not significantly different from the control group (Figure 7a). Treatment with EE_2 resulted in consistently lower *CmIGF1-1* mRNA levels than controls throughout ten weeks, reaching the lowest level at 10 weeks and being significantly lower ($F_{4,8} = 62.65$, $p < 0.01$) than that in the control group (Figure 7a). Hormone treatment (EE_2 and MT) led to the maintenance of low transcription levels of *CmIGF1-2*, which were significantly lower than that in the control group ($F_{2,8} = 69.249$, $p < 0.01$) (Figure 7b). *CmIGF2* remained at a low expression level in the control group, while *CmIGF2* mRNA levels were considerably boosted ($F_{2,8} = 73.017$, $p < 0.01$) after the administration of MT and EE_2, with the highest levels detected at 6 weeks and 8 weeks ($F_{4,8} = 74.372$, $p < 0.01$), respectively, which were much higher than that in the control group (Figure 7c). Finally, the body weight and length of fish in the control and treated groups were measured at 10 weeks post-injection. Females treated with MT achieved an average body length of 21.4 ± 1.1 cm and a weight of 240.1 ± 9.2 g. Compared to the control group, which presented an average body length of 26.7 ± 1.8 cm and a weight of 331.5 ± 9.3 g, the growth rates were reduced by 19.8% in length and 27.4% in weight. Females treated with EE_2 reached an average body length of 19.2 ± 1.5 cm and a weight of 207.5 ± 8.7 g, indicating the reductions in growth rates of 28.0% in length and 37.3% in weight compared to the control group.

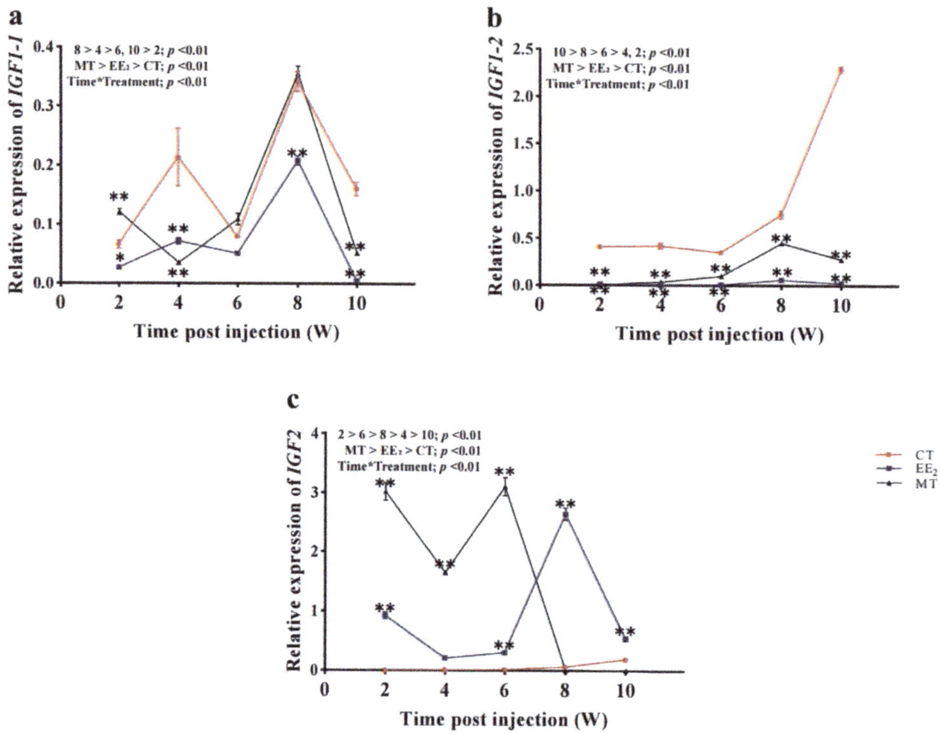

Figure 7. Effects of long-term hormone treatment on *CmIGFs* in female blotched snakehead. (**a**) *CmIGF1-1*; (**b**) *CmIGF1-2*; (**c**) *CmIGF2*. *β-actin* and *EF1α* were used as the internal controls for qPCR. Data are presented as mean ± S.E.M (n = 3). Asterisks represent significant differences between treatments and control at the corresponding time point for each dataset. * $p < 0.05$, and ** $p < 0.01$.

3.8. Effects of Long-Term Hormone Treatment on CmIGFs in Males

As shown in Figure 8a,b, the expression levels of *CmIGF1-1* ($F_{2,8} = 57.041$, $p < 0.01$) and *CmIGF1-2* ($F_{2,8} = 62.436$, $p < 0.01$) remained low in males after the administration of EE_2, being lower than those in the control group. Upon the administration of MT, the transcription of *CmIGF1-1* ($F_{4,8} = 78.874$, $p < 0.01$) and *CmIGF1-2* ($F_{4,8} = 43.802$, $p < 0.01$) remained at low levels until 6 weeks, then increased sharply and reached their peaks at 8 and 10 weeks, respectively, which were significantly higher than those in the control group (Figure 8a,b). EE_2 and MT treatments up-regulated ($F_{2,8} = 69.567$, $p < 0.01$) the expression levels of *CmIGF2* compared to the control group at 2 weeks (Figure 8c). After that, *CmIGF2* expression displayed a decreasing trend and achieved the lowest level at 10 weeks ($F_{4,8} = 58.721$, $p < 0.01$) (Figure 8c). In contrast, *CmIGF2* transcripts dramatically increased from 4 weeks and reached the peak ($F_{2,8} = 69.567$, $p < 0.01$) at 8 weeks with EE_2 administration and subsequently rapidly declined to the lowest level at 10 weeks ($F_{4,8} = 58.721$, $p < 0.01$) (Figure 8c). At 10 weeks post-treatment, males treated with EE_2 achieved an average body length of 24.3 ± 1.6 cm and a body weight of 330.6 ± 7.4 g. Compared to the control group, which presented an average body length of 28.6 ± 1.3 cm and a body weight of 370.5 ± 8.8 g, the growth rates were reduced by 15.1% in length and 10.8% in weight. Males treated with MT reached an average body length of 33.7 ± 1.4 cm and a body weight of 406.2 ± 10.3 g, indicating the increased growth rates of 15.1% in length and 15.1% in weight compared to the control group.

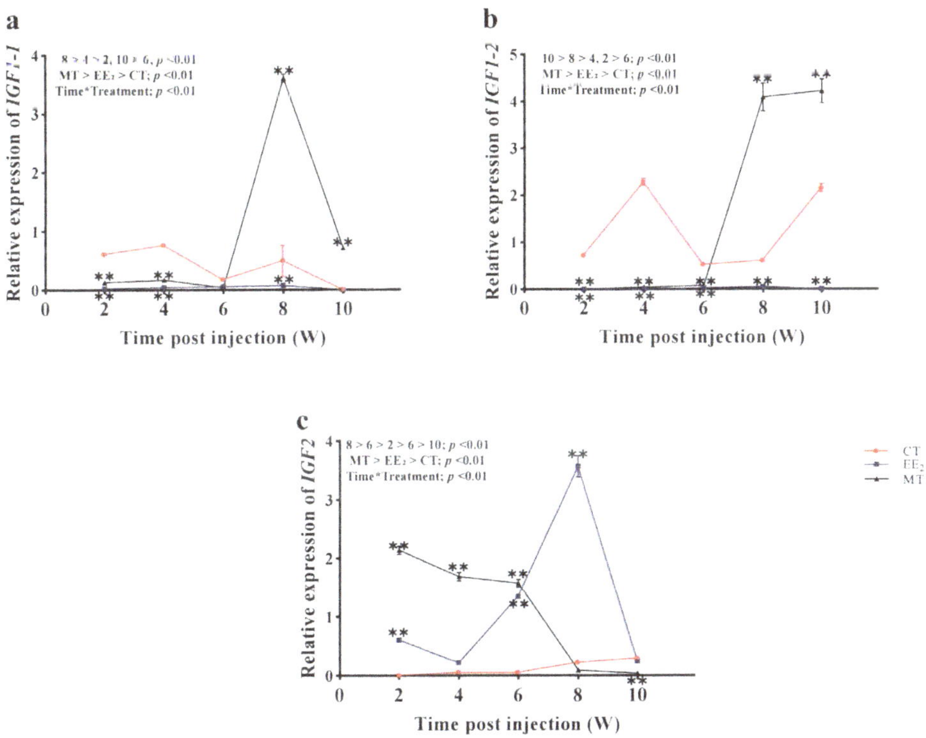

Figure 8. Effects of long-term hormone treatment on *CmIGFs* in male blotched snakehead. (**a**) *CmIGF1-1*; (**b**) *CmIGF1-2*; (**c**) *CmIGF2*. *β-actin* and *EF1α* were used as the internal controls for qPCR. Data are presented as mean ± S.E.M (n = 3). Asterisks represent significant differences between treatments and control at the corresponding time point for each dataset. ** $p < 0.01$.

4. Discussion

Growth, one of the most crucial economic traits in farmed fishes, plays a vital role in the advancement of aquaculture. In the present study, *IGFs*, which are known to be involved in growth regulation, were characterized in blotched snakehead (*C. maculata*). Consistent with findings in zebrafish (*D. rerio*) [9] and giant grouper (*E. lanceolatus*) [16], two copies of *IGF1* were identified in blotched snakehead, designated as *CmIGF1-1* and *CmIGF1-2*, located on chromosomes 15 and 10, respectively [30]. The *IGF1* gene in many fish species, including giant grouper (*E. lanceolatus*) [16], ussuri catfish (*Pseudobagrus ussuriensis*) [34], and zebrafish (*D. rerio*) [9], typically consists of five exons and four introns, with intron 2 being the longest. In our study, *CmIGF1-1* also consisted of five exons and four introns, with intron 2 being the longest at 8800 bp. In contrast, *CmIGF1-2* manifested a more concise structure, with solely three exons and two introns. The structural differences between the two gene products may potentially stem from divergent splicing methodologies [35]. The *IGF2* gene differs from *IGF1* in exon–intron composition in ussuri catfish (*P. ussuriensis*) [34] and yellow catfish (*P. fulvidraco*) [11], typically consisting of four exons and three introns. In this study, *CmIGF2* comprised four exons and three introns. The alignment of amino acid sequences revealed conserved IGF domains in both IGF1-1 and IGF1-2 across species. Furthermore, IGF2 exhibited conserved IGF and IGF2_C domains, which play a vital role in protein folding, maintaining proper secondary spatial structure, and facilitating normal physiological functions [36]. Evolutionary analysis revealed that IGF1 and IGF2 clustered together among all fish species, diverging from those of tetrapods, which encompass mammals, birds, and reptiles. Our findings are consistent with both systematic evolutionary analysis and traditional taxonomy.

Two distinct variant forms have been identified in *CmIGF1*, whereas *CmIGF2* was expressed as a single form, consistent with observations in giant grouper (*E. lanceolatus*) [16] and olive flounder (*P. olivaceus*) [37]. In contrast, in zebrafish (*D. rerio*) [9], both *IGF1* and *IGF2* present two variants, a disparity that may stem from gene duplication. This suggests that the retention of gene duplication among species may be selective, reflecting the diversity of evolutionary histories and genomic adaptabilities [38]. *CmIGF1-1* and *CmIGF1-2* exhibited notable expression levels in liver, consistent with the previous findings in spotted scat (*S. argus*) [25], giant grouper (*E. lanceolatus*) [16], and Japanese sea bass (*Lateolabrax japonicus*) [39], indicating the significant role of liver cells in IGF1 secretion. *CmIGF2* was broadly distributed across tissues, although it showed predominant expression in the liver. A high expression of *IGF2* has been observed in the gonads of Chinese sturgeon (*Acipenser dabryanus*) [40] and the gills of redbanded seabream (*Pagrus Auriga*) [24]. The observed variation in the predominant tissue for *IGF2* mRNA expression among different fish species may be attributed to interspecific differences. Sex-biased differences in the expression of *CmIGFs* were observed, with higher levels in male livers than in females, consistent with reports in yellow catfish (*P. fulvidraco*) [11] where *IGFs* expression levels were higher in the livers of males with faster growth rates than female individuals. Sexual dimorphism in growth has been observed in European eels (*Anguilla anguilla*) [41] and tongue sole (*C. semilaevis*) [42], with females demonstrating faster growth rates than males. In the late juvenile stage of European eels (*A. anguilla*) [41] and the larval phase of tongue sole (*C. semilaevis*) [10], the expression levels of *IGF1* and *IGF2* are significantly higher in females compared to males. In addition to *IGF1* and *IGF2*, a gonad-specific IGF family gene (*IGF3*) has been isolated in the gonads of turbot (*Scophthalmus maximus*) [4], zebrafish (*D. rerio*) [9], orange-spotted grouper (*Epinephelus coioides*) [43], and common carp (*Cyprinus carpio*) [44], uniquely expressed in ovaries and testes. *IGF3* commences at the early stage of sex determination and differentiation and remains high in expression throughout gonadal development, offering insights into fish gonadal development, particularly in economically valuable species. However, the expression profiles of *CmIGF1-1*, *CmIGF1-2*, and *CmIGF2* across various tissues did not reveal any gonad-specific expression patterns like *IGF3*. Additionally, the comparison of the amino acid sequences of IGF3 from zebrafish (*D. rerio*) retrieved from the NCBI database (https://www.ncbi.nlm.nih.gov/, accessed on 20 January 2024) with those of *Cm*IGF1-1, *Cm*IGF1-2, and *Cm*IGF2 showed low similarity levels, at only 25.6%, 22.3%, and 43.1%, respectively. Furthermore, the *IGF3* gene was not identified in the genome of blotched snakehead (*C. maculata*) (SRA Accession No. PRJNA730430) [30], suggesting the potential absence of IGF3 in blotched snakehead (*C. maculata*), warranting further investigation.

CmIGFs were detected in all tissues sampled from 45 to 365 dah, with global expression levels higher in males than in females. The expression of *CmIGF1-1* and *CmIGF1-2* gradually increased with growth and culminated in a minor peak at 105 dah, similar to observations in yellow catfish (*P. fulvidraco*) [11] and spotted scat (*S. argus*) [25]. Subsequently, a gradual decrease in the expression of *CmIGF1-1* and *CmIGF1-2* was observed from 135 dah, potentially due to a substantial portion of energy from feed being used for gonadal development [28,45]. *CmIGF1-1* and *CmIGF1-2* revealed the highest expression levels at 365 dah, corresponding to sexual maturity in blotched snakehead (*C. maculata*), with energy primarily directed toward growth. Furthermore, *CmIGF1-1* and *CmIGF1-2* exhibited elevated expression levels in male snakehead compared to females, possibly due to deferred sexual maturity in males and preferential energy allocation toward growth [46]. Similar expression patterns of *IGF1* have observed in Nile tilapia (*O. niloticus*) at different developmental stages [47], showing temporal and spatial variations with the gradual elevation of hepatic expression levels during the initial growth stages, followed by a minor peak, rapid decline, and resurgence to peak levels in the late sampling period. *CmIGF2* expression was high at 45 dah and decreased to a minimum at 135 dah and then gradually increased to peak at 365 dah, suggesting a crucial role of *IGF2* in early growth and maturation in blotched snakehead (*C. maculata*). Similar results have been reported in aucha perch (*Siniperca chuatsi*) [48] and dark sleeper (*Odontobutis potamophila*) [49]. However, *IGF2* exhibits different expression profiles in mammals, where it is highly expressed during early embryonic development and subsequently diminishes [5], suggesting a broader role of IGF2 in fishes beyond embryonic growth and development. Our results suggest that differences in growth between male and female individuals may be connected with sexually dimorphic expression levels of *IGFs*, warranting further investigation.

Sex steroid hormones are assimilated by fish and influence the expression of genes associated with growth, thereby regulating protein synthesis, cellular proliferation, and overall growth [50,51]. Previous studies have suggested interactions between sex steroid hormones and components of the GH/IGF axis [11,18,21,25,27]. However, these interactions are varied, and the reason for the variety among these studies remains unclear. In this study, both *CmIGF1-1* and *CmIGF1-2* were down-regulated in response to EE_2 treatment in both sexes, consistent with findings in rainbow trout (*O. mykiss*) where E_2 treatment reduces the hepatic expression levels of *IGF1* [21]. Additionally, EE_2 inhibits *GH* expression in the pituitary of blotched snakehead (*C. maculata*) [27], indicating that EE_2 down-regulates hepatic *CmIGF1* transcription levels in both sexes, probably by diminishing the expression of *CmGH*. After EE_2 stimulation, the expression of *CmIGF2* was up-regulated in both sexes, which differs slightly from the results in Nile tilapia (*O. niloticus*), where E2 stimulation increased *IGF2* expression in females but had no significant effect on *IGF2* expression in males [18]. Conversely, in spotted scat (*S. argus*), E2 treatment increased the transcription of both *IGF1* and *IGF2* in males, while in females, *IGF1* was up-regulated and *IGF2* was down-regulated [25]. These findings indicate that the transcriptional levels of *IGFs* are influenced by sex steroid hormones and exhibit variation among different fishes, potentially attributable to differences in experimental methods (such as administration method, concentration, and exposure duration) and developmental stages.

The ability of androgens to regulate the GH/IGF axis is supported by several observations of this study. In females, *CmIGF1-1* expression initially increased after MT treatment for up to 2 weeks, but eventually, its expression was inhibited, while the expression of *CmIGF1-2* remained inhibited. Additionally, *CmIGF2* expression was significantly up-regulated in females before 6 weeks, but eventually, its expression was inhibited. Comparable results were observed in other fish species. In female yellow catfish (*P. fulvidraco*), *IGF-1* and *IGF-2* were up-regulated in the liver after 2–3 weeks of MT treatment and then declined, significantly falling to a lower level than the control group by 4 weeks [11]. Similarly, in female Nile tilapia (*O. niloticus*), following T treatment, *IGF1* expression initially increased, then significantly dropped below the control group by 3 weeks [18]. Such fluctuations may be due to the resistance of endogenous estrogen to exogenous androgen, thereby regulating *IGFs* expression [35,52]. *CmIGF1-1* and *CmIGF1-2* expression levels were significantly up-regulated in males treated with MT after 6 weeks. It is possible that the anabolic effect of MT leads to an increase in food intake in males, thus enhancing energy allocation for growth in males [53]. Similar findings have been observed in rainbow trout (*O. mykiss*), where T directly increases steady-state *IGF-1* and *IGF-2* expression levels [21]. Likewise, *IGF1* and *IGF2* are provoked by T treatment in male Nile tilapia (*O. niloticus*), with expression increasing 7–21 days after injection [18]. However, the responses of *CmIGF1-1* and *CmIGF1-2* to MT stimulation differed after 8 weeks, suggesting that MT regulates the expression of GH/IGF axis genes in a time-dependent manner. Additionally, MT treatment significantly up-regulated *CmIGF2* expression compared to the control group before 6 weeks, and subsequently, it gradually decreased to below the control level. We suggest that exogenous sex steroid hormones stimulate IGFs via IGF-1R in blotched snakehead

(*C. maculata*), resulting in a compensatory increase in *CmIGF2* expression in response to low *CmIGF1* concentration, as IGF-1R has higher binding affinity for IGF1 than for IGF2 [54]. The specific pathways through which MT regulates the expression of *CmIGFs*, thereby affecting the growth of blotched snakehead, are currently unclear. Perhaps in the future, we can conduct a deeper investigation by using liquid chromatography–mass spectrometry (LC–MS) to detect changes in sex steroid hormones in both males and females during hormone treatment processes.

Sexual dimorphism in growth is governed by a complex interplay of factors, including the differential expression of GH/IGF axis genes, the influence of exogenous sex steroid hormones, and various other elements such as environmental conditions, the abundance of IGF receptors, interactions among growth axis genes, and the modulation of endogenous hormones. Future research projects will focus on delving deeper into the mechanisms underlying growth dimorphism between male and female snakehead.

5. Conclusions

The full-length sequences of *CmIGFs* were cloned, and their sequence characteristics and gene expression patterns were thoroughly scrutinized. The investigation unveiled that *CmIGFs* closely resembled those documented in bony fishes, exhibiting the typical IGF domain of the IGF family. Gene expression analyses showed the predominant expression of *CmIGFs* in the male liver, signifying that the liver serves as the primary site for the synthesis and secretion of IGFs. Furthermore, the study showed that exogenous sex steroid hormones, EE_2 and MT, exerted significant modulatory effects on the expression of *CmIGFs* genes. Specifically, EE_2 was observed to suppress the expression of *CmIGF1-1* and *CmIGF1-2* while promoting the expression of *CmIGF2*. In females, MT up-regulated the expression of *CmIGF1-1* and *CmIGF2* in a time-dependent manner, while consistently inhibiting the expression of *CmIGF1-2*. Interestingly, MT was found to promote the expression of *CmIGFs* in males in a time-dependent manner, and *CmIGF1* and *CmIGF2* may exhibit a complementary relationship, with a compensatory increase in *CmIGF2* expression in response to low *CmIGF1* concentration. These findings lay the groundwork for future investigations aimed at unraveling the molecular mechanism underlying the growth dimorphism between female and male blotched snakeheads.

Supplementary Materials: The following supporting information can be downloaded at: https://www.mdpi.com/article/10.3390/fishes9040120/s1. Figure S1: Nucleotide and putative amino sequences of *Cm*IGF1-1 (a), *Cm*IGF1-2 (b), and *Cm*IGF2 (c). The sequence numbers of nucleotide (lower row) and putative amino acid (upper row) are shown on the left. The translation initiation codons and stop codons are in bold. The motif-associated mRNA instability (ATTTA) is shown as a double underscore. The poly-adenylation signal sequence (AATAA) is shown as a wavy line. The signal peptide is shown as an underscore. The IGF domain is marked with a green background. The IGF2_C domain is marked with a yellow background. The box represents the IGF-1R recognition sequence. Figure S2: Multiple protein alignments of *Cm*IGF1-1 (a), *Cm*IGF1-2 (b), and *Cm*IGF2 (c) in various species. The amino acid sequences of IGF1s from typical organisms were aligned using the Clustal X 2.1 program. The black shade represents 100% identity, dark gray represents 80% identity, and the IGF domain is marked by a green box. *At*IGF1-1 stands for IGF1-1 protein in *A. testudineus* (Protein ID. XP_026218654.1), *At*IGF1-2 stands for IGF1-2 protein in *A. testudineus* (Protein ID. XP_026218655.1), *At*IGF2 stands for IGF2 protein in *A. testudineus* (Protein ID. XP_026221051.1), *Bt*IGF1 stands for IGF1 protein in *B. taurus* (Protein ID. NP_001071296.1), *Bt*IGF2 stands for IGF2 protein in *B. taurus* (Protein ID. XP_005227329.1), *Ca*IGF1a stands for IGF1a protein in *C. argus* (Protein ID. KAF3696413.1), *Ca*IGF2 stands for IGF2 protein in *C. argus* (Protein ID. KAF3689021.1), *Cc*IGF1-1 stands for IGF1-1 protein in *C. carpio* (Protein ID. XP_018948511.1), *Cc*IGF1-2 stands for IGF1-2 protein in *C. carpio* (Protein ID. XP_018948527.1), *Cc*IGF2 stands for IGF2 protein in *C. carpio* (Protein ID. XP_018955405.1), *Cm*IGF1-1 stands for IGF1-1 protein in *C. maculata* (Protein ID. UTQ11150.1), *Cm*IGF1-2 stands for IGF1-2 protein in *C. maculata* (Protein ID. UTQ11151.1), *Cm*IGF2 stands for IGF2 protein in *C. maculata* (Protein ID. UTQ11152.1), *Ci*GF2a stands for IGF2a protein in *C. idella* (Protein ID. AJG05606.1), *Ci*GF2b stands for IGF2b protein in *C. idella* (Protein ID. AJG05607.1), *Cj*IGF2 stands for IGF2 protein in *C. japonica*

(Protein ID. XP_015719328.1), *Dr*IGF1 stands for IGF1 protein in *D. rerio* (Protein ID. AAI14263.1), *Dr*IGF2a stands for IGF2a protein in *D. rerio* (Protein ID. NP_571508.1), *Dr*IGF2b stands for IGF2b protein in D. rerio (Protein ID. NP_001001815.1.), *El*IGF1a stands for IGF1a protein in *E. lanceolatus* (Protein ID. ABZ10840.1), *El*IGF1b stands for IGF1b protein in *E. lanceolatus* (Protein ID. ABZ10841.1), *Gg*IGF1 stands for IGF1 protein in *G. gallus* (Protein ID. NP_001004384.1), *Gg*IGF2 stands for IGF2 protein in *G. gallus* (Protein ID. NP_001025513.1), *Hs*IGF1a stands for IGF1a protein in *H. sapiens* (Protein ID. AAA52538.1), *Hs*IGF1b stands for IGF1a protein in *H. sapiens* (Protein ID. AAA52537.1), *Ip*IGF2 stands for IGF2 protein in *I. punctatus* (Protein ID. ADO29240.1), *Mm*IGF2 stands for IGF2 protein in *M. musculus* (Protein ID. NP_001116208.1), *Po*IGF1a stands for IGF1a protein in *P. olivaceus* (Protein ID. CAA09267.1), *Po*IGF1b stands for IGF1b protein in *P. olivaceus* (Protein ID. CAA09268.1), *Sa*IGF1 stands for IGF1 protein in *S. aurata* (Protein ID. ABQ52656.1), *Sm*IGF1a stands for IGF1a protein in *S. maximus* (Protein ID. XP 035474340.1), *Sm*IGF1b stands for IGF1b protein in *S. maximus* (Protein ID. XP 035474347.1), *Tr*IGF2 stands for IGF2 protein in *T. rubripes* (Protein ID. ADO29240.1), *Xl*IGF1 stands for IGF1 protein in *X. laevis* (Protein ID. AAA70330.1), *Xl*IGF2a stands for IGF2a protein in *X. laevis* (Protein ID. AAL11445.1), *Xl*IGF2b stands for IGF2b protein in *X. laevis* (Protein ID. AAH72153.1), *Xt*IGF1 stands for IGF1 protein in *X. tropicalis* (Protein ID. XP_002936875.1), and *Xt*IGF2 stands for IGF2 protein in *X. tropicalis* (Protein ID. AAI56000.1).

Author Contributions: Conceptualization, X.Z. and M.O.; data curation, X.Z. and Y.W.; formal analysis, X.Z., J.Z.(Jin Zhang) and Y.Z.; funding acquisition, K.C., J.Z. (Jian Zhao) and M.O.; investigation, X.Z., Y.W., H.L., Q.L. and S.F.; project administration, K.C., J.Z. (Jian Zhao) and M.O.; visualization, K.C., Q.L., X.Z. and J.Z. (Jian Zhao); writing—original draft, X.Z. and Y.W.; writing—review and editing, J.Z. (Jian Zhao), K.C. and M.O. All authors have read and agreed to the published version of the manuscript.

Funding: This work was supported by the China Agriculture Research System of Ministry of Finance and Ministry of Agriculture and Rural Affairs (CARS-46), the Central Public-interest Scientific Institution Basal Research Fund of the Chinese Academy of Fishery Sciences (2023XT0202, 2023TD37), the National Natural Science Foundation of China (32373127), the Guangdong Province Rural Revitalization Strategy Special Fund (2022-SPY-00-016), and the National Freshwater Genetic Resource Center (FGRC18537).

Institutional Review Board Statement: All fish experiments in the present study were approved by the Pearl River Fisheries Research Institute and the Chinese Academy of Fishery Sciences under contract LAEC-PRFRI-2021-01-04, and the experimental process complied with protocols of international guidelines for the ethical use of animals in research.

Informed Consent Statement: Not applicable.

Data Availability Statement: All the data related to this project are available from the corresponding author and will be provided upon request.

Conflicts of Interest: The authors declare that they have no known competing financial interests or personal relationships that could have influenced the work reported in this paper.

References

1. Dixit, M.; Poudel, S.B.; Yakar, S. Effects of GH/IGF axis on bone and cartilage. *Mol. Cell. Endocrinol.* **2021**, *519*, 111052. [CrossRef] [PubMed]
2. Ranke, M.B.; Wit, J.M. Growth hormone—Past, present and future. *Nat. Rev. Endocrinol.* **2018**, *14*, 285–300. [CrossRef]
3. Sun, C.; Sun, H.; Dong, J.; Tian, Y.; Hu, J.; Ye, X. Correlation analysis of mandarin fish (*Siniperca chuatsi*) growth hormone gene polymorphisms and growth traits. *J. Genet.* **2019**, *98*, 58. [CrossRef]
4. Zhao, C.; Zheng, S.; Dang, Y.; Wang, M.; Ren, Y. Identification of a new insulin-like growth factor 3 (igf3) in turbot (*Scophthalmus maximus*): Comparison and expression analysis of IGF system genes during gonadal development. *Fishes* **2023**, *8*, 240. [CrossRef]
5. Baral, K.; Rotwein, P. The Insulin-like The insulin-like growth factor 2 gene in mammals: Organizational complexity within a conserved locus. *PLoS ONE* **2019**, *14*, e0219155. [CrossRef]
6. Cao, Q.; Duguay, S.J.; Plisetskaya, E.; Steiner, D.F.; Shu, J. Nucleotide sequence and growth hormone-regulated expression of salmon insulin-like growth factor I mRNA. *Mol. Endocrinol.* **1989**, *3*, 2005–2010. [CrossRef] [PubMed]
7. Bersin, T.V.; Cordova, K.L.; Journey, M.L.; Beckman, B.R.; Lema, S.C. Food deprivation reduces sensitivity of liver Igf1 synthesis pathways to growth hormone in juvenile gopher rockfish (*Sebastes carnatus*). *Gen. Comp. Endocrinol.* **2024**, *346*, 114404. [CrossRef]

8. Caelers, A.; Berishvili, G.; Meli, M.L.; Eppler, E.; Reinecke, M. Establishment of a real-time RT-PCR for the determination of absolute amounts of IGF-I and IGF-II gene expression in liver and extrahepatic sites of the tilapia. *Gen. Comp. Endocrinol.* **2004**, *137*, 196–204. [CrossRef] [PubMed]
9. Zou, S.; Kamei, H.; Modi, Z.; Duan, C. Zebrafish IGF genes: Gene duplication, conservation and divergence, and novel roles in midline and notochord development. *PLoS ONE* **2009**, *4*, e7026. [CrossRef]
10. Ma, Q.; Liu, S.; Zhuang, Z.; Sun, Z.; Liu, C.; Su, Y.; Tang, Q. Molecular cloning, expression analysis of insulin-like growth factor I (IGF-I) gene and IGF-I serum concentration in female and male tongue sole (*Cynoglossus semilaevis*). *Comp. Biochem. Physiol. Part B Biocchem. Mol. Biol.* **2011**, *160*, 208–214. [CrossRef]
11. Ma, W.; Wu, J.; Zhang, J.; He, Y.; Gui, J.; Mei, J. Sex differences in the expression of GH/IGF axis genes underlie sexual size dimorphism in the yellow catfish (*Pelteobagrus fulvidraco*). *Sci. China Life Sci.* **2016**, *59*, 431–433. [CrossRef] [PubMed]
12. Zhang, K.; Chen, H.; Jiang, D.; Deng, S.; Zhu, C.; Wu, T.; Li, G. Insulin-like growth factors 1 and 2 in spotted scat (*Scatophagus argus*): Molecular cloning and differential expression during embryonic development. *J. Guangdong Ocean. Uni.* **2018**, *38*, 7–14. (In Chinese) [CrossRef]
13. Shamblott, M.J.; Chen, T.T. Identification of a second insulin-like growth factor in a fish species. *Proc. Natl. Acad. Sci. USA* **1992**, *89*, 8913–8917. [CrossRef]
14. Yuan, Y.; Hong, Y. Medaka insulin-like growth factor-2 supports self-renewal of the embryonic stem cell line and blastomeres in vitro. *Sci. Rep.* **2017**, *7*, 78. [CrossRef]
15. White, Y.A.R.; Kyle, J.T.; Wood, A.W. Targeted gene knockdown in zebrafish reveals distinct intraembryonic functions for insulin-like growth factor II signaling. *Endocrinology* **2009**, *150*, 4366–4375. [CrossRef] [PubMed]
16. Dong, H.; Zeng, L.; Duan, D.; Zhang, H.; Wang, Y.; Li, W.; Lin, H. Growth hormone and two forms of insulin-like growth factors I in the giant grouper (*Epinephelus lanceolatus*): Molecular cloning and characterization of tissue distribution. *Fish Physiol. Biochem.* **2010**, *36*, 201–212. [CrossRef]
17. Yuan, X.; Jiang, X.; Pu, J.; Li, Z.; Zou, S. Functional conservation and divergence of duplicated insulin-like growth factor 2 genes in grass carp (*Ctenopharyngodon idellus*). *Gene* **2011**, *470*, 46–52. [CrossRef] [PubMed]
18. Yue, M.; Zhao, J.; Tang, S.; Zhao, Y. Effects of estradiol and testosterone on the expression of growth-related genes in female and male *Nile tilapia, Oreochromis niloticus*. *J. World Aquac. Soc.* **2018**, *49*, 216–228. [CrossRef]
19. Nipkow, M.; Wirthgen, E.; Luft, P.; Rebl, A.; Hoeflich, A.; Goldammer, T. Characterization of igf1 and igf2 genes during maraena whitefish (*Coregonus maraena*) ontogeny and the effect of temperature on embryogenesis and igf expression. *Growth Horm. Igf. Res.* **2018**, *40*, 32–43. [CrossRef]
20. Venken, K.; Boonen, S.; Kopchick, J.; Coschigano, K.; Moverare, S.; Bouillon, R.; Ohlsson, C.; Vanderschueren, D. Growth without growth hormone receptor: Estradiol is a major growth-hormone independent regulator of hepatic insulin-like growth factor-I synthesis. *J. Bone Miner. Res.* **2004**, *19*, S43. [CrossRef]
21. Norbeck, L.A.; Sheridan, M.A. An in vitro model for evaluating peripheral regulation of growth in fish: Effects of 17β-estradiol and testosterone on the expression of growth hormone receptors, insulin-like growth factors, and insulin-like growth factor type 1 receptors in rainbow trout (*Oncorhynchus mykiss*). *Gen. Comp. Endocrinol.* **2011**, *173*, 270–280. [CrossRef] [PubMed]
22. Ma, X.; Zhang, Y.; Chen, Y.; Zhou, L. Steroid hormones (E_2 and MT) displayed difference in sex for *Nile tilapia Oreochromis niloticus*. *Oceanol. Limnol. Sin.* **2015**, *46*, 1487–1493. (In Chinese)
23. Shen, M.; Zhu, J.; Wu, T.; Ren, T.; Zhang, Z.; Li, X.; Liu, F.; Zheng, S. Cloning and expression analysis of insunlin-like growth factor-I gene before and after reproduction in *Zacco platypus*. *Acta Hydrobiol. Sin.* **2020**, *44*, 1182–1190. [CrossRef]
24. Ponce, M.; Infante, C.; Funes, V.; Manchado, M. Molecular characterization and gene expression analysis of insulin-like growth factors I and II in the redbanded seabream, *Pagrus auriga*: Transcriptional regulation by growth hormone. *Comp. Biochem. Physiol. Part B Biocchem. Mol. Biol.* **2008**, *150*, 418–426. [CrossRef]
25. Zhang, K.; Wu, T.; Chen, H.; Jiang, D.; Zhu, C.; Deng, S.; Zhang, Y.; Li, G. Estradiol-17 beta regulates the expression of insulin-like growth factors 1 and 2 via estradiol receptors in spotted scat (*Scatophagus argus*). *Comp. Biochem. Physiol. Part B Biocchem. Mol. Biol.* **2019**, *237*, 110328. [CrossRef]
26. Ou, M.; Chen, K.; Gao, D.; Wu, Y.; Luo, Q.; Liu, H.; Zhao, J. Characterization, expression and cpg methylation analysis of dmrt1 and its response to steroid hormone in blotched snakehead (*Channa maculata*). *Comp. Biochem. Physiol. Part B Biocchem. Mol. Biol.* **2022**, *257*, 110672. [CrossRef] [PubMed]
27. Ou, M.; Chen, K.; Gao, D.; Wu, Y.; Chen, Z.; Luo, Q.; Liu, H.; Zhao, J. Comparative transcriptome analysis on four types of gonadal tissues of blotched snakehead (*Channa maculata*). *Comp. Biochem. Physiol. Part D Genom. Proteom.* **2020**, *35*, 100708. [CrossRef]
28. Gao, D.; Ou, M.; Wu, Y.; Chen, K.; Liu, H.; Luo, Q.; Zhao, J. Gene cloning and expression analysis of growth hormonr gene from blotched snakehead (*Channa maculata*). *J. Agric. Biotech.* **2021**, *29*, 2328–2341. Available online: http://journal05.magtech.org.cn/Jwk_ny/EN/10.3969/j.issn.1674-7968.2021.12.007 (accessed on 18 January 2024). (In Chinese).
29. Zhao, J.; Ou, M.; Wang, Y.; Liu, H.; Luo, Q.; Zhu, X.; Chen, B.; Chen, K. Breeding of YY super-male of blotched snakehead (*Channa maculata*) and production of all-male hybrid (*Channa argus* ♀× *C. maculata* ♂). *Aquaculture* **2021**, *538*, 736450. [CrossRef]
30. Ou, M.; Huang, R.; Yang, C.; Gui, B.; Luo, Q.; Zhao, J.; Li, Y.; Liao, L.; Zhu, Z.; Wang, Y.; et al. Chromosome-level genome assemblies of *Channa argus* and *Channa maculata* and comparative analysis of their temperature adaptability. *Gigascience* **2021**, *10*, giab070. [CrossRef]

31. Tamura, K.; Peterson, D.; Peterson, N.; Stecher, G.; Nei, M.; Kumar, S. MEGA5: Molecular evolutionary genetics analysis using maximum likelihood, evolutionary distance, and maximum parsimony methods. *Mol. Biol. Evol.* **2011**, *28*, 2731–2739. [CrossRef] [PubMed]
32. Mao, H.; Chen, K.; Zhu, X.; Luo, Q.; Zhao, J.; Li, W.; Wu, X.; Xu, H. Identification of suitable reference genes for quantitative real-time PCR normalization in blotched snakehead *Channa maculata*. *J. Fish Biol.* **2017**, *90*, 2312–2322. [CrossRef] [PubMed]
33. Livak, K.J.; Schmittgen, T.D. Analysis of relative gene expression data using real-time quantitative pcr and the 2(T)(-Delta Delta C) method. *Methods* **2001**, *25*, 402–408. [CrossRef] [PubMed]
34. Wei, M.; Wang, M.; Ning, J.; Fan, F.; Zhu, C. Cloning and association analysis with growth traits of igf gene in ussuri catfish *Pseudobagrus ussuriensis*. *Fish Sci.* **2022**, *41*, 738–748. (In Chinese) [CrossRef]
35. Shamblott, M.J.; Chen, T. Age-related and tissue-specific levels of five forms of insulin-like growth factor mRNA in a teleost. *Mol. Mar. Biol. Biotechnol.* **1993**, *2*, 351–361.
36. Ndandala, C.B.; Dai, M.S.; Mustapha, U.F.; Li, X.; Liu, J.; Huang, H.; Li, G.; Chen, H. Current research and future perspectives of GH and IGFs family genes in somatic growth and reproduction of teleost fish. *Aquac. Rep.* **2022**, *26*, 101289. [CrossRef]
37. Sun, P.; Wu, Z.; You, F.; Li, J. Annual cycle change of sex steroid hormones in cultured *Paralichthys olivaceus*. *Mar. Fish* **2013**, *35*, 34–37. [CrossRef]
38. Parey, E.; Louis, A.; Montfort, J. An atlas of fish genome evolution reveals delayed rediploidization following the teleost whole-genome duplication. *Genome Res.* **2022**, *32*, 1685–1697. [CrossRef]
39. Qian, K.; Wen, H.; Chi, M.; Ni, M.; Zhang, D.; Ding, Y. Solation of full-length cDNA of Insulin-like growth factor-1 (IGF-1) gene of Japanese sea bass (*Lateolabrax japonicas*) and its expression analysis. *J. Ocean Univ. China* **2014**, *44*, 27–34. (In Chinese) [CrossRef]
40. Chen, Y.; Liu, Y.; Lai, J.; Song, M.; Gong, Q. Molecular cloning of insulin-like growth factor 1, 2 in *Acipenser dabryanus* and their expression level during starvation stress. *Southwest China J. Agric. Sci.* **2020**, *33*, 447–455. (In Chinese) [CrossRef]
41. Degani, G.; Tzchori, I.; Yom-Din, S.; Goldberg, D.; Jackson, K. Growth differences and growth hormone expression in male and female European eels [*Anguilla anguilla* (L.)]. *Gen. Comp. Endocrinol.* **2003**, *134*, 88–93. [CrossRef] [PubMed]
42. Ma, Q.; Liu, S.; Zhuang, Z.; Lin, L.; Sun, Z.; Liu, C.; Ma, H.; Su, Y.; Tang, Q. Genomic structure, polymorphism and expression analysis of the growth hormone (GH) gene in female and male half-smooth tongue sole (*Cynoglossus semilaevis*). *Gene* **2012**, *493*, 92–104. [CrossRef]
43. Yang, H.; Chen, H.; Zhao, H.; Liu, L.; Xie, Z.; Xiao, L.; Li, S.; Zhang, Y.; Lin, H. Molecular cloning of the insulin-like growth factor 3 and difference in the expression of igf genes in orange-spotted grouper (*Epinephelus coioides*). *Comp. Biochem. Phys. B* **2015**, *186*, 68–75. [CrossRef] [PubMed]
44. Song, F.; Wang, L.; Zhu, W.; Fu, J.; Dong, J.; Dong, Z. A novel igf3 gene in common carp (*Cyprinus carpio*): Evidence for its role in regulating gonadal development. *PLoS ONE* **2016**, *7*, e1002447. [CrossRef] [PubMed]
45. Wu, Y.; Ou, M.; Gao, D.; Chen, K.; Luo, Q.; Liu, H.; Zhao, J. Molecular cloning, expression and respone of foxl2 gene induced by sex steroid hormones in blotched snakehead *Channa maculata*. *J. Dalian Ocean Uni.* **2022**, *37*, 49–60. (In Chinese) [CrossRef]
46. Li, X.; Mei, J.; Ge, C.; Liu, X.; Gui, J. Sex determination mechanisms and sex control approaches in aquaculture animals. *Sci. China Life Sci.* **2022**, *65*, 1091–1122. [CrossRef]
47. Hu, Y.; Zou, Z.; Zhu, J.; Li, D.; Xiao, W.; Han, Y.; Le, Y.; Wang, T.; Yang, H. The expression analysis of IGF1 gene during different development stages in *Nile tilapia*. *Chin. Agric. Sci. Bull.* **2014**, *30*, 107–111. (In Chinese)
48. Liu, J.; Zhao, J.; Zhang, M.; Dai, W. Cloning and expression of full-length cDNA of insulin-like growth factor-II in mandarin fish *Siniperca chuatsi*. *J. Dalian Ocean Uni.* **2012**, *27*, 495–501. (In Chinese) [CrossRef]
49. Zhu, W.; Zhang, H.; Wang, T.; Wang, D.; Zhang, H.; Yin, S.; Chen, S.; Zhou, G. Cloning and temporal expression analysis of GHR and IGF-2 in *Odontobutis potamophila*. *Mar. Fish* **2019**, *41*, 421–433. (In Chinese) [CrossRef]
50. James, M.O. Steroid catabolism in marine and freshwater fish. *J. Steroid. Biochem. Mol. Biol.* **2011**, *127*, 167–175. [CrossRef]
51. Shved, N.; Berishvili, G.; D'Cotta, H.; Baroiller, J.F.; Segner, H.; Eppler, E.; Reinecke, M. Ethinylestradiol differentially interferes with IGF-I in liver and extrahepatic sites during development of male and female bony fish. *J. Endocrinol.* **2007**, *195*, 513–523. [CrossRef] [PubMed]
52. Yuan, C.; Wu, T.; Zhang, Y.; Gao, J.; Yang, Y.; Qin, F.; Liu, S.; Zheng, Y.; Wang, Z. Responsiveness of four gender-specific genes, figla, foxl2, scp3 and sox9a to 17α-ethinylestradiol in adult rare minnow *Gobiocypris rarus*. *Gen. Comp. Endocrinol.* **2014**, *200*, 44–53. [CrossRef]
53. Liu, S.; Lv, W.; Lv, X. Effect of 17α-methyltestosterone (MT) on mRNA expression of vtg gene in liver of *Brachydanio rerio*. *Anim. Husb. Feed Sci.* **2016**, *3*, 9–11. (In Chinese) [CrossRef]
54. Fruchtman, S.; McVey, D.C.; Borski, R.J. Characterization of pituitary IGF-I receptors: Modulation of prolactin and growth hormone. *Am. J. Physiol. Regul. Integr. Comp. Physiol.* **2002**, *283*, R468–R476. [CrossRef] [PubMed]

fishes

MDPI

Article

Comparative Analysis of Enzymatic Activities and Transcriptional Profiles of Various Hepatic Enzymes between Male and Female Yellowfin Tuna (*Thunnus albacares*)

Dongge Liu [1,2,†], Hao Yang [1,†], Shuisheng Li [3], Hai Huang [4], Guangli Li [1] and Huapu Chen [1,4,*]

[1] Guangdong Research Center on Reproductive Control and Breeding Technology of Indigenous Valuable Fish Species, Guangdong Provincial Key Laboratory of Pathogenic Biology and Epidemiology for Aquatic Economic Animals, Fisheries College, Guangdong Ocean University, Zhanjiang 524088, China; pieappleliu@hotmail.com (D.L.); lxq02221406@163.com (H.Y.); guangligdou@163.com (G.L.)
[2] Guangdong Havwii Agriculture Group Co., Ltd., Zhanjiang 524266, China
[3] State Key Laboratory of Biocontrol, and the Guangdong Province Key Laboratory for Aquatic Economic Animals, Sun Yat-sen University, Guangzhou 510275, China; lshuish@mail.sysu.edu.cn
[4] Key Laboratory of Utilization and Conservation for Tropical Marine Bioresources of Ministry of Education, Hainan Key Laboratory for Conservation and Utilization of Tropical Marine Fishery Resources, Yazhou Bay Innovation Institute, Hainan Tropical Ocean University, Sanya 572022, China; huanghai74@126.com
* Correspondence: chenhp@gdou.edu.cn; Tel.: +86-188-2070-6692; Fax: +86-759-2382459
† These authors contributed equally to this work.

Abstract: Yellowfin tuna (*Thunnus albacares*) is a valuable pelagic migratory fish with potential for aquaculture. Despite this, there is limited understanding of the biological and physiological characteristics of this species, particularly regarding sex differences in growth performance. The liver, a crucial organ for digestion and metabolism, plays a significant role in regulating fish growth. This study aimed to compare liver enzyme activities and transcriptome profiles between female and male yellowfin tuna to uncover the molecular mechanisms underlying difference between the sexes. The results revealed that female yellowfin tuna exhibited higher amylase and lipid metabolism enzyme activities, while male yellowfin tuna showed higher glucose-6-phosphate dehydrogenase and antioxidant enzyme activities. Additionally, through Illumina sequencing technology, the study generated 37.74 Gb of clean data and identified 36,482 unique genes (UniGenes) in the liver transcriptome. A total of 2542 differentially expressed genes were found, with enriched Gene Ontology terms and pathways related to metabolic processes, particularly lipid metabolism and transport. These findings suggest that female yellowfin tuna have superior digestive enzyme activities and lipid metabolism, while male yellowfin tuna excel in sugar metabolism, ATP production, and antioxidant defense. This study provides valuable insights into sex differences in yellowfin tuna and could aid in advancing full-cycle aquaculture practices for this species.

Keywords: *Thunnus albacares*; liver; metabolism; transcriptome; enzyme activity

Key Contribution: Female and male yellowfin tuna differ in digestion, lipid, and sugar metabolism.

1. Introduction

Yellowfin tuna (*Thunnus albacares*) is a highly sought-after marine species globally, belonging to the family Mackerelidae and the genus Tuna [1]. This pelagic migratory fish is predominantly found in tropical and subtropical waters of the Pacific, Atlantic, and Indian Oceans [2]. Yellowfin tuna can grow rapidly, reaching lengths of up to 200 cm, with growth rates influenced by sex and body size. Males tend to be larger than females, and growth accelerates after reaching a body length of over 63 cm [3,4]. Due to its distinct flavor and high nutritional value, yellowfin tuna has gained popularity as a premium seafood in the international market [5]. Global landings of yellowfin tuna have averaged approximately 1.25 million tons annually over the last decade, making it the second largest tuna species worldwide [6]. Studies have indicated a decline in wild spawning of yellowfin tuna since the 1970s, coupled with an increase in fishing mortality among adults and juveniles. The wild populations of yellowfin tuna are currently fully exploited, particularly in the Central and Western Pacific [7]. However, the quality and food safety of yellowfin tuna are susceptible to various environmental and biological factors, including overfishing and climate change [8–10]. Therefore, the establishment of aquaculture-based production systems for yellowfin tuna is crucial. This approach will not only ensure a consistent supply of yellowfin tuna without being limited by seasonal or geographical factors, but also alleviate pressure on wild stocks and promote sustainable resource management.

Currently, yellowfin tuna is farmed mainly through capture-based aquaculture, where wild juveniles or subadults are captured and fattened up in nets for several months before being harvested [11,12]. This practice has been adopted in countries such as Mexico, Panama, and Indonesia [13,14]. In China, the artificial culture of yellowfin tuna is still at an early stage. The Deep-sea Aquaculture Technology and Species Development Innovation Team of the Chinese Academy of Fisheries Sciences has realized the indoor recirculating water and offshore deep-water net-pen culture of yellowfin tuna in Lingshui Lizu autonomous county of Hainan province [15].

To develop the full-cycle aquaculture of yellowfin tuna, it is necessary to understand the biological and physiological characteristics of the species, such as reproduction, nutrition, and metabolism [16]. The liver plays a key role in fish nutrition by receiving and distributing large amounts of dietary nutrients through the portal vein, which is directly connected to the digestive tract [17]. The liver also performs essential metabolic functions such as processing and storing nutrients, synthesizing enzymes and other cofactors, forming and secreting bile, and metabolizing xenobiotic compounds [18]. The liver plays a central role in growth regulation and has, therefore, been extensively studied to reveal the genetic and metabolic mechanisms that lead to differences in growth rates in fishes [19–21], and these findings have contributed to a better understanding of the regulation of growth.

By studying changes in liver enzyme activities, we can provide optimized feeding strategies to improve the growth and health of yellowfin tuna in aquaculture. Transcriptome sequencing technology is a powerful tool to analyze the type, structure, and expression level of all transcription products of a specific tissue or cell under different conditions, which can reveal the molecular regulatory mechanisms of specific biological processes. This technique has been widely used in studies of fish growth and metabolism [22,23].

In this study, we compared the physiological indices of male and female yellowfin tuna and performed comparative transcriptome analysis to identify responsive genes in the liver. Our aim was to gain a comprehensive understanding of sex differences in liver functions. These data will help to add biometric data to yellowfin tuna culture.

2. Materials and Methods

2.1. Fish and Sample Preparation

Wild yellowfin tuna were caught with baited lines in the South China Sea (17°24′ N, 110°36′ E) in mid-May 2023, and the sex of the fish was determined by gonadal morphology observation. Three females and three males of mature individuals (body length: 100–125 cm, body weight: 13–26.5 kg) were selected for sampling. Liver samples were rapidly frozen in liquid nitrogen and then stored at −80 °C until RNA extraction and biochemical analysis. Meanwhile, small pieces of gonadal and liver tissues from each fish were fixed in Bouin's solution for histological analysis.

2.2. Histological Procedures for Liver and Gonad Tissues

Tissue specimens were fixed in Bouin's solution for 24 h and then transferred to 70% ethanol for dehydration. Next, the tissues were further dehydrated in a gradient ethanol series (75–100%), and then cleared with xylene and embedded in molten paraffin. Finally, serial sections were made at 6–8 μm. The sections were stained with hematoxylin and eosin (H&E) and evaluated histomorphologically using a light microscope (Nikon IQ50, Tokyo, Japan).

2.3. Measurement of Enzyme Activities

The liver samples were weighed and added to ice-cold 0.85% physiological saline in a sample to saline mass ratio of 1:10. The samples were then homogenized using a tissue cell crusher in an ice bath. After centrifugation at 4 °C and 900× g for 10 min, the supernatant was collected as the enzyme source solution. The total protein content of this solution was determined using the BCA method [24]. Enzyme activities related to carbohydrate and lipid metabolism, as well as oxidative stress, were assessed using commercial assay kits following the manufacturer's instructions. The activities were measured by using the absorbance of reaction products or substrates at specific wavelengths with a photometric microplate reader (multiscan MK3, Thermo Fisher Scientific, Chelmsford, MA, USA). The enzymes analyzed and their corresponding assay kits are detailed in Table S1, and include amylase (AMS), lipase (LPS), malondialdehyde (MDA), pyruvate kinase (PK), acyl-CoA oxidase (ACO), malic enzyme (NADP-ME), fatty acid synthase (FAS), glucose-6-phosphate dehydrogenase (G-6-PD), acetyl-CoA carboxylase (ACC), carnitine-acylcarnitine translocase (CACT), lipoprotein lipase (LPL), superoxide dismutase (SOD), glutathione peroxidase (GPX), and catalase (CAT).

2.4. RNA-Seq and Bioinformatics Analysis

Fish sex and gonadal stages were determined by histological methods. Six liver samples (three of each sex) were used for preparing transcriptome (RNA-Seq) sequencing libraries. The RNA-Seq process was carried out as previously described [25], with the following steps: isolation of total RNA using a TRIzol kit (Invitrogen, Carlsbad, CA, USA); quantification and integrity assessment of RNA using a Nanodrop 2000c spectrophotometer and Agilent 2100 Bioanalyzer system; and construction of cDNA libraries following Illumina RNA sequencing protocol. Sequencing was performed using the Illumina HiSeq™ 2000 platform, which generated paired-end (PE) reads of 125 bp. The sequencing data were submitted to the NCBI Sequence Read Archive (SRA) under BioProject number PRJNA1013240.

The bioinformatics analysis included read quality control, assembly, annotation, and differential expression analysis. The sequences were further processed using the bioinformatics pipeline tool BMKCloud (www.biocloud.net, accessed on 19 July 2023) online platform. First, the quality control of raw sequencing data was performed using an in-house perl script to remove adaptor sequences, reads containing poly-N sequences, and low-quality reads. Clean RNA-Seq data were then assembled using Trinity Assembler with default parameters [26]. Gene expression levels were quantified in fragments per kilobase per million reads (FPKM) using RSEM v1.2.21 [27]. Gene function was annotated using the NR (NCBI non-redundant protein sequences), Pfam (Protein family), KOG/COG/eggNOG (Clusters of Orthologous Groups of proteins), Swiss-Prot (manually annotated and reviewed protein sequence database), KEGG (Kyoto Encyclopedia of Genes and Genomes), and GO (Gene Ontology) databases. All subsequent analyses were conducted with males regarded as the control group.

2.5. Real-Time Quantitative PCR (RT-qPCR) Validation

The liver samples were homogenized and total RNA was extracted using TRIzol reagent (Invitrogen, Carlsbad, CA, USA) according to the manufacturer's instructions. The RNA was reverse-transcribed to cDNA using the RevertAid First-Strand CDNA Synthesis Kit (Thermo Scientific, Waltham, MA, USA). cDNA was used for RT-qPCR to quantify the expression levels of target genes using the SYBR Premix Ex Taq II (TaKaRa Bio Inc., Shiga, Japan) on a LightCycler 480 system (Roche, Basel, Switzerland). Each sample was analyzed in three biological replicates and three technical replicates. The reference gene β-actin was used as an internal control to standardize the mRNA levels. In addition, the amplification efficiency and correlation coefficient (R2) were determined by standard curves of 10-fold dilutions (1, 1/10 1/100 1/1000 and 1/10,000) of cDNA template. The primer pairs for the target genes are listed in Table S2. Relative gene expression levels were calculated using the $2^{-\Delta\Delta Ct}$ method.

2.6. Statistical Analysis

Data are presented as mean ± standard error of the mean (SEM). One-way ANOVA was performed using SPSS Statistics 24.0 software (SPSS Inc., Chicago, IL, USA) to test the differences among groups. A p-value of less than 0.05 was considered statistically significant.

3. Results

3.1. Histological Observation

All tuna individuals analyzed in this study were sexually mature (Figure S1). Figure 1 illustrates male and female yellowfin tuna liver cells, which exhibited a flattened and compact texture. Although no significant differences were detected in the liver cells between the male and female yellowfin tuna, the male hepatocytes showed a deeper basophilic reaction compared to the female hepatocytes, suggesting higher metabolic activity. Additionally, vacuolization was more pronounced in female hepatocytes, possibly associated with increased lipid metabolism.

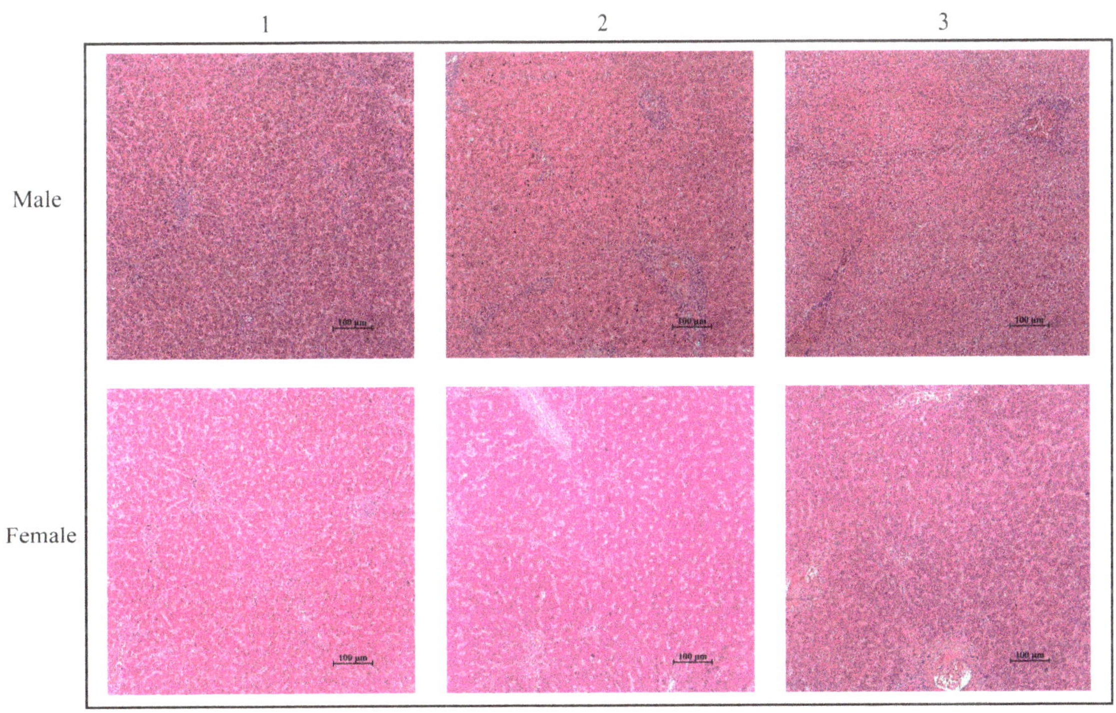

Figure 1. Histological observation of the liver of male and female tuna. Scale bar: 100 μm.

3.2. Differences in Liver Metabolic Enzymes between Male and Female Yellowfin Tuna

Differences in digestive enzyme activities, energy metabolism, lipid metabolism, and antioxidant enzyme activities in the liver of male and female yellowfin tuna were tested. In terms of digestive enzyme activities (Figure 2A), the AMS activity ($p < 0.01$) and LPS ($p < 0.05$) activity were significantly higher in females than in males. LPL did not differ significantly between the sexes. In terms of energy metabolism (Figure 2B), the hepatic

G6PD activity was significantly higher in males than in females ($p < 0.05$). A similar result was observed for NADP-ME ($p < 0.01$), while no significant difference was found for PK. The activity of hepatic ACO was significantly higher in females than in males ($p < 0.05$). There were no statistically significant differences in the liver for CACT ($p > 0.05$), ACC ($p > 0.05$), or FAS ($p > 0.05$). For antioxidant enzymes (Figure 2D), namely, GSH-PX, males exhibited higher activity ($p < 0.05$).

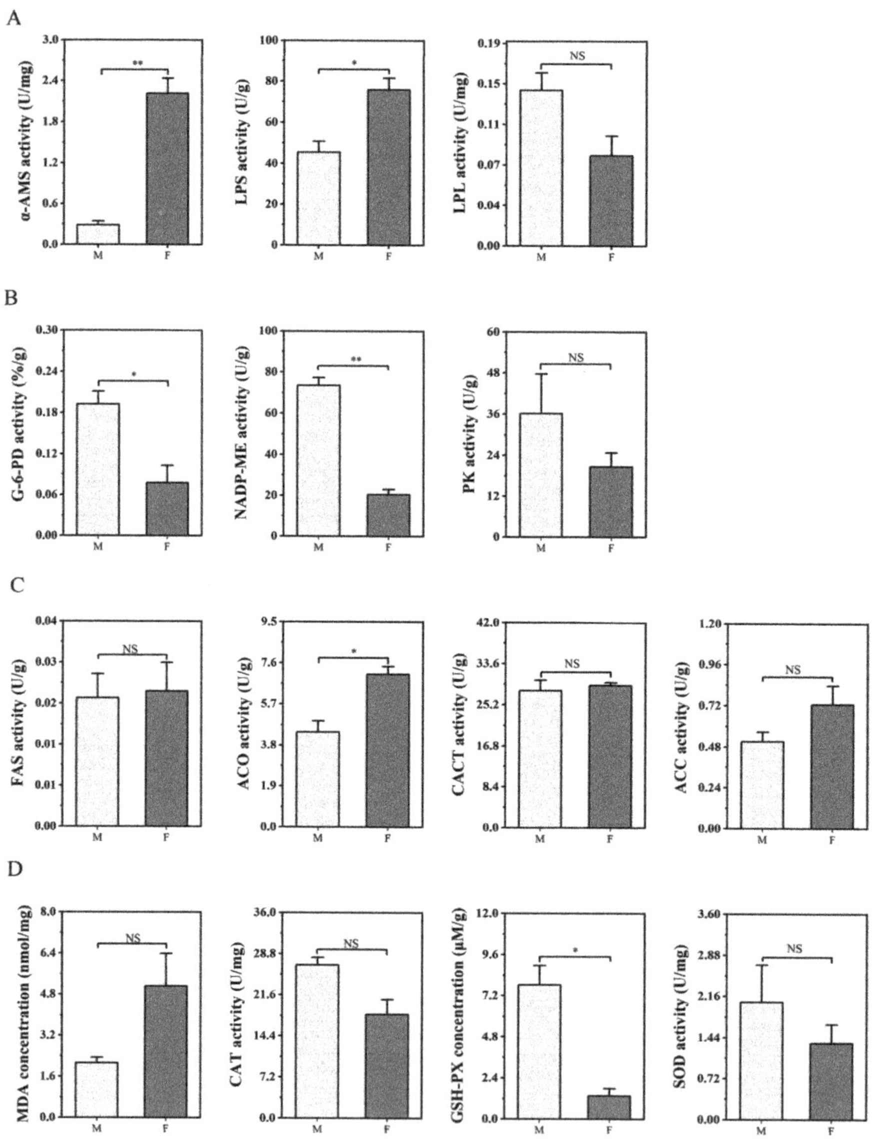

Figure 2. Comparative assessment of liver enzyme activities related to (**A**) digestion, (**B**) energy metabolism, (**C**) lipid metabolism, and (**D**) antioxidant defense in male and female yellowfin tuna. Error lines indicate SEM ($n = 3$). Asterisks indicate significant differences between the two groups (* $p < 0.05$, ** $p < 0.01$). NS: not significant ($p > 0.05$).

3.3. RNA-Seq of the Liver Transcriptome

RNA from the liver tissues from adult female and male yellowfin tuna was sequenced using synthetic sequencing (SBS) technology on the Illumina HiSeq high-throughput platform. Raw reads from F-1, F-2, F-3, M-1, M-2, and M-3 samples resulted in 37.74 Gb of clean data after quality control. The number of clean reads for each sample ranged from 19,904,230 to 23,628,924. The Q30 base percentages for all samples exceeded 93.57%, with GC content ranging from 43.62% to 49.18% (see Table 1).

Table 1. Statistics for evaluation of sample sequencing data.

ID	Clean Reads	GC Content	% ≥ Q30
F-1	21,116,430	48.53%	93.71%
F-2	21,116,430	48.72%	94.01%
F-3	23,628,924	49.18%	94.17%
M-1	20,375,537	46.30%	93.57%
M-2	19,904,230	43.62%	93.98%
M-3	20,183,521	47.51%	93.67%

F represents female yellowfin tuna, M represents male yellowfin tuna, and the suffixed number is the sample number.

The de novo assembly yielded 36,482 UniGenes with an average length of 1128 bp. Among these, 7460 (20.45%) UniGenes had lengths between 1000 bp and 2000 bp, while 6879 (18.86%) UniGenes had lengths exceeding 2000 bp (Table S3). The UniGene sequences underwent annotation by comparing them with various databases including COG, GO, KEGG, KOG, Pfam, Swiss-Prot, TrEMBL, eggNOG, and NR (Table 2). A total of 19,440 UniGenes were annotated using BLAST with an E-value of 1×10^{-5} and HMMER with an E-value of 1×10^{-10}. Across the database comparisons, the percentages of genes annotated were as follows: COG (16.55%), GO (84.69%), KOG (59.35%), Pfam (66.89%), Swiss-Prot (46.18%), TrEMBL (98.53%), eggNOG (85.63%), and NR (97.69%). These results demonstrate the robustness and credibility of the assembly.

Table 2. UniGene annotation statistics.

Anno_Database	Number Annotated_	300 ≤ Length < 1000	Length ≥ 1000	Percentage
COG_Annotation	3217	481	2589	16.55
GO_Annotation	16,464	4328	9749	84.69
KEGG_Annotation	15,549	4046	9279	79.98
KOG_Annotation	11,537	2689	7503	59.35
Pfam_Annotation	13,003	2881	8979	66.89
Swissprot_Annotation	8977	1998	6041	46.18
TrEMBL_Annotation	19,155	5279	10,932	98.53
eggNOG_Annotation	16,647	4429	9752	85.63
nr_Annotation	18,991	5205	10,868	97.69
All_Annotated	19,440	5416	10,994	100.00

3.4. Differential Expression Analysis

A total of 2542 genes were identified in this study as being differentially expressed between male and female groups based on expression levels. Among these genes, 1195 were down-regulated and 1347 were up-regulated in females (Figure 3A). Heatmaps of hierarchical clusters of DEGs were used to visually represent the overall gene expression pattern between the sexes. The male samples were first clustered separately and then grouped with the female branch, indicating significant differences in transcription patterns between the male and female livers (Figure 3B). The differentially expressed genes were annotated, with a total of 1894 genes annotated across various databases including COG, GO, KEGG, KOG, NR, Pfam, Swiss-Prot, and eggNOG (Table 3). Additionally, 10 DEGs were selected and validated using RT-qPCR (Figure 3C), confirming the reliability and accuracy of gene expression levels determined through transcriptomics analysis.

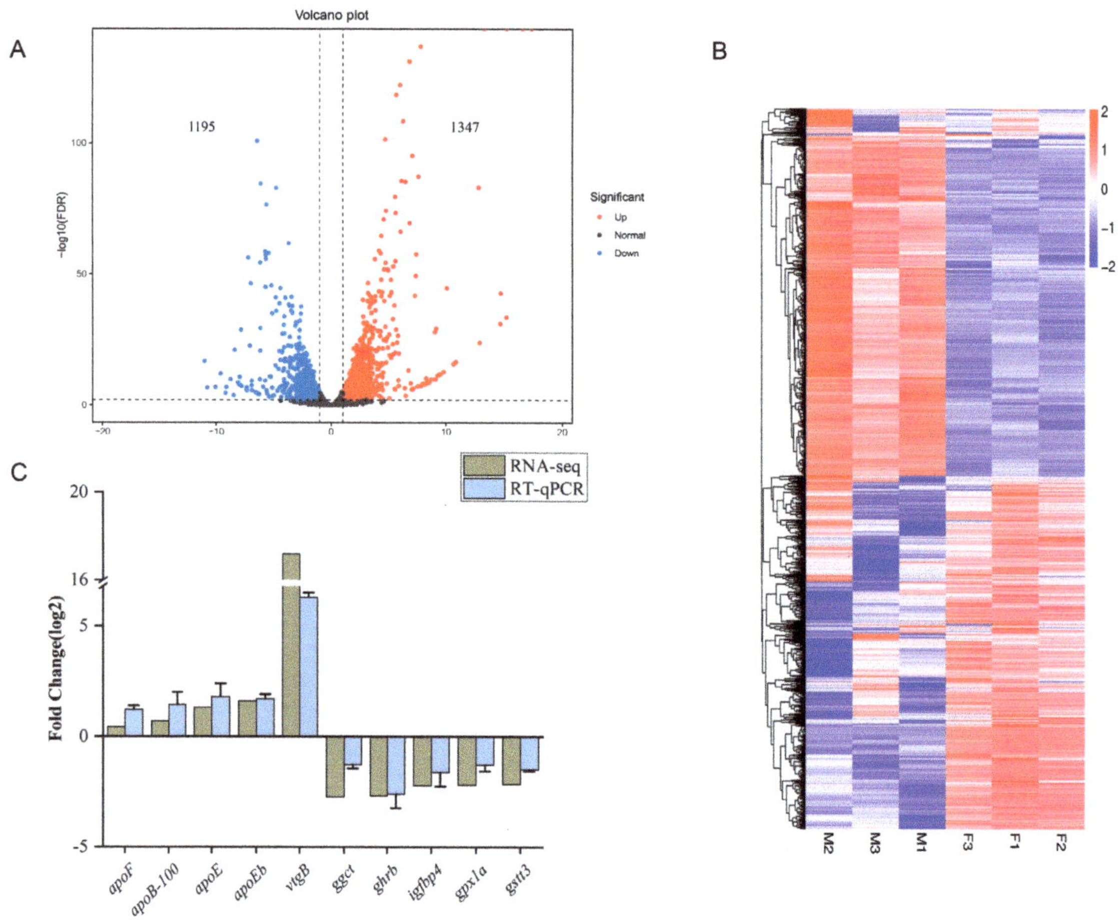

Figure 3. Transcriptome differential gene validation and its clustering analysis. (**A**) Male and female yellowfin tuna liver basal differential genes. (**B**) Clustering diagram of differential gene expression patterns for illustrating the overall pattern of gene expression in different liver samples. (**C**) RT-qPCR differential expression validation. Error lines indicate SEM ($n = 3$).

Table 3. Statistics on the number of differentially expressed genes annotated with various databases.

DEG Set	Total	COG	GO	KEGG	KOG	NR	Pfam	Swiss-Prot	eggNOG
MvsF	1894	519	1681	1600	1308	1883	1582	1063	1697

3.5. Enriched GO Terms and KEGG Pathways

Based on the GO enrichment analysis, the DEGs were categorized into three main functional classes: cellular components (CCs), molecular functions (MFs), and biological processes (BPs) (Figure 4). The top three GO terms involved in biological processes included serine family amino acid metabolism (GO:0009069), cellular amino acid metabolism (GO:0006520), and lipid transport (GO:0006869). The cellular-component-class DEGs included integral component of membrane (GO:0016021), endoplasmic reticulum membrane (GO:0005789), and proteasome regulatory particle (GO:0005838). The top three molecular function terms were lipid transporter activity (GO:0005319), transmembrane transporter activity (GO:0022857), and amino acid binding (GO:0016597). 0016597). Additionally, we noted that a number of lipid-related GO terms were enriched, including fatty acid transport (GO:0015908), lipid localization (GO:0010876), lipid catabolic process (GO:0016042), and triglyceride metabolic process (GO:0006641).

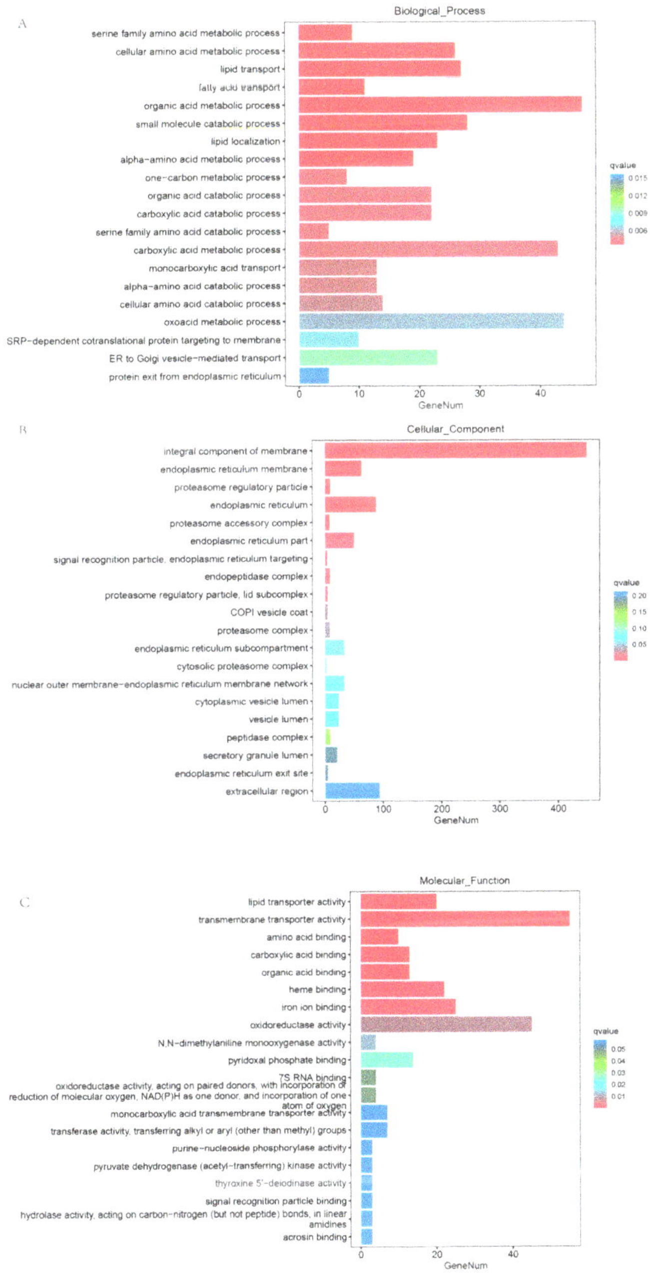

Figure 4. Functional categorization of differentially expressed genes in liver GO enrichment. (**A**) Biological process, (**B**) molecular function, and (**C**) cellular component.

The annotation of differentially expressed genes using KEGG identified a total of 205 enriched pathways. These results were further categorized based on pathway types (Figure 5A). The liver transcriptome of yellowfin tuna exhibited involvement in six pathway types, including cellular processes, environmental information processing, genetic information processing, human diseases, and metabolism. Notably, pathways such as protein processing in the endoplasmic reticulum, endocytosis, MAPK signaling pathway, and FoxO signaling pathway showed a high percentage of DEGs. The up- and down-regulation results, along with the numbers of enriched genes, for the top 20 significant pathways are depicted in Figure 5B. Functionally, these pathways and the genes within them primarily relate to metabolism, encompassing lipid metabolism (steroid hormone biosynthesis, glycerophospholipid metabolism, fatty acid degradation, arachidonic acid metabolism, and steroid biosynthesis), carbohydrate metabolism (pentose and glucuronate interconversions, glyoxylate and dicarboxylate metabolism), and amino acid metabolism (glycine, serine, and threonine metabolism; arginine and proline metabolism; tryptophan metabolism; and alanine, aspartate, and glutamate metabolism). These functional categories offer insights into the differences in liver metabolism between male and female yellowfin tuna. Table S4 showcases the gene expression profiles related to these enriched pathways.

A

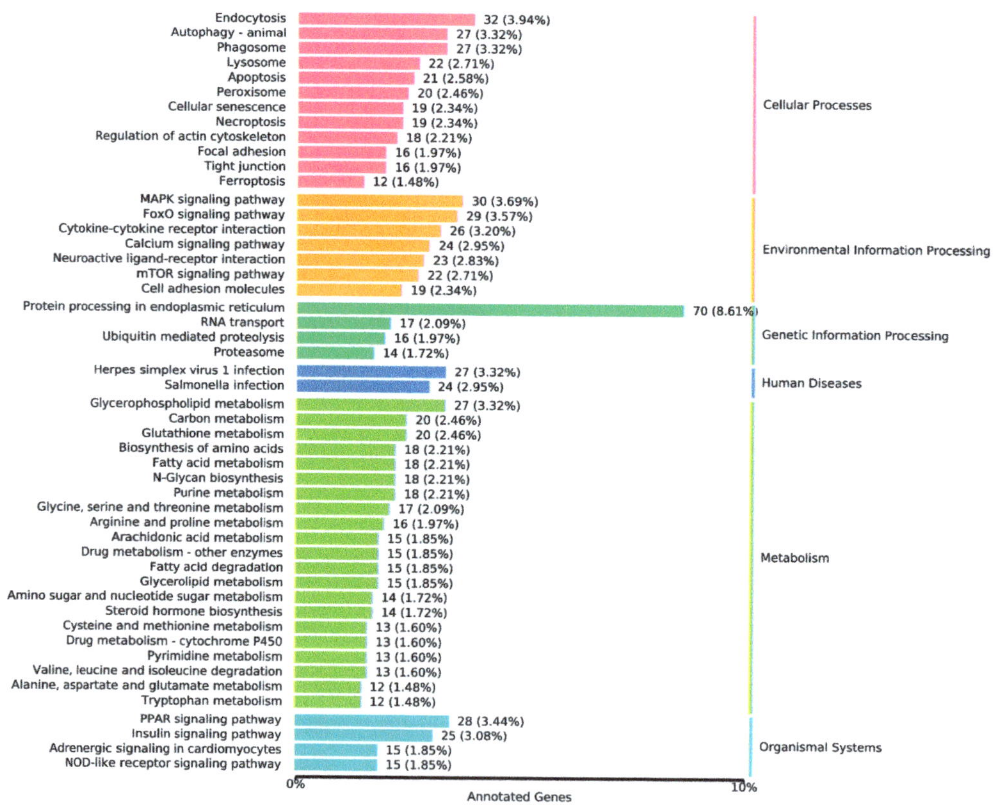

Figure 5. *Cont.*

B

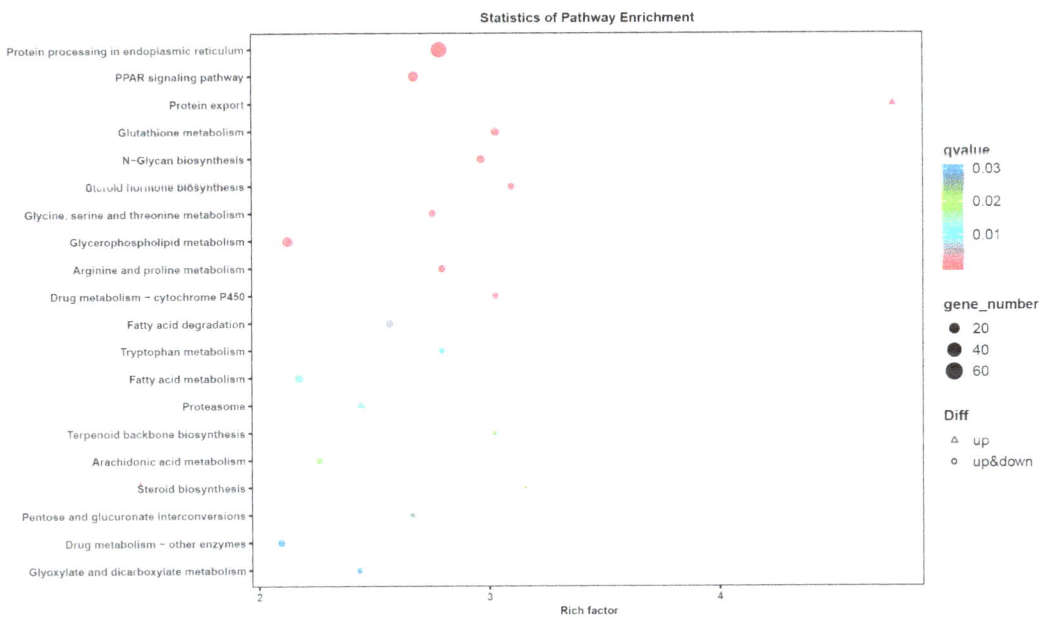

Figure 5. (**A**) KEGG categorization diagram of differentially expressed genes, with metabolic pathway names on the vertical axis, pathway names on the left, and corresponding categorization categories on the right. Colors in the same column indicate the same category. (**B**) Scatter plot of KEGG pathway enrichment for differentially expressed genes, with the pathway names on the vertical axis, enrichment factor on the horizontal axis, color representing the q value (*p*-value corrected by multiple hypothesis testing), and circle size indicating the number of enriched genes in the pathway. In the gene expression pathways context, 'up' signifies that all genes in the pathway are up-regulated, while 'up&down' indicates a mix of up-regulated and down-regulated genes.

4. Discussion

4.1. Digestive Differences

Fish liver contains a rich enzyme system that plays a crucial role in metabolic processes, particularly in growth and development. Among them, digestive enzymes are the most important class of enzymes in fish liver, which can break down large molecules such as proteins, fats and starches in food, releasing small molecules such as amino acids, fatty acids, and monosaccharides for absorption and utilization [28,29]. We compared the differences in digestive enzyme activity and related gene expression in the liver between female and male yellowfin tuna. The results showed that female yellowfin tuna had higher amylase (α-AMS) and lipase (LPS) activities, and at the same time, the transcriptome results showed that genes relate to glyceride metabolism were significantly up-regulated in females. These genes include *lipc*, *plpp1*, *pipp2d*, and *plpp5*. Lipc encodes hepatic lipase, one of the three members of the triglyceride lipase family, which can promote the degradation and uptake of vascular lipoproteins and participate in the hydrolysis of triglycerides [30]. *Plpp1*, *pipp2d*, and *plpp5* belong to the PAP2-like superfamily, which are transmembrane phospholipid phosphatases that can convert phospholipids into diacylglycerol, thereby regulating lipid metabolism [31]. These results indicate that female yellowfin tuna have higher digestive capacity, which can improve food utilization efficiency.

4.2. Energy Metabolism Differences

Energy metabolism plays a crucial role in fish, particularly in species with seasonal variations in reproductive activity. The energy allocation pattern shifts throughout the reproductive cycle and can be influenced by hormone levels. For instance, in *Oncorhynchus masou*, enzyme activities like pyruvate kinase, lactate dehydrogenase, and malate dehydrogenase decrease as the spawning season nears [32]. An important indicator of energy metabolism is the activity of G-6-PD, the primary and rate-limiting enzyme of the pentose phosphate pathway. The products of this pathway, NADPH and 5-phosphoribose, are crucial for fatty-acid synthesis and glutathione reduction [33,34]. Our study compared G-6-PD and NADP-ME activities in the liver of male and female yellowfin tuna. The results revealed higher G-6-PD and NADP-ME activities in male yellowfin tuna, suggesting increased NADPH production. Transcriptome analysis also showed the enrichment of energy metabolism pathways in males, such as glycolysis/gluconeogenesis (ko00010) and citrate cycle (ko00020), which are linked to ATP synthesis. These findings indicate that male yellowfin tuna exhibit enhanced glucose metabolism, ATP production, and overall activity in May.

4.3. Lipid Metabolism Differences

Lipid metabolism is a crucial process for fish, especially for those with high reproductive demands, such as yellowfin tuna. The liver is the main site of lipid synthesis, storage, and transport in fish [35]. We investigated the sex differences in lipid metabolism of yellowfin tuna liver by measuring the enzyme activity and gene expression of key enzymes and transporters involved in fatty acid synthesis, degradation, and transport. We found that female yellowfin tuna had a higher lipid metabolism capacity than male yellowfin tuna, which may be needed to meet their reproductive needs.

One of our main findings was that female yellowfin tuna had significantly higher ACO activity than male yellowfin tuna. ACO is the rate-limiting enzyme of fatty acid β oxidation [36]. This finding indicates that female yellowfin tuna can oxidize more fatty acids to produce energy. In addition, we observed a significant increase in the expression of carnitine palmitoyltransferase 1 (*cpt1*) and carnitine palmitoyltransferase 2 (*cpt2*) genes in female yellowfin tuna. These genes encode CPT1 and CPT2, which are key enzymes for transporting fatty acids to mitochondria for β oxidation [37]. These results indicate that female yellowfin tuna have a stronger fatty-acid degradation pathway.

The female yellowfin tuna also had higher expression levels for genes involved in fatty acid synthesis, such as acetyl-CoA carboxylase 1 (*acc1*) and fatty acid synthase (*fas*). ACC1 is a multifunctional enzyme that catalyzes the formation of malonyl-CoA, the precursor of fatty acid synthesis, and also prevents the transfer of acyl groups from acyl-CoA to carnitine [38]. FAS is a homodimeric protein that synthesizes fatty acids from acetyl-CoA and malonyl-CoA [39]. FAS is expressed in most tissues, but mainly in the liver and adipose tissue, and plays a role in lipid metabolism and energy balance [39]. Although we did not detect significant differences in the activities of FAS, CACT, or ACC between male and female yellowfin tuna, the up-regulation of *acc1* and *fas* genes indicates that female yellowfin tuna have higher potential for fatty acid synthesis.

In addition, we also found that female yellowfin tuna had higher expression of genes related to lipid transport and absorption, such as apolipoprotein F (*apoF*), apolipoprotein B-100 (*apoB-100*), apolipoprotein E (*apoE*), and apolipoprotein Eb (*apoEb*). These genes encode apolipoproteins, which are the protein components of lipoproteins that carry lipids in blood and tissues [40]. Apolipoprotein E is particularly important for lipid metabolism, as it can interact with various lipoprotein receptors and mediate the clearance of lipoproteins from the blood circulation [41]. Apolipoprotein E is also involved in oocytes' uptake of yolk [42]. Moreover, we observed a simultaneous up-regulation of vitellogenin B (*vtgb*) and vitellogenin C (*vtgc*) genes in the female yellowfin tuna. These genes encode vitellogenins, which are yolk proteins synthesized in the liver and transported to the ovary [43]. These results indicate that female yellowfin tuna have a more active lipid transport, possibly to provide lipids for oocyte development and maturation.

4.4. Antioxidant Differences

The antioxidant capacity of fish is closely related to their health status. Antioxidant capacity is the ability of fish to scavenge reactive oxygen species (ROS) and protect cells from oxidative damage. ROS are generated as by-products of normal metabolism or under environmental stress, such as temperature changes, hypoxia, or pollution. ROS can cause lipid peroxidation, which is a common oxidative stress reaction that causes damage to cell membranes and lipophilic substances. Lipid peroxidation can be measured by MDA content, which is an important indicator of lipid peroxidation [44]. We compared the differences in antioxidant enzyme activity and related gene expression in the liver between female and male yellowfin tuna. While there was no notable disparity in MDA content in the livers of female and male yellowfin tuna, there was a discernible trend towards an increase, potentially linked to females' comparatively lower antioxidant enzyme activities. Antioxidant enzymes are a class of enzymes that can scavenge reactive oxygen species and protect cells from oxidative damage, including GSH-Px, CAT, and SOD.

We found that male yellowfin tuna liver had higher GSH-Px activity than female yellowfin tuna liver, while CAT and SOD activities also showed an upward trend. GSH-Px can remove harmful substances such as lipid hydroperoxides [45], while CAT and SOD can decompose ROS such as hydrogen peroxide and superoxide anion, respectively [46].

At the same time, the transcriptome also detected the expression level of genes related to antioxidants in the liver and found that genes related to glutathione metabolism were significantly up-regulated in males, including *gpx1a*, *gstt3*, and *ggct*. These genes encode GSH-Px, glutathione S-transferase, and γ-glutamyl cycle transferase, respectively, which are involved in the regulation of glutathione metabolism. Other studies have also found sex-related differences in antioxidant defense. For example, in tilapia (*Oreochromis mossambicus*), males had higher SOD and glutathione S-transferase activities than females [47]; in brown trout (*Salmo trutta*), females had higher SOD and CAT activities than males [48]. These differences may be related to the different reproductive strategies, energy metabolism, and environmental adaptability of different fishes [49]. In yellowfin tuna, we suggest that the lower antioxidant level of females may be due to their more efficient lipid metabolism, because lipid metabolism produces more peroxides, and excessive saturated fatty acids or other lipids (such as phospholipids) increase the susceptibility of cell membranes to oxidations [50].

4.5. Expression Differences of Growth-Hormone-Axis-Related Genes

One of the factors regulating fish growth is the growth hormone axis, which includes the pituitary secretion of growth hormone (GH) and liver production of GH-dependent insulin-like growth factor I (IGF-I). GH binds to receptors (GHRs) on various target tissues, activating intracellular signaling pathways that affect growth, metabolism, and immunity [51,52]. IGF-I can stimulate cell proliferation, differentiation, and metabolism, thereby promoting body development and growth. IGF-I also can bind to its binding proteins (IGFBPs), thereby regulating the bioavailability and biological effects of IGF-I. We found that the transcription levels of two isoforms of GHR, *ghra* and *ghrb*, in male liver were higher than those in female liver. This result indicates that male fish have a stronger ability to respond to GH stimulation and activate downstream signaling pathways. We also observed that the expression level of the IGFBP4 gene in male liver was higher than that in female liver. IGFBP4 is a regulatory factor of IGF-I activity, as it can inhibit the binding of IGF-I to its receptor and shorten its half-life. The increase in IGFBP4 expression in male fish may be a feedback mechanism to prevent the excessive transmission of the IGF-I signal and maintain homeostasis [53,54]. Differences in growth axis function may account for the more rapid growth of males. However, growth is a complex biological process with continuous changes, so more research is needed to elucidate the molecular differences between male and female yellowfin tuna in growth axis function.

5. Conclusions

This study compared the liver enzyme activities and transcriptome profiles of female and male yellowfin tuna to investigate sex-related differences at a molecular level. The findings highlighted differences in digestion, energy metabolism, lipid metabolism, antioxidant defense, and growth axis function between the two sexes. Female yellowfin tuna exhibited superior digestive and lipid metabolism capabilities, likely to support reproductive needs and enhance food efficiency. Conversely, male yellowfin tuna showed higher sugar metabolism, ATP production, and antioxidant defense, possibly due to their more active lifestyle and need to combat oxidative stress. Moreover, males displayed a stronger response to growth hormone and insulin-like growth factor I, indicating differences in growth axis function contributing to the faster growth of males. Transcriptomic analysis revealed significant liver differences between the female and male yellowfin tuna, reflecting distinct physiological states and metabolic demands. Overall, these insights contribute valuable biometric data for yellowfin tuna aquaculture and lay a foundation for the further exploration of sex-related differences at a molecular level.

Supplementary Materials: The following supporting information can be downloaded at https://www.mdpi.com/article/10.3390/fishes9050184/s1: Table S1: Product numbers of commercial assay kits; Table S2: Primers used for detection of DEGs; Table S3: Assembly result statistics; Table S4: Gene expression of partially enriched pathways; Figure S1: Histological identification of male and female yellowfin tuna gonads.

Author Contributions: Data curation, D.L. and H.Y.; Formal analysis, H.Y. and D.L.; Funding acquisition, H.C.; Investigation, H.Y., D.L. and S.L.; Methodology, H.Y. and H.H.; Project administration, G.L. and H.C.; Resources, H.C.; Supervision, H.C.; Writing—original draft, D.L.; Writing—review and editing, H.C. All authors have read and agreed to the published version of the manuscript.

Funding: This article was funded by grants from the Natural Science Foundation of China (32273131), the Science and Technology Plan of Guangdong province (2023B0202010016), the Youth Science and Technology Innovation Talent of Guangdong TeZhi plan talent (2023TQ07A888), the Science and Technology Plan of Zhanjiang City (2022A01214), and the Science and Technology Plan of Yangjiang City (2022011 and SDZX2023027).

Institutional Review Board Statement: Animal handling techniques strictly adhered to the Guidelines for the Care and Use of Laboratory Animals. The Animal Research and Ethics Committee of Guangdong Ocean University authorized this protocol (NIH Pub. No. 85-23, revised 1996).

Data Availability Statement: Sequencing data were submitted to the NCBI Sequence Read Archive (SRA) under BioProject number PRJNA1013240. Data are contained within the article and Supplementary Materials.

Acknowledgments: We wish to thank the anonymous reviewers and the editor of the journal for their valuable comments.

Conflicts of Interest: Dongge Liu was employed by the company Guangdong Havwii Agriculture Group Co., Ltd. The remaining authors declare that the research was conducted in the absence of any commercial or financial relationships that could be construed as a potential conflict of interest.

References

1. Báez, J.C.; Czerwinski, I.A.; Ramos, M.L. Climatic Oscillations Effect on the Yellowfin Tuna (*Thunnus albacares*) Spanish Captures in the Indian Ocean. *Fish. Oceanogr.* **2020**, *29*, 572–583. [CrossRef]
2. Tridjoko; Hutapea, J.H.; Setiadi, A.; Gunawan; Selamet, B. Maintenance and Spawning on Yellowfin Tuna Broodstock Reared in Floating Net Cage. *IOP Conf. Ser. Earth Environ. Sci.* **2021**, *890*, 012039. [CrossRef]
3. Dortel, E.; Massiot-Granier, F.; Rivot, E.; Million, J.; Hallier, J.-P.; Morize, E.; Munaron, J.-M.; Bousquet, N.; Chassot, E. Accounting for Age Uncertainty in Growth Modeling, the Case Study of Yellowfin Tuna (*Thunnus albacares*) of the Indian Ocean. *PLoS ONE* **2013**, *8*, e60886. [CrossRef] [PubMed]
4. Gascuel, D.; Fonteneau, A.; Capisano, C. Modélisation d'une Croissance En Deux Stances Chez l'albacore (*Thunnus albacares*) de l'Atlantique Est. *Aquat. Living Resour.* **1992**, *5*, 155–172. [CrossRef]

5. Zhang, Y.; Ma, X.; Dai, Z. Comparison of Nonvolatile and Volatile Compounds in Raw, Cooked, and Canned Yellowfin Tuna (*Thunnus albacores*). *J. Food Process. Preserv.* **2019**, *43*, e14111. [CrossRef]
6. Pacicco, A.E.; Brown-Peterson, N.J.; Murie, D.J.; Allman, R.J.; Snodgrass, D.; Franks, J.S. Reproductive Biology of Yellowfin Tuna (*Thunnus albacares*) in the Northcentral U.S. Gulf of Mexico. *Fish. Res.* **2023**, *261*, 106620. [CrossRef]
7. Liu, H.; Fu, Z.; Yu, G.; Ma, Z.; Zong, H. Effects of Acute High-Temperature Stress on Physical Responses of Yellowfin Tuna (*Thunnus albacares*). *J. Mar. Sci. Eng.* **2022**, *10*, 1857. [CrossRef]
8. Nicol, S.; Lehodey, P.; Senina, I.; Bromhead, D.; Frommel, A.Y.; Hampton, J.; Havenhand, J.; Margulies, D.; Munday, P.L.; Scholey, V.; et al. Ocean Futures for the World's Largest Yellowfin Tuna Population Under the Combined Effects of Ocean Warming and Acidification. *Front. Mar. Sci.* **2022**, *9*, 816772. [CrossRef]
9. Pecoraro, C.; Zudaire, I.; Bodin, N.; Murua, H.; Taconet, P.; Díaz-Jaimes, P.; Cariani, A.; Tinti, F.; Chassot, E. Putting All the Pieces Together: Integrating Current Knowledge of the Biology, Ecology, Fisheries Status, Stock Structure and Management of Yellowfin Tuna (*Thunnus albacares*). *Rev. Fish Biol. Fish.* **2017**, *27*, 811–841. [CrossRef]
10. Wu, Y.-L.; Lan, K.-W.; Evans, K.; Chang, Y.-J.; Chan, J.-W. Effects of Decadal Climate Variability on Spatiotemporal Distribution of Indo-Pacific Yellowfin Tuna Population. *Sci. Rep.* **2022**, *12*, 13715. [CrossRef]
11. Bramantya, B.; Gunawan; Sari, L.A. Spawning Technique of Yellowfin Tuna (*Thunnus albacares*) Infloating Nets Cage. *IOP Conf. Ser. Earth Environ. Sci.* **2021**, *679*, 012029. [CrossRef]
12. Margulies, D.; Scholey, V.P.; Wexler, J.B.; Stein, M.S. Chapter 5—Research on the Reproductive Biology and Early Life History of Yellowfin Tuna *Thunnus albacares* in Panama. In *Advances in Tuna Aquaculture*; Benetti, D.D., Partridge, G.J., Buentello, A., Eds.; Academic Press: San Diego, CA, USA, 2016; pp. 77–114. ISBN 978-0-12-411459-3.
13. Benetti, D.D.; Partridge, G.J.; Stieglitz, J. Chapter 1—Overview on Status and Technological Advances in Tuna Aquaculture Around the World. In *Advances in Tuna Aquaculture*; Benetti, D.D., Partridge, G.J., Buentello, A., Eds.; Academic Press: San Diego, CA, USA, 2016; pp. 1–19. ISBN 978-0-12-411459-3.
14. Estess, E.E.; Klinger, D.H.; Coffey, D.M.; Gleiss, A.C.; Rowbotham, I.; Seitz, A.C.; Rodriguez, L.; Norton, A.; Block, B.; Farwell, C. Bioenergetics of Captive Yellowfin Tuna (*Thunnus albacares*). *Aquaculture* **2017**, *468*, 71–79. [CrossRef]
15. Zhou, S.; Zhang, N.; Fu, Z.; Yu, G.; Ma, Z.; Zhao, L. Impact of Salinity Changes on the Antioxidation of Juvenile Yellowfin Tuna (*Thunnus albacares*). *J. Mar. Sci. Eng.* **2023**, *11*, 132. [CrossRef]
16. Mourente, G.; Tocher, D.R. Tuna Nutrition and Feeds: Current Status and Future Perspectives. *Rev. Fish. Sci.* **2009**, *17*, 373–390. [CrossRef]
17. Roques, S.; Deborde, C.; Richard, N.; Skiba-Cassy, S.; Moing, A.; Fauconneau, B. Metabolomics and Fish Nutrition: A Review in the Context of Sustainable Feed Development. *Rev. Aquac.* **2020**, *12*, 261–282. [CrossRef]
18. Wolf, J.C.; Wolfe, M.J. A Brief Overview of Nonneoplastic Hepatic Toxicity in Fish. *Toxicol. Pathol.* **2005**, *33*, 75–85. [CrossRef]
19. Reinecke, M. Influences of the Environment on the Endocrine and Paracrine Fish Growth Hormone–Insulin-like Growth Factor-I System. *J. Fish Biol.* **2010**, *76*, 1233–1254. [CrossRef]
20. Lozano, A.R.; Borges, P.; Robaina, L.; Betancor, M.; Hernández-Cruz, C.M.; García, J.R.; Caballero, M.J.; Vergara, J.M.; Izquierdo, M. Effect of Different Dietary Vitamin E Levels on Growth, Fish Composition, Fillet Quality and Liver Histology of Meagre (*Argyrosomus regius*). *Aquaculture* **2017**, *468*, 175–183. [CrossRef]
21. Mourente, G.; Bell, J.G. Partial Replacement of Dietary Fish Oil with Blends of Vegetable Oils (Rapeseed, Linseed and Palm Oils) in Diets for European Sea Bass (*Dicentrarchus labrax* L.) over a Long Term Growth Study: Effects on Muscle and Liver Fatty Acid Composition and Effectiveness of a Fish Oil Finishing Diet. *Comp. Biochem. Physiol. B Biochem. Mol. Biol.* **2006**, *145*, 389–399. [CrossRef]
22. Wu, N.; Song, Y.-L.; Wang, B.; Zhang, X.-Y.; Zhang, X.-J.; Wang, Y.-L.; Cheng, Y.-Y.; Chen, D.-D.; Xia, X.-Q.; Lu, Y.-S.; et al. Fish Gut-Liver Immunity during Homeostasis or Inflammation Revealed by Integrative Transcriptome and Proteome Studies. *Sci. Rep.* **2016**, *6*, 36048. [CrossRef]
23. Caballero-Solares, A.; Xue, X.; Parrish, C.C.; Foroutani, M.B.; Taylor, R.G.; Rise, M.L. Changes in the Liver Transcriptome of Farmed Atlantic Salmon (*Salmo salar*) Fed Experimental Diets Based on Terrestrial Alternatives to Fish Meal and Fish Oil. *BMC Genom.* **2018**, *19*, 796. [CrossRef] [PubMed]
24. Olson, B.J.S.C. Assays for Determination of Protein Concentration. *Curr. Protoc. Pharmacol.* **2016**, *73*, A.3A.1–A.3A.32. [CrossRef] [PubMed]
25. Ren, X.; Liu, J.; Ndandala, C.B.; Li, X.; Guo, Y.; Li, G.; Chen, H. Physiological Effects and Transcriptomic Analysis of sbGnRH on the Liver in Pompano (*Trachinotus ovatus*). *Front. Endocrinol.* **2022**, *13*, 869021. [CrossRef] [PubMed]
26. Grabherr, M.G.; Haas, B.J.; Yassour, M.; Levin, J.Z.; Thompson, D.A.; Amit, I.; Adiconis, X.; Fan, L.; Raychowdhury, R.; Zeng, Q.; et al. Full-Length Transcriptome Assembly from RNA-Seq Data without a Reference Genome. *Nat. Biotechnol.* **2011**, *29*, 644–652. [CrossRef] [PubMed]
27. RSEM: Accurate Transcript Quantification from RNA-Seq Data with or without a Reference Genome | BMC Bioinformatics. Available online: https://link.springer.com/article/10.1186/1471-2105-12-323 (accessed on 15 September 2023).
28. Friedman, I.S.; Fernández-Gimenez, A.V. State of Knowledge about Biotechnological Uses of Digestive Enzymes of Marine Fishery Resources: A Worldwide Systematic Review. *Aquac. Fish.* 2023; *in press*. [CrossRef]

29. Zambonino Infante, J.L.; Cahu, C.L. Dietary Modulation of Some Digestive Enzymes and Metabolic Processes in Developing Marine Fish: Applications to Diet Formulation. *Aquaculture* **2007**, *268*, 98–105. [CrossRef]
30. Holmes, R.S.; VandeBerg, J.L.; Cox, L.A. Vertebrate Hepatic Lipase Genes and Proteins: A Review Supported by Bioinformatic Studies. *Open Access Bioinform.* **2011**, *3*, 85–95. [CrossRef] [PubMed]
31. Smith, S.W.; Weiss, S.B.; Kennedy, E.P. The enzymatic dephosphorylation of phosphatidic acids. *J. Biol. Chem.* **1957**, *228*, 915–922. [CrossRef]
32. Leonard, J.B.K.; Iwata, M.; Ueda, H. Seasonal Changes of Hormones and Muscle Enzymes in Adult Lacustrine Masu (*Oncorhynchus masou*) and Sockeye Salmon (*O. nerka*). *Fish Physiol. Biochem.* **2001**, *25*, 153–163. [CrossRef]
33. Ciftci, M.; Turkoglu, V.; Coban, T.A. Effects of Some Drugs on Hepatic Glucose 6-Phosphate Dehydrogenase Activity in Lake Van Fish (*Chalcalburnus Tarischii pallas*, 1811). *J. Hazard. Mater.* **2007**, *143*, 415–418. [CrossRef]
34. Martins, D.A.; Valente, L.M.P.; Lall, S.P. Apparent Digestibility of Lipid and Fatty Acids in Fish Oil, Poultry Fat and Vegetable Oil Diets by Atlantic Halibut, *Hippoglossus hippoglossus* L. *Aquaculture* **2009**, *294*, 132–137. [CrossRef]
35. Yuan, X.; Liang, X.-F.; Liu, L.; Fang, J.; Li, J.; Li, A.; Cai, W.; Xue, M.; Wang, J.; Wang, Q. Fat Deposition Pattern and Mechanism in Response to Dietary Lipid Levels in Grass Carp, Ctenopharyngodon Idellus. *Fish Physiol. Biochem.* **2016**, *42*, 1557–1569. [CrossRef]
36. Kondo, T.; Kishi, M.; Fushimi, T.; Kaga, T. Acetic Acid Upregulates the Expression of Genes for Fatty Acid Oxidation Enzymes in Liver To Suppress Body Fat Accumulation. *J. Agric. Food Chem.* **2009**, *57*, 5982–5986. [CrossRef]
37. Weil, C.; Lefèvre, F.; Bugeon, J. Characteristics and Metabolism of Different Adipose Tissues in Fish. *Rev. Fish Biol. Fish.* **2013**, *23*, 157–173. [CrossRef]
38. Tong, L. Acetyl-Coenzyme A Carboxylase: Crucial Metabolic Enzyme and Attractive Target for Drug Discovery. *Cell. Mol. Life Sci. CMLS* **2005**, *62*, 1784–1803. [CrossRef]
39. Tomacha, J.; Dokduang, H.; Padthaisong, S.; Namwat, N.; Klanrit, P.; Phetcharaburanin, J.; Wangwiwatsin, A.; Khampitak, T.; Koonmee, S.; Titapun, A.; et al. Targeting Fatty Acid Synthase Modulates Metabolic Pathways and Inhibits Cholangiocarcinoma Cell Progression. *Front. Pharmacol.* **2021**, *12*, 696961. [CrossRef] [PubMed]
40. Fang, X.; Wei, Y.; Liu, Y.; Wang, J.; Dai, J. The Identification of Apolipoprotein Genes in Rare Minnow (*Gobiocypris rarus*) and Their Expression Following Perfluorooctanoic Acid Exposure. *Comp. Biochem. Physiol. Part C Toxicol. Pharmacol.* **2010**, *151*, 152–159. [CrossRef]
41. St Clair, R.W.; Beisiegel, U. What Do All the Apolipoprotein E Receptors Do? *Curr. Opin. Lipidol.* **1997**, *8*, 243. [CrossRef] [PubMed]
42. Schippers, E.F.; Berbée, J.F.P.; van Disseldorp, I.M.; Versteegh, M.I.M.; Havekes, L.M.; Rensen, P.C.N.; van Dissel, J.T. Preoperative Apolipoprotein CI Levels Correlate Positively with the Proinflammatory Response in Patients Experiencing Endotoxemia Following Elective Cardiac Surgery. *Intensive Care Med.* **2008**, *34*, 1492–1497. [CrossRef]
43. Poupard, G.; André, M.; Durliat, M.; Ballagny, C.; Boeuf, G.; Babin, P.J. Apolipoprotein E Gene Expression Correlates with Endogenous Lipid Nutrition and Yolk Syncytial Layer Lipoprotein Synthesis during Fish Development. *Cell Tissue Res.* **2000**, *300*, 251–261. [CrossRef] [PubMed]
44. Moore, K.; Roberts, L.J. Measurement of Lipid Peroxidation. *Free Radic. Res.* **1998**, *28*, 659–671. [CrossRef]
45. Olsvik, P.A.; Kristensen, T.; Waagbø, R.; Rosseland, B.O.; Tollefsen, K.-E.; Baeverfjord, G.; Berntssen, M.H.G. mRNA Expression of Antioxidant Enzymes (SOD, CAT and GSH-Px) and Lipid Peroxidative Stress in Liver of Atlantic Salmon (*Salmo salar*) Exposed to Hyperoxic Water during Smoltification. *Comp. Biochem. Physiol. Part C Toxicol. Pharmacol.* **2005**, *141*, 314–323. [CrossRef] [PubMed]
46. Farombi, E.O.; Adelowo, O.A.; Ajimoko, Y.R. Biomarkers of Oxidative Stress and Heavy Metal Levels as Indicators of Environmental Pollution in African Cat Fish (*Clarias gariepinus*) from Nigeria Ogun River. *Int. J. Environ. Res. Public Health* **2007**, *4*, 158–165. [CrossRef] [PubMed]
47. Figueiredo-Fernandes, A.; Fontaínhas-Fernandes, A.; Peixoto, F.; Rocha, E.; Reis-Henriques, M.A. Effects of Gender and Temperature on Oxidative Stress Enzymes in Nile Tilapia *Oreochromis niloticus* Exposed to Paraquat. *Pestic. Biochem. Physiol.* **2006**, *85*, 97–103. [CrossRef]
48. Parolini, M.; Iacobuzio, R.; De Felice, B.; Bassano, B.; Pennati, R.; Saino, N. Age- and Sex-Dependent Variation in the Activity of Antioxidant Enzymes in the Brown Trout (*Salmo trutta*). *Fish Physiol. Biochem.* **2019**, *45*, 145–154. [CrossRef] [PubMed]
49. Costantini, D. Meta-Analysis Reveals That Reproductive Strategies Are Associated with Sexual Differences in Oxidative Balance across Vertebrates. *Curr. Zool.* **2018**, *64*, 1–11. [CrossRef]
50. Sreejai, R.; Jaya, D.S. Studies on the Changes in Lipid Peroxidation and Antioxidants in Fishes Exposed to Hydrogen Sulfide. *Toxicol. Int.* **2010**, *17*, 71–77. [CrossRef] [PubMed]
51. Yada, T. Growth Hormone and Fish Immune System. *Gen. Comp. Endocrinol.* **2007**, *152*, 353–358. [CrossRef] [PubMed]
52. Pérez-Sánchez, J.; Simó-Mirabet, P.; Naya-Català, F.; Martos-Sitcha, J.A.; Perera, E.; Bermejo-Nogales, A.; Benedito-Palos, L.; Calduch-Giner, J.A. Somatotropic Axis Regulation Unravels the Differential Effects of Nutritional and Environmental Factors in Growth Performance of Marine Farmed Fishes. *Front. Endocrinol.* **2018**, *9*, 687. [CrossRef] [PubMed]

53. Ding, H.; Wu, T. Insulin-Like Growth Factor Binding Proteins in Autoimmune Diseases. *Front. Endocrinol.* **2018**, *9*, 499. [CrossRef]
54. Song, F.; Zhou, X.-X.; Hu, Y.; Li, G.; Wang, Y. The Roles of Insulin-Like Growth Factor Binding Protein Family in Development and Diseases. *Adv. Ther.* **2021**, *38*, 885–903. [CrossRef]

fishes

Article

Construction of a Growth Model and Screening of Growth-Related Genes for a Hybrid Puffer (*Takifugu obscurus* ♀× *Takifugu rubripes* ♂)

Chaoyu Wang [1], Yan Shi [1,*], Yuanye Gao [1], Shuo Shi [1], Mengmeng Wang [1], Yunlong Yao [1], Zhenlong Sun [2], Yaohui Wang [2] and Zhe Zhao [1,*]

1 Jiangsu Province Engineering Research Center for Marine Bio-Resources Sustainable Utilization, College of Oceanography, Hohai University, Nanjing 210098, China; w_chaoyu@163.com (C.W.)
2 Jiangsu Zhongyang Group Company Limited, Haian 226600, China
* Correspondence: yshi@hhu.edu.cn (Y.S.); zhezhao@hhu.edu.cn (Z.Z.)

Abstract: The obscure puffer (*Takifugu obscurus*) is a popular cultured species and accounts for around 50% of the total pufferfish production in China. A hybrid puffer was generated by crossing a female obscure puffer with a male tiger puffer (*T. rubripes*). Its growth model has not been developed and the genetic basis underlying its growth superiority has not yet been fully investigated. In this study, the growth model and morphological traits of the hybrid puffer were explored. The results indicated that the hybrid puffer exhibited a significant growth advantage compared to the obscure puffer. There were also significant differences in their morphological traits. We conducted genotyping-by-sequencing (GBS) on hybrid and obscure puffer groups, identifying 215,288 high-quality single nucleotide polymorphisms (SNPs) on 22 chromosomes. Subsequently, a total of 13 growth-related selection regions were identified via a combination of selection signatures and a genome-wide association study (GWAS); these regions were mainly located on chromosomes 10 and 22. Ultimately, the screened regions contained 13 growth-related genes, including *itgav*, *ighv3-43*, *ighm*, *atp6v1b2*, *pld1*, *xmrk*, *inhba*, *dsp*, *dsg2*, and *dsc2*, which regulate growth through a variety of pathways. Taken together, the growth models and candidate genes used in this study will aid our understanding of production characteristics and the genetic basis of growth rates. The hybrid will also be of great significance for the genome-assisted breeding of pufferfish in the future.

Keywords: hybrid puffer; growth; selection signatures; non-linear growth models; SNPs

Key Contribution: In this study, a growth model of how hybrid puffer (*Takifugu obscurus* ♀× *Takifugu rubripes* ♂) grows and develops is constructed, and the candidate genes identified in this study will help our understanding of the culture characteristics and the genetic basis of the growth rate. The result will also be of great significance for genome-assisted breeding of pufferfish in the future.

1. Introduction

Pufferfishes, especially members of the genus *Takifugu*, are economically important aquaculture species in East Asia owing to their desirable taste and high nutritional quality. In China, due to their tetrodotoxin (TTX) content, only two *Takifugu* species have been approved for artificial cultivation and consumed, namely, tiger puffer (*Takifugu rubripes*) and obscure puffer (*T. obscurus*), since 2016. Their total output increased to 31,060 tons in 2022 [1]. Normally, the tiger puffer exhibits a very fast growth rate in a farming environment and has the largest body size within the *Takifugu* genus [2]; however, this species resides obligately in seawater and is only allowed to be farmed in coastal areas. The obscure puffer is a migratory species that has acclimated to completely freshwater regions [3,4], and it is gradually becoming one of the main *Takifugu* species used for production. Nevertheless, compared with the tiger puffer, the obscure puffer has a smaller body size and a slower growth rate.

Growth is one of the most significant traits in the aquaculture industry, and it is directly related to yield and economic efficiency [5]. At the same time, growth rate is a crucial factor that optimizes feeding rates and breeding density. Therefore, constructing a growth model to estimate or predict fish weight is essential for achieving better control of the output of an aquaculture system [6]. Hybridization, a significant breeding method, has been widely used in aquatic breeding since the 1950s [7]. It can be effective to combine the advantageous traits of both parental species to rapidly improve the traits of the offspring, especially growth [8–10]. For example, in terms of growth and stress resistance, a high growth rate and high spot disease resistance were induced by the distant hybridization of *Percocypris pingi* (♂) and *Schizothorax wangchiachii* (♀) [11]; further, a high growth rate and hypoxia-tolerance traits of hybrid *Pelteobagrus fulvidraco* (♀) × *Leiocassis longirostris* (♂) were observed [12]. Nevertheless, the enhancement of growth increases the difficulty of aquaculture prediction. For example, hybrid Jinhu grouper (*Epinephelus fuscoguttatus* ♀× *E. tukula* ♂) and Hulong grouper (*E. fuscoguttatus* ♀× *E. lanceolatus* ♂) groups showed significantly higher growth than the parental *E. fuscoguttatus* group in one study, and there were different growth models between the groups, all of which showed allometric relationships between body weight and body length ($W = aL^b$, $b < 3$) [13]. Overall, the hybrid group exhibited a significantly improved growth rate, accompanied by alterations in growth models and morphological traits. These observations offer insights into the growth advantages of hybrid groups

Extensive research has been conducted on hybrid puffers to study factors such as the slow growth rate and diverse osmoregulatory abilities of the pufferfishes. The exploration of hybrid puffers began in the 1960s [14]. Early studies focused on the embryonic development [14], morphological characteristics [15], growth performance, and osmoregulation of the hybrid puffer, with results showing that the hybrid puffer (*T. flavidus* ♀× *T. obscurus* ♂) cannot be hatched in freshwater [16] and that the hybrid puffer (*T. flavidus* ♀× *T. rubripes* ♂) exhibits different morphological characteristics from its parents [15]. In recent years, with the development of molecular biology, research on the hybrid puffer has involved various aspects, such as mitochondrial DNA sequences [17], gut microbiota [18], body composition [19], and screening for salinity regulation genes [20]. However, until now, for the hybrid puffer (*T. obscurus* ♀× *T. rubripes* ♂), there is still a lack of a long-term growth models, growth-related SNPs, and growth-related gene studies.

Recently, numerous studies have demonstrated that variations in growth and morphological traits are associated with genetic diversity among groups [7]. Additionally, genomic analysis methods, such as selection signature analysis [10] and genome-wide association studies (GWASs) [21], have demonstrated that the molecular mechanisms of growth traits can be explored and that the growth-related selection regions, genes, and single nucleotide polymorphisms (SNPs) can be further investigated. For example, regarding the Australasian Snapper (*Chrysophrys auratus*), selection signature analysis was used to identify growth genes associated with fast-growing groups [22]. Similarly, the growth-related genes of mandarin fish (*Siniperca chuatsi*), including *rnf213*, *mkk6*, and *nck2*, were screened using a genome-wide association study (GWAS) [23]. Nevertheless, there is still a paucity of analyses of growth performance and growth models for hybrid pufferfish, and their growth-related molecular regulatory mechanisms are still unknown.

In this study, growth models and morphological traits were analyzed between the hybrid and obscure pufferfishes. Subsequently, individuals with different growth traits were genotyped. Concurrently, selection signatures and a GWAS were employed to investigate the growth-related SNPs and selected regions to identify the genes underlying the genetic adaptation responsible for the rapid growth observed in hybrid puffers. Taken together, this study will improve comprehension of the genetic mechanism of pufferfish growth and provide important theoretical support for genome-assisted breeding.

2. Materials and Methods

2.1. Construction of Hybrid Group

A hybridization experiment was conducted on 22 February 2022, at Zhongyang Aquatic Co., Ltd. (Nantong, China), to generate hybrid puffers between randomly selected parental female *T. obscurus* and male *T. rubripes* groups (Figure 1). The parental fishes were approximately 3–4 years old, with body weights ranging from 700 to 1500 g. Following a month of conditioning, sexually mature individuals, including randomly selected groups of 10 male obscure puffers, 20 female obscure puffers, and 10 male *T. rubripes* individuals, were used for breeding under identical conditions (400 m^2, 18.0 ± 1 °C). Their body weights were 777.63 ± 186.37 g, 859.19 ± 183.33 g, and 1195.50 ± 149.22 g, respectively (Table S1). The selected fish were injected with human chorionic gonadotropin (HCG) and luteinizing hormone-releasing hormone (LHRH-A2) [24,25]. The two-injection method [26] was employed in this study; the first injected dose administered to the females was 200–300 IU/kg of HCG + 50–100 μg/kg of LRHA-A2, while the males were not injected. The second injected dose administered to females was 200–300 IU/kg of HCG + 100–300 μg/kg of LRHA-A2, while the males were injected with half this dose [11]. Following an interval of 24–48 h, the abdomens of the mature parental fish were gently pressed to obtain approximately 500 g of mature eggs and 20 mL of sperm; then, hybrid and obscure puffer groups were created (Figure 1).

Figure 1. Phenotypes of the obscure puffer (*T. obscurus*), tiger puffer (*T. rubripes*), and hybrid puffer. (**A**) Parental obscure and tiger puffers. (**B**) Experimental offspring of the hybrid and obscure puffer groups.

Individuals of obscure and hybrid puffers were selected at 0 days post-hatching (dph) after the hatching of fertilized eggs. These puffers were reared at different culture stages (0–270 days post-hatching, dph) for growth comparison. During the initial 30-day period (0–30 dph), two 20 m^2-size tanks were used and each pond contained 5000 hybrid puffers and 5000 obscure puffers, with sterilized groundwater (25 ± 1 °C, 0‰) that was changed every 10 days. The fish in both groups were fed rotifers (*Brachionus calyciflorus*) and brine shrimp (*Artemia salina*) five times daily at specific intervals (6:00, 9:00, 12:00, 15:00, and 17:00) [27], maintaining a rotifer density of 5000–6000/L and a brine shrimp density of 20–50/L in the water column with each feeding. After 30 days, considering the loss of fish and the breeding density, 3000 healthy individuals each of hybrid and obscure puffers were randomly selected and transferred to same 444 m^2 tanks, with the same water and temperature. The fish were fed commercial feed (Zhongyang brand, pufferfish variety, containing 40% crude protein) three times daily at specific intervals (8:00, 14:00, and 18:00), the feeding situation was observed for 60 min after each feeding, and the remaining food was removed in time if no fish were eating. Freshwater (25 ± 1 °C and 7–8 mg/L dissolved oxygen) was supplied, and replaced every 10 days.

2.2. Growth and Morphological Traits Comparison

During the experiment, electronic scales and vernier calipers, with precisions of 0.01 g and 0.01 mm, respectively, were utilized to measure the growth traits and morphological traits.

The body weight (BW), body length (BL), and total length (TL) of 100 randomly selected individuals from each group were measured at 30, 60, 90, 120, 150, 180, 210, and 270 dph. The weight gain ratio (WGR), absolute growth rate (AGR), specific growth rate (SGR), and condition factor (CF) were calculated according to the following equations [28–30]:

$$\text{WGR } (\%) = (W_i - W_0)/W_0 \times 100\% \quad (1)$$

$$\text{AGR } (\text{g d}^{-1}) = (W_i - W_0)/(t_i - t_0) \quad (2)$$

$$\text{SGR } (\%\ \text{d}^{-1}) = (ln_{Wi} - \ln_{W0})/(t_i - t_0) \times 100\% \quad (3)$$

$$\text{CF } (\%) = (W_i/L_i^3) \times 100\% \quad (4)$$

where t is the experiment duration (d), W_i is the weight of t_i, W_0 is the weight of t_0, and L_i is the length of W_i.

The morphological traits of 100 randomly selected samples from each group were measured at 270 dph. The sex of each fish was determined based on a sex-linked SNP in the *amhr2* gene [31]. Fifteen morphological traits were measured [30]: BW, TL, BL, caudal peduncle height (CH), head length (HL), snout length (SL), head-behind length (HBL), eye length (EL), eye spacing (ES), nostril spacing (NS), outlet hole spacing (OS), snout cleft (SC), chest length (CL), abdominal length (AL), and caudal girth (CG) (Figure 2). To reduce the effects of excessive individual weight differences, the data obtained were corrected using BL in analyses of morphological traits. The Spearman correlation coefficient was utilized to conduct a correlation analysis of morphological traits among two populations using SPSS (version 26.0). The correlation threshold was established as follows: $|r| < 0.4$ indicated a low correlation; $0.4 \leq |r| \leq 0.7$ indicated a moderate correlation; and $|r| > 0.7$ indicated a high correlation. Based on the results of a normal distribution analysis, a principal component analysis (PCA) of the morphological traits was performed using SPSS (version 26.0). Simultaneously, 75 samples were randomly selected from each group. Additionally, a cluster analysis of two groups was performed using OriginPro (version 2019b) [32] based on the unweighted pair-group method with arithmetic means (UPGMA) and morphological traits.

The functional relationship between BW and BL was calculated according to the following equation [27]:

$$y = \text{a}\, x^{\text{b}} \quad (5)$$

where x is body length; y is body weight, a is the intercept on the y-axis, and b is the allometric growth index.

The von Bertalanffy growth models were calculated according to the following equations [33,34]:

$$W_t = \text{W}_\infty\, (1 - \text{e}^{-\text{k}(t-\text{t}_0)}) \quad (6)$$

$$L_t = \text{L}_\infty\, (1 - \text{e}^{-\text{k}(t-\text{t}_0)}) \quad (7)$$

where W_t and L_t are body weight and body length at age t, respectively; L_∞ is the asymptotic maximum body length; W_∞ is the asymptotic maximum body weight; k is the Brody coefficient (or "growth constant"), describing how fast W_∞ and L_∞ are approached; and t_0 is the theoretical age at size 0 in the von Bertalanffy growth model.

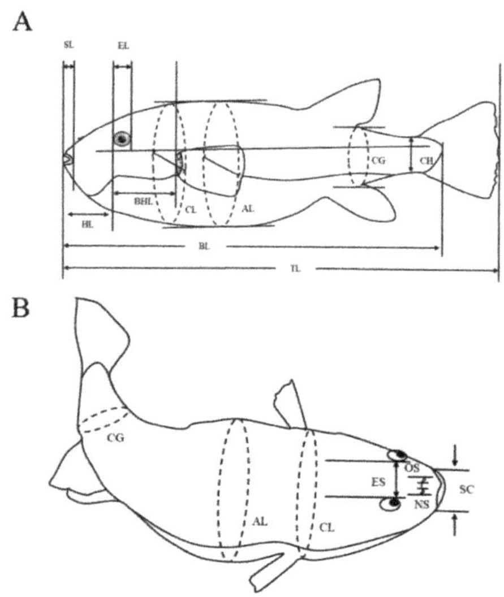

Figure 2. Diagram depicting growth index measurement. (**A**) Lateral view. (**B**) Dorsal view. BW, body weight; TL, total length; BL, body length; CH, caudal peduncle height; HL, head length; SL, snout length; HBL, head-behind length; EL, eye length; ES, eye spacing; NS, nostril spacing; OS, outlet hole spacing; SC, snout cleft; CL, chest length; AL, abdominal length; CG, caudal girth.

The inflection point age of growth was calculated according to the following equation:

$$T_t = ln_b/\mathrm{k} + \mathrm{t}_0 \tag{8}$$

where T_t is the time taken to reach the growth inflection point; b is the allometric growth index; k is the Brody coefficient (or "growth constant"), describing how fast W_∞ and L_∞ are approached; and t_0 is the theoretical age at size 0 in the von Bertalanffy growth model.

Finally, the Fisher and Bayes discriminant equations for the two groups were calculated using stepwise discriminant analysis based on Wilks' lambda method [32] and different morphological traits.

2.3. Sample Collection and Sequencing

At the age of 270 days, 100 individuals were taken from both the obscure and hybrid puffer groups. These individuals were then categorized as either fast-growing or slow-growing based on their body weights. Then, 20 fast-growing individuals and 20 slow-growing individuals were selected from each group. A total of 80 pufferfish samples were collected (Table S2). Their growth traits (BW, BL, TL, AL, and CL) were measured, and the fins of each fish were clipped and preserved in 95% ethanol at −20 °C. DNA was extracted using the cetyltrimethylammonium bromide (CTAB) method, while DNA quality was measured using a Qubit fluorometer (Thermo Fisher Scientific, Carlsbad, CA, USA) and a NanoDrop spectrophotometer (Thermo Fisher Scientific).

Based on the size and GC content of the *T. rubripes* genome (https://www.ncbi.nlm.nih.gov/datasets/genome/GCF_901000725.2/, accessed on 15 July 2023), appropriate endonucleases for genomic fragmentation were predicted. The qualified genomic DNA was digested using restriction enzymes, followed by end repair, the A-tail was added, and an Illumina sequencing connector was added by the NEBNext Ultra DNA Library Prep Kit (NEB, Ipswich, MA, USA). DNA fragments (300–400 bp) were amplified and enriched using PCR. Finally, the PCR products were purified using the AMPure XP system (Beckman Coulter, Brea, CA, USA). Sequencing libraries were examined using an Agilent 2100 Bioanalyzer (Agilent Technologies, Santa Clara, CA, USA) and quantified using qPCR.

Sequencing was performed by employing a NovaSeq 6000 sequencer using the PE 150 sequencing strategy.

2.4. Quality Control and Genotyping

The reads were subjected to quality control using FASTP software (version 0.18.0) [23,35], Low-quality reads were removed, and low quality was defined as follow: reads aligned to the barcode adaptor; reads with connectors; reads including $\geq$ 10% unidentified nucleotides (N); and reads for which $\geq$ 50% of the bases had a Phred quality score $\leq$ 20. The filtered reads were mapped against the reference genome using the MEM algorithm in BWA (version 0.7.12) [35,36], and the comparison parameter was–k 32 –M. After comparison, the potential PCR duplicates were tagged and removed using PICARD (version 1.129) (Picard: http://sourceforge.net/projects/picard/, accessed on 10 August 2023). SNPs were identified and filtered using GATK (version 3.4-46) [37] Variant Filtration with proper standards (–Window 4, –filter "QD < 4.0 | | FS > 60.0 | | MQ < 40.0", –G_filter "GQ < 20"), and functional annotation of the detected variants was performed using ANNOVAR (version 2) [38]. In addition, to enhance the accuracy of this study, we applied stringent filtering criteria to the SNPs using PLINK2 (version 2.0) [39]. Stringent filtering conditions were set as follows: for selection signatures, the missing rate was < 50%, and for GWAS, the filtered SNPs were all indels and non-biallelic SNPs, SNPs with a minor allele frequency (MAF) < 0.05, SNPs with a call rate < 0.9, individuals with a missing data rate > 0.5, and heterozygosity rate > 0.8. Finally, these analyses of SNPs were used in the subsequent GWAS and selection signatures analyses.

2.5. Analysis of Population Structure and Kinship

After SNP quality control, linkage disequilibrium (LD) pruning was performed according to the r^2 value. LD pruning was conducted with a window size of 50, a step of 5, and an r^2 threshold of 0.2 [40,41]. According to the filtered SNPs, principal component analysis (PCA) and kinship analysis were performed on the experimental population using GCTA software (v1.93.2) [42]. Simultaneously, an NJ tree (model: *p*-distance; bootstrap replications 1000) was constructed using MEGA-X software (version 10) [43], and the population structure was constructed using Admixture software (version 1.3) [44]. Since the experimental population comprises individuals with different growth rates and hybrid individuals, it was hypothesized that the samples belong to multiple subgroups. Therefore, the number of subgroups (K) for the samples was assumed to range from 1 to 9. Subsequently, the optimal number of clusters was determined based on the cross-validation error rate (CV error).

2.6. Screening for Selection Signatures and a Genome-Wide Association Study

The F_{st} (a measure of population differentiation, varies from 0 to 1) and π ratio were used to identify the candidate selected region. PopGenome software (in R package) [45,46] was used to perform sliding-window analysis based on the filtered SNPs, using a window size of 100 kb and a step size of 10 kb according to physical length. This step facilitated a comparative evaluation of intra- and inter-population diversity, encompassing analyses of F_{st} [47] and the π ratio [48]. The selected regions were determined by comparing the top 1% of results from the F_{st} and π ratio, ranked by the degree of genetic differentiation.

A GWAS was used to investigate the genetic architecture of the growth-related traits. According to a modified Bonferroni correction method [23,49], the significance and suggestive association thresholds were set at 0.05/N and 1/N, respectively, where N is the number of SNPs used in the association analysis. SNP density maps, Manhattan plots, and quantile-quantile (Q-Q) plots were generated using CMplot in R [23,50].

2.7. Identification and Functional Annotation of Candidate Genes

SNPs were mapped against the *T. rubripes* genome (https://www.ncbi.nlm.nih.gov/datasets/genome/GCF_901000725.2/, accessed on 22 August 2023). In the selection signatures, the growth-related candidate genes were identified by scanning candidate selection regions; in the GWAS analysis, the genes containing significant SNPs were scanned, and the genes of ± 300 kb of significant and suggestive SNPs were scanned. Subsequently, to enhance the accuracy of those analysis, we performed an intersection analysis of those genes.

All candidate genes were functionally annotated based on the reference genome of *T. rubripes* posted by the National Center for Biotechnology Information (NCBI) (https://www.ncbi.nlm.nih.gov/). Gene Ontology (GO) and Kyoto Encyclopedia of Genes and Genomes (KEGG) enrichment analyses were performed using OmicShare (www.omicshare.com/tools) and Metascape (https://metascape.org/gp/index.html#/reportfinal/tm8hy07mo, accessed on 30 August 2023), respectively [40]. The calculated p values were subjected to FDR correction, with a threshold of FDR $\leq$ 0.05. GO terms that met this criterion were considered significantly enriched. Similarly, the calculated p values underwent FDR correction using the same threshold, and pathways meeting this condition were defined as significantly enriched.

2.8. Statistical Analysis

The data were analyzed using SPSS (version 26.0) [32]. Morphological traits were corrected for body weight. The data were checked for the normality of distribution using the Kolmogorov–Smirnov test. Bartlett's test was used to determine the homogeneity of variance among different groups. Significant differences were analyzed via a t-test, in which $p < 0.01$ and $p < 0.05$ were considered highly significant and significant differences, respectively. Data are expressed as means ± standard deviation.

3. Results

3.1. The Disparities in Growth Performance

At 270 dph, the BW, BL, and TL of the hybrid were significantly higher than those of the obscure puffer (Table S3, Figure S1, $p < 0.01$); the BW and TL of the hybrid reached 266.13 ± 63.90 g and 22.00 ± 1.6 cm, respectively. At 120 dph, due to the health condition of the experimenters, the hybrid puffer did not meet the specified measurement target and therefore was excluded from this analysis. The subsequent analysis remains unaffected. The hybrid achieved a maximum AGR of 1.09 g/d, which is 1.7 times higher than that of the obscure puffer, and the maximum SGR reached 2.11%/d, indicating that the hybrid exhibited a significant growth advantage. The AGR and SGR results indicate that the growth rate decreased with increasing age in both groups, while the hybrid still exhibited a significantly higher growth rate than the obscure puffer (Table 1, $p < 0.01$). In addition, the CF of the hybrid was lower than that of the obscure puffer prior to 60 dph ($p > 0.05$), a finding we hypothesize is related to the growth model differences between the hybrid and obscure puffers. In conclusion, the hybrid exhibited a significant growth advantage over the obscure puffer.

Table 1. Growth traits of hybrid and obscure puffer groups.

Stage		Obscure Puffer				Hybrid Puffer			
Family	DPH	WGR	AGR	SGR	CF	WGR	AGR	SGR	CF
Larval	60	10.48	0.17	4.07%	4.97	17.93	0.3	4.90%	3.56
	90	50.56	0.56	4.38%	4.24	67.22	0.88	4.69%	4.2
Juvenile	120	75.1	0.63	3.61%	3.87	-	-	-	-
	150	92.77	0.62	3.03%	3.85	122.31	0.82	3.08%	4.09
	180	123.99	0.69	2.68%	3.63	184.23	1.02	2.90%	4.32
	210	137.85	0.66	2.35%	3.54	203.16	0.97	2.53%	3.74
Young	270	173.31	0.64	1.91%	3.62	294.29	1.09	2.11%	3.97

DPH, day post-fertilization; WGR, weight gain ratio; AGR, absolute growth rate; SGR, specific growth rate; CF, condition factor.

3.2. The Variations in Morphological Characteristics

To reduce the effects of excessive individual weight differences, the data were corrected using BL in the analyses of morphological traits. These standardized traits were comparatively analyzed (Table S4), and the results showed that 13 traits were significantly different (BW/BL, TL/BL, CH/BL, SL/BL, HBL/BL, EL/BL, ES/BL, OS/BL, SC/BL, CL/BL, AL/BL, CL/BL, and CF/BL; $p < 0.01$), while two traits were not significantly different (HL/BL and NS/BL; $p > 0.05$), between the hybrid and obscure puffers. Spearman correlation analysis of the morphological traits and sex (Figure S2, Table S5) showed that in the obscure puffer, the CH/BL ($|r| = 0.50$), CL/BL ($|r| = 0.46$), AL/BL ($|r| = 0.53$), CG/BL ($|r| = 0.47$), and CF/BL ($|r| = 0.61$) were significantly positively correlated with BW/BL ($p < 0.01$); the EL/BL ($|r| = 0.30$) was significantly negatively correlated with BW/BL ($p < 0.01$); the BW/BL ($|r| = 0.3$) was significantly positively correlated with sex ($p < 0.01$); the HL/BL ($|r| = 0.34$) was significantly negatively correlated with sex ($p < 0.01$); and the SL/BL ($|r| = 0.25$) and HBL/BL ($|r| = 0.28$) were significantly negatively correlated with sex ($p < 0.05$). In the hybrid puffer, CL/BL ($|r| = 0.43$), AL/BL ($|r| = 0.51$), CG/BL ($|r| = 0.49$) and CF/BL ($|r| = 0.71$) showed a significantly positively correlated with BW/BL ($p < 0.01$); the CH/BL ($|r| = 0.26$) was significantly positively correlated with BW/BL ($p < 0.05$); and the EL/BL ($|r| = 0.45$) was significantly negatively correlated with BW/BL ($p < 0.01$); only ES/BL ($|r| = 0.26$) showed a significant positive correlation with sex ($p < 0.05$); there were no growth-related traits that had a significant correlation with sex ($p > 0.05$). In conclusion, the above results indicate that there is a significant difference in morphological traits between the hybrid and obscure puffers, and that the growth of the obscure puffer was significantly associated with sex. These morphological trait differences may allow distinction between these two groups, which are difficult to distinguish by appearance, through morphological features. Additionally, the differences in morphological traits also interact with the growth patterns, indicating the existence of different growth models between the obscure and hybrid puffers.

Based on the morphological traits, a principal component analysis (PCA) was conducted on a mixed group of 150 individuals with a normal distribution (75 hybrid and 75 obscure puffers). The results showed that the morphological traits were grouped into four categories (Figure 3). The cumulative contribution rate was 80.97%. The contribution rate of the first principal component was 39.53%, including AL/BL, CL/BL, CG/BL, BW/BL, CF/BL, CH/BL, and EL/BL, which reflected a pufferfish body shape. The second principal component was 15.48%, including HL/BL and HBL/BL, which reflected the anterior torso ratio. The third principal component accounted for 15.34%, including OS/BL, ES/BL, and SC/BL, which reflected the morphological structure of the head. According to the analysis of PCA1, PCA2, and PCA3, the obscure and hybrid puffers from PCA1 can be better distinguished in terms of body shape. From the aspect of anterior torso ratio in PCA2, the two groups are difficult to distinguish due to a large amount of overlap. Although there are certain differences in the morphological structure of the head from the perspective of PCA3, there were still several individuals that were difficult to clearly classify. In addition, even from the perspective of body shape in PCA1, the two groups are distributed on both sides and can be distinguished well, but there are still a few individuals that overlap. Similarly, there are overlapping individuals in the cluster tree results (Figure S3). We analyzed these overlapping individuals and the results show that the overlapping individuals in the PCA and cluster tree analysis are largely the same individuals. Then, analysis of the morphological traits of overlapping individuals showed that these individuals are not only affected by body weight but also other morphological traits, such as body length and chest length. Among overlapping individuals, the hybrid puffer usually exhibits weight gain and rapid elongation compared to other individuals in the same group. This result further suggests that there are different growth models between the obscure and hybrid puffer groups. In addition, the different growth models were genetically controlled; therefore, it can be speculated that the morphological traits of hybrid puffers among overlapping individuals may be influenced by genetics. The relationship between genetic and morphological traits is analyzed in the following section, "3.5 Analysis of population structure". In general, there are multiple morphological differences between obscure and hybrid puffers, and the two groups can be well distinguished using morphological traits related to body shape.

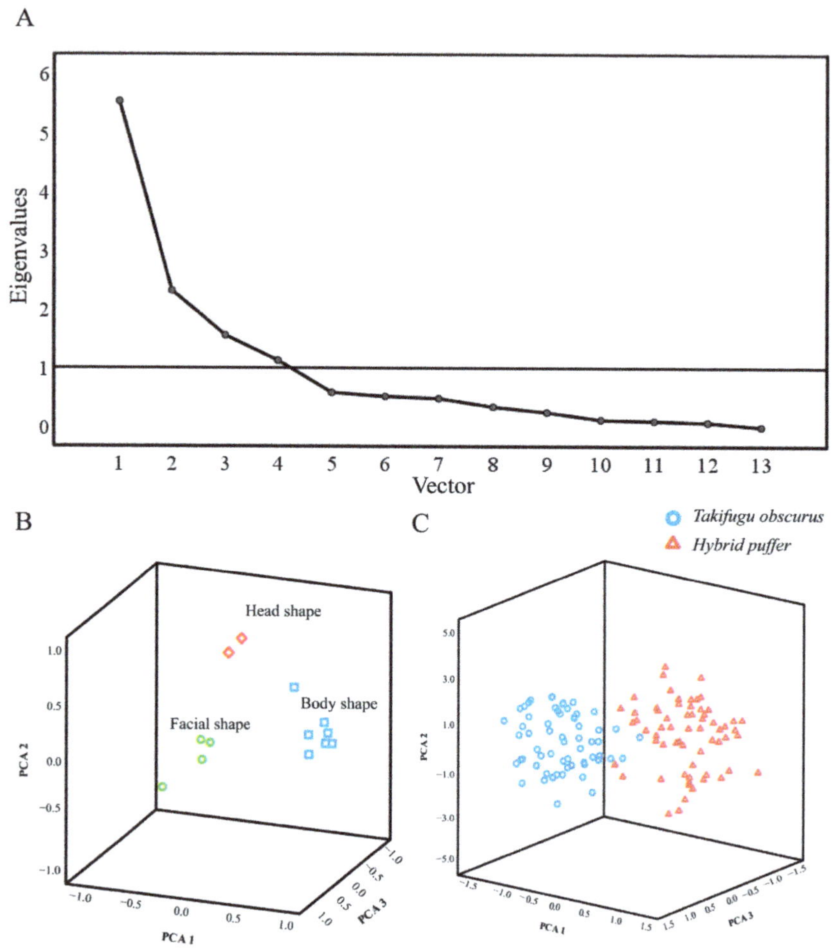

Figure 3. Principal component analysis of the morphological traits of the pufferfish. (**A**) Morphological trait principal component analysis scree plot. (**B**) Principal component analysis of 15 morphological traits. (**C**) Principal component analysis of 150 experimental individuals based on morphological traits.

3.3. The Growth Model and Discriminant Analysis

The relationship between BL and BW was calculated. The results showed that b = 2.759 and $R^2 = 0.99$ for the obscure puffers and b = 3.072 and $R^2 = 0.99$ for the hybrid puffers (Figure 4), indicating that the obscure puffer ($b < 3$) showed negative allometric growth, while the hybrid ($b \approx 3$) showed uniform growth. Growth is determined by the regulation of genes. Therefore, we inferred that the differences in growth models are related to genetics, and these different growth models lead to increased morphological trait differences in body shape between the two groups during the growth and development process.

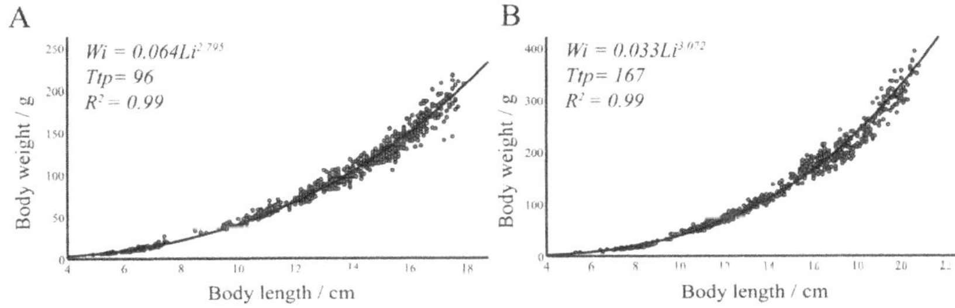

Figure 4. The relationship between body weight and body length. (**A**) The relationship in the obscure puffer group. (**B**) The relationship in the hybrid group. Wi represents the formula for the body length weight relationship; Ttp represents the time of taken to reach the inflection point for growth.

The von Bertalanffy growth models were also calculated (Figure 5). The result showed that the growth inflection point age of obscure puffer was 96 dph, and the corresponding BW and BL were 48.04 g and 10.69 cm, respectively. Meanwhile, the theoretical maximum BW and BL at 270 dph were 165.64 g and 16.64 cm ($R^2 = 0.99$), respectively. Similarly, the growth inflection point age of hybrid was 167 dph, and the corresponding BW and BL were 145.93 g and 15.37 cm, respectively; the theoretical maximum BW and BL at 270 dph were 489.23 g and 22.79 cm ($R^2 = 0.99$), respectively, in the hybrid.

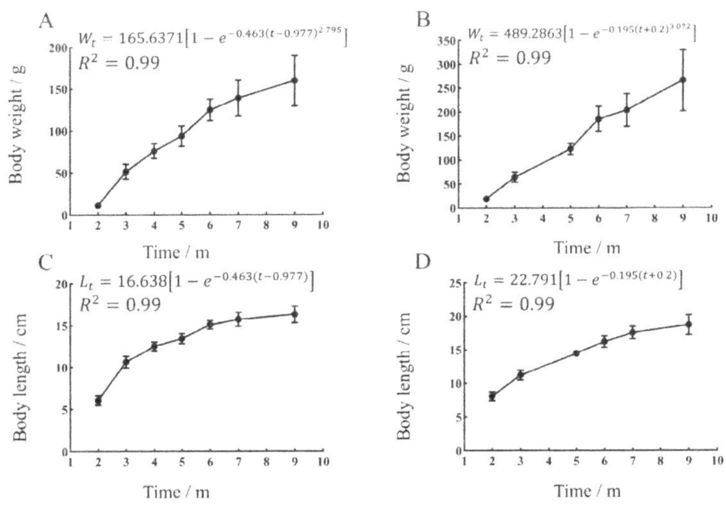

Figure 5. The von Bertalanffy growth models of the obscure puffer and hybrid. (**A**) The von Bertalanffy growth model of body weight for the obscure puffer. (**B**) The von Bertalanffy growth model of body weight for the hybrid puffer. (**C**) The von Bertalanffy growth model of body length for the obscure puffer. (**D**) The body length von Bertalanffy growth model of body length for the hybrid. *Wt* represents the von Bertalanffy growth model for body weight; *Lt* represents the von Bertalanffy growth model for body length.

Based on morphological disparities, stepwise discriminant analysis was conducted using Fisher and Bayes discriminant equations. Stepwise discriminant analysis demonstrated that the BW/BL, EL/BL, SC/BL, AL/BL, CG/BL, and CF/BL ratios exhibited the highest discriminatory power. According to Fisher's principle, the discriminant equation created was as follows:

$$Y = 0.886 * BW/BL + 437.567 * EL/BL - 39.976 * SC/BL + 15.646 * AL/BL + 35.454 * CG/BL - 5.744 * CF/BL - 36.175 \quad (9)$$

where Y is the type of group; Y > 0 indicates the hybrid puffer. Y < 0 indicates the obscure puffer.

According to Bayes' theorem, the discriminant equations were derived as follows:

$$\text{Obscure puffer} = 25.088 * BW/BL + 19385.331 * EL/BL + 2570.616 * SC/BL + 1001.791 * AL/BL + 1264.575 * CG/BL - 244.468 * CF/BL - 933.933 \quad (10)$$

$$\text{Hybrid puffer} = 32.477 * BW/BL + 23035.524 * EL/BL + 2237.132 * SC/BL + 1132.311 * AL/BL + 1560.337 * CG/BL - 292.387 * CF/BL - 1233.482 \quad (11)$$

Individual morphological traits were measured and calculated; if obscure puffer > hybrid, the individual was an obscure puffer and vice versa for a hybrid.

Furthermore, an additional 50 individuals who were not included in the analysis (25 obscure puffers and 25 hybrid puffers) were randomly selected to further validate the equations. The identification rate for both was 96%.

3.4. Quality Control of the Sequencing Reads

A total of 556.50 M reads (79.02 Gbp) were obtained, with 297.02 M reads (42.18 Gbp) and 259.47 M reads (36.85 Gbp) for the hybrid and obscure puffer groups, respectively. The sequencing depths were 96 X and 110 X, respectively. The proportions of high-quality bases (Q score > 20) were 95.90% and 95.73%, respectively. The average GC contents were 45.57% and 45.50%, respectively. After processing and screening, 215,788 and 182,925 high-quality SNPs distributed across 22 chromosomes were included in the selection signatures and GWAS, respectively (Figure 6). The SNPs involved in the GWAS comprised 114,665 transitions and 68,260 transversions. In the transitions, C -> T occupied a maximum of 34,235. In the transversions, C -> A occupied a maximum of 9639. However, the distribution of these SNPs in the genome was not uniform. Following annotation, 23.48% (48,714) of the variants were identified in intergenic regions, 52.30% (108,498) were identified in intronic regions, and only 6.15% (12,759) were identified in exonic regions (Table S6). Among them, 57.02% of the SNPs in the exon region were synonymous mutations, while 40.60% were nonsense mutations and missense mutations (Table S7).

3.5. Analysis of Population Structure

Based on SNPs, the results regarding kinship showed that the genetic relationship coefficient ranged from −0.2 to 0.4, and the two groups contained subgroups with close genetic relationships (Figure S2). The results of the genetic PCA showed that PCA1 and PCA2 accounted for 27.55% of the total phenotypic variation (Figure 6). Specifically, PCA1 explained 19.75% of the total genetic variation and PCA2 explained 7.80%. Meanwhile, the results showed that there is a notable distinction between the obscure and hybrid puffer groups. This is the same as the PCA results for both of their morphological traits. Moreover, the genetic PCA results showed that both groups exhibited subgroup structures (Figure 6). Similarly, the NJ tree result also showed that there are sister groups within the two groups (Figure 6). We combined morphological traits and kinship to analyze the subgroup structures, and the results showed that the subgroup structure was related to morphological traits and kinship. Among the subgroups, one subgroup showed similar growth performance amongst its members, while the growth performance between different subgroups was significantly different. In addition, the analysis of population structure showed that the optimal number of subgroups is K = 4, and there were two subgroups within each of the two groups (Figure S4). In the different clustering results, the obscure puffer can be clearly distinguished, whereas there are few hybrid puffers with the same bloodline as the obscure puffer, indicating that the hybrid puffer and the obscure puffer share a common ancestor. Analysis of the hybrid puffer with an overlapping background shows that most of the individuals with an overlapping background grow slowly and have morphological traits similar to the obscure puffer. Combining the PCA results regarding morphological traits, the results showed that these hybrid puffers were the individuals that were overlapping with the obscure puffer. Therefore, this finding indicates that genetics significantly influence the growth performance of hybrid puffers, and the genetic traits of fast-growing individuals differ significantly from those of slow-growing individuals. Researching these genetics provides support for the analysis of growth mechanisms. Interestingly, when K = 5 or K = 6, the cross-validation error values are close to K = 4. We interpret that this is because the experimental group is built from a smaller parental group. The small parental groups increase the probability of producing multiple lineage groups during the fertilization process. However, excessive subgroup numbers result in an uneven distribution of individuals across different subgroups, which is not conducive to GWAS analysis. Consequently, this study was not adopted. Overall, the results imply that morphological traits are interrelated with genetics, and there are significant genetic differences between the obscure and hybrid puffers leading to differences in their growth and morphological traits. Both experimental groups have certain kinship and subgroup structures; therefore, to improve the accuracy of the GWAS, a mixed linear model (MLM) was employed using GEMMA (version 0.98.1) [51].

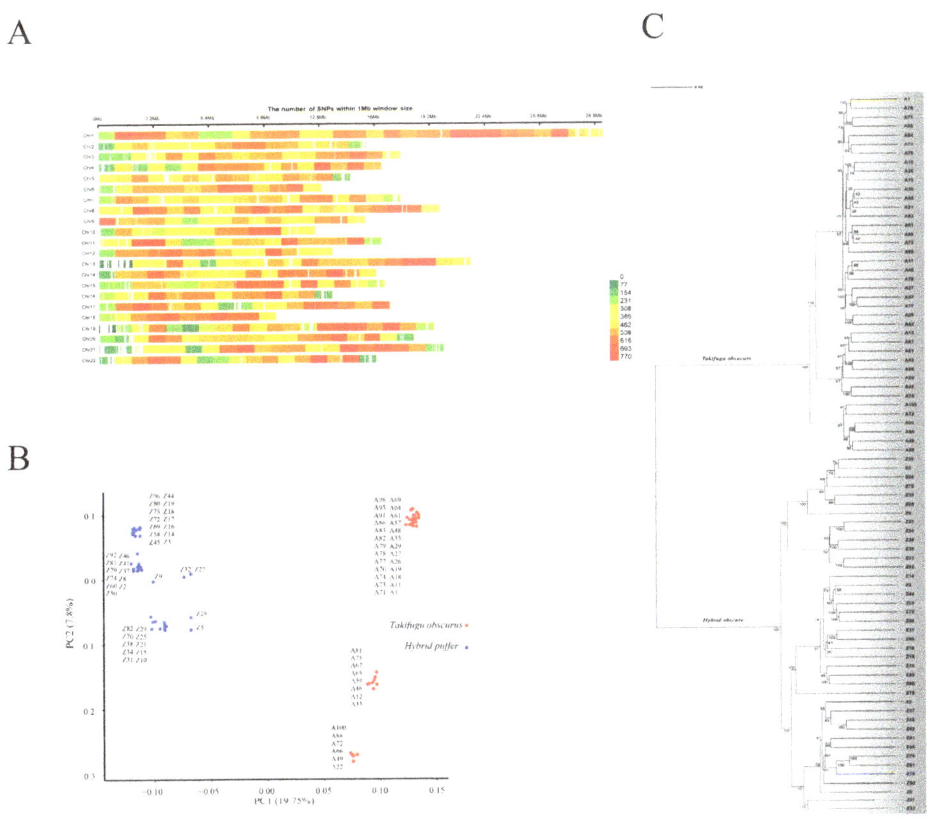

Figure 6. The distribution of SNPs and the experimental groups' population structures. (**A**) The distribution of SNPs on 22 chromosomes. (**B**) Principal component analysis of 80 experimental individuals based on SNPs. (**C**) The experimental phylogenetic tree. A1–A100 represent the individual IDs of the obscure puffer; Z1–Z100 represent the individual IDs of the hybrid puffer.

3.6. Detection of Genome Selection Signatures

Based on the constructed population pairs exhibiting extreme differences in growth-related phenotypic traits, we used the F_{st} and π ratio methods to identify the genomic signatures of selection in the hybrid and obscure puffer groups. We obtained a set of 1720 genes from a total of 860 candidate selected regions in the top 1% of selection regions ($F_{st} \geq 0.60$, π ratio ≥ 1.05, Figure 7) by combining the F_{st} and π ratio.

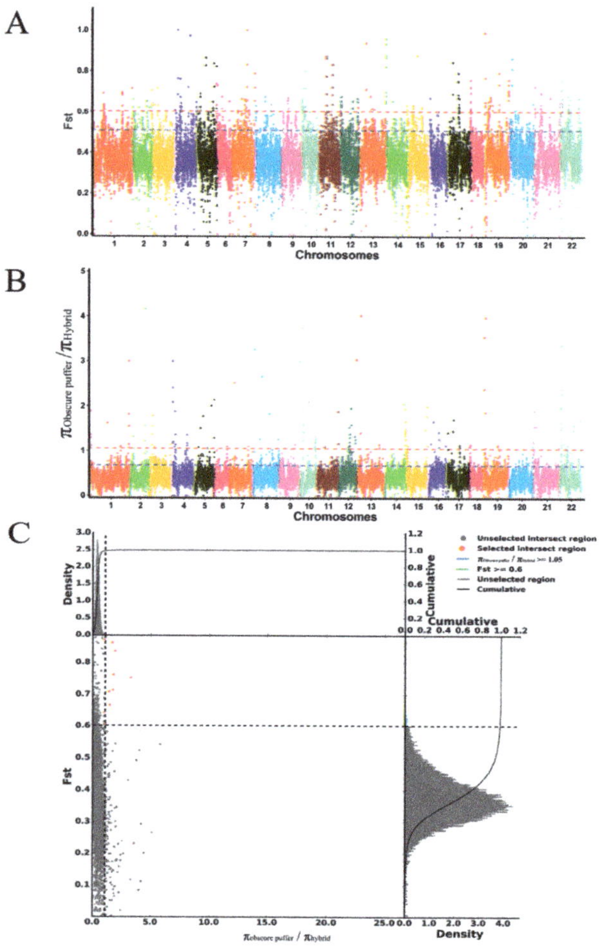

Figure 7. Candidate selection regions of the hybrid and obscure puffer groups. (**A**,**B**) Candidate selection regions detected using F_{st} (**A**) and π ratio (**B**) statistics are plotted across the genome. The *y*-axis of the Manhattan plots displays the F_{st} values and π ratio scores calculated in 100 kb with steps of 50 kb. The red horizontal dashed line represents the top 1% threshold in the F_{st} value (0.60) and π ratio scores (1.05). (**C**) The candidate selection region intersection of the F_{st} and π ratio.

The F_{st} method identified a total of 445 candidate selected regions encompassing 823 genes (Table S8), which were mainly enriched on chromosomes 4, 10, and 22, while the π ratio method identified 415 candidate selected regions encompassing 944 genes (Table S8), which were mainly enriched on chromosomes 10 and 22. In addition, a total of 30 intersection candidate regions, encompassing 47 genes, were identified (Table S8).

3.7. GO Term and KEGG Pathway Analysis

GO and KEGG analyses were conducted on the candidate genes within the selected regions to provide insights for subsequent investigations. The results obtained from the GO analysis of the intersection of F_{st} and π ratio analyses (Figure S5) indicated that

57.45% of the genes (n = 27) were categorized into the biological process (BP) category, with 633 enriched terms, of which 118 were significant; 55.32% of the genes (n = 26) were categorized as cellular components (CC), with 93 enriched terms, of which 11 were significant; and 59.57% of the genes (n = 28) were categorized as molecular function (MF), with 95 enriched terms, of which 14 were significant (Table S9). The top 20 enriched GO terms are shown in Figure 8, ranked by p value. In biological processes, the most abundant terms included regulation of actin filament-based movement (GO:1903115), regulation of muscle contraction (GO:0006937), regulation of blood circulation (GO:1903522), and regulation of muscle system processes (GO:0090257), all of which involved an average of eight genes ($p < 0.01$). In the cellular components, the cell–cell junction (GO:0005911), cell–cell adherens junction (GO:0005913), and proton-transporting V-type ATPase complex (GO:0033176) all contained an average of four genes ($p < 0.05$). In terms of molecular function, protein binding was involved in cell adhesion (GO:0098631), insulin-like growth factor binding (GO:0005520), and inorganic cation transmembrane transporter activity (GO:0022890), all of which involved an average of two genes ($p < 0.05$).

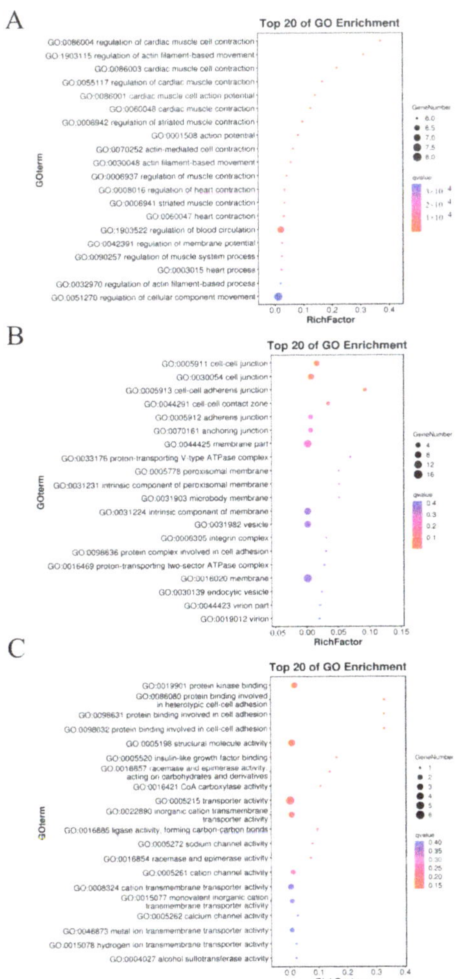

Figure 8. Top 20 enriched GO terms of genes identified under selection regions. (**A**) Biological process. (**B**) Cellular component. (**C**) Molecular function.

A KEGG pathway enrichment analysis of the candidate genes, ranked by *p* value, was conducted (Figure 9). The top 20 enriched KEGG terms in the F_{st} analysis are presented; the growth-related pathways specifically enriched included hematopoietic cell lineage (ko04640), mannose type O-glycan biosynthesis (ko00515), and vitamin digestion and absorption (ko04977) (Table S10). The π ratio analysis (Figure S6) included cell adhesion molecules (ko04514) and cellular senescence (ko04218) (Table S10). In addition, the result of the analysis of all candidate genes indicated that the human diseases category exhibited the highest abundance, and the other pathways are mainly enriched in organismal systems and metabolism (Figure S6).

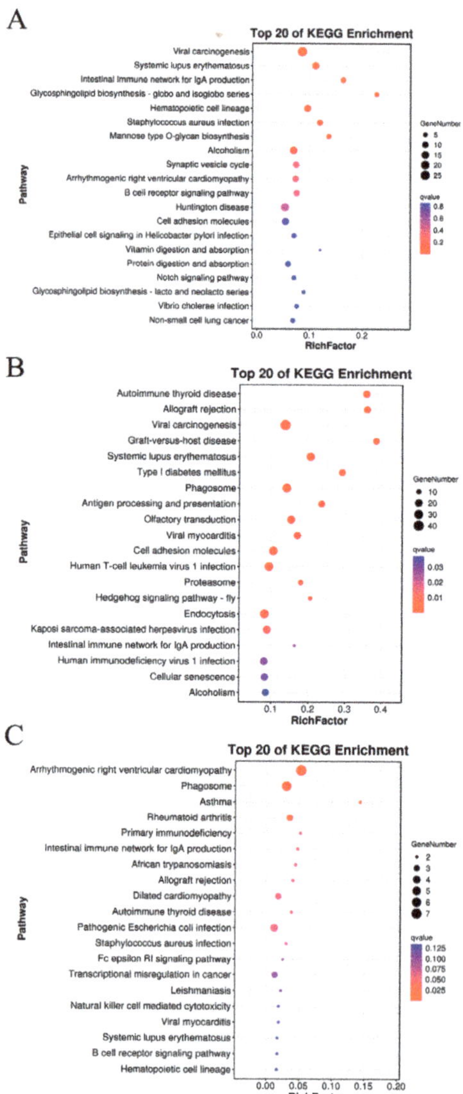

Figure 9. Top 20 enriched KEGG terms of genes identified under selection regions. (**A**) KEGG enrichment analysis results of the F_{st} screening region genes. (**B**) KEGG enrichment analysis results of π ratio screening region genes. (**C**) KEGG enrichment analysis results of genes in the intersection region screened using the F_{st} and π ratio.

3.8. Genome-Wide Association Analysis of Growth Traits

The original phenotypic data for each sample were measured and recorded. The growth-related traits exhibited predominantly non-normally distributed patterns according to the Kolmogorov–Smirnov test; therefore, these traits were transformed to achieve normality. Subsequently, a GWAS was conducted on five growth-related traits (BW, TL, BL, CL, and AL) using GEMMA software (version 0.98.1), encompassing 182,925 SNPs.

For the obscure puffer, the GWAS showed 32 SNPs significantly associated ($-\log_{10} > 4.7$) (Figure S7, Table S11) and 163 suggestively associated ($-\log_{10} > 3.4$) with growth traits. These SNPs were located mainly on chromosomes 1, 21, and 22. Among them, 12 SNPs were associated with multiple traits (Table 2): LOC1:2211857 and LOC1:3193101 were associated with BW and AL, with SNP effects (beta) ranging from −37.33 to 29.33; LOC22:5379885 was associated with BW and TL, with SNP effects of 32.05 and 1.20, respectively; LOC1:3253768, LOC1:3253792, LOC1:3327200, LOC1:3403622, LOC1:3479403, and LOC1:3479441 were associated with TL and AL, with SNP effects ranging from −1.37 to 1.47; and LOC3:3685861, LOC3:3685871, and LOC12:5883873 were associated with TL and BL, with SNP effects ranging from −2.80 to 1.38. Although no significant SNPs related to BW and CL were found, 21 and 55 suggestive SNPs related to them also had certain reference values. In addition, LOC1:3193101 (*rnf213*) and LOC12:5883873 (*LOC115251859*) were both significant SNPs ($-\log_{10} > 4.7$) and associated with multiple traits. A QQ plot of the p values is shown in Figure S8.

Table 2. The SNPs associated with multiple traits in GWAS analysis in pufferfishes.

Family	Trait	SNP	Chr	Position	Beta	p Value	Region	Gene Name
Obscure puffer	BW and AL	LOC1:2211857	1	2,211,857	−37.33243	1.83×10^{-4}	Intergenic	NA
		LOC1:3193101	1	3,193,101	29.33228	2.79×10^{-4}	Exonic	*rnf213*
	BW and TL	LOC22:5379885	22	5,379,885	32.05076	1.04×10^{-4}	Intergenic	NA
	TL and AL	LOC1:3253768	1	3,253,768	−1.273611	2.08×10^{-4}	Intronic	*baiap2*
		LOC1:3253792	1	3,253,792	−1.273611	2.08×10^{-4}	Intronic	*baiap2*
		LOC1:3327200	1	3,327,200	−1.273611	2.08×10^{-4}	Intronic	*wapl*
		LOC1:3403622	1	3,403,622	−1.273611	2.08×10^{-4}	Intronic	*grid1*
		LOC1:3479403	1	3,479,403	1.435881	1.04×10^{-4}	Intronic	*ccser2*
		LOC1:3479441	1	3,479,441	1.435881	1.04×10^{-4}	Intronic	*ccser2*
	TL and BL	LOC3:3685861	3	3,685,861	1.141095	3.91×10^{-4}	Intronic	*cfap74*
		LOC3:3685871	3	3,685,871	−1.141095	3.91×10^{-4}	Intronic	*cfap75*
		LOC12:5883873	12	5,883,873	−2.1259	2.70×10^{-4}	ncRNA intronic	*LOC115251859*
Hybrid puffer	BW and CL	LOC1:5146711	1	5,146,711	78.12444	4.94×10^{-5}	Intronic	*vclb*
		LOC20:7178403	20	7,178,403	−58.05161	3.70×10^{-5}	Intronic	*lingo3*
	BW and TL	LOC21:8126695	21	8,126,695	118.4649	5.24×10^{-5}	Intergenic	NA

BW, body weight; BL, body length; TL, total length; CL, chest length; AL, abdominal length.

Meanwhile, for the hybrid group, the GWAS showed that 1 SNP significantly associated ($-\log_{10} > 5.3$) (Figure S7, Table S11) and 40 SNPs suggestively associated ($-\log_{10} > 4.0$) with the growth traits. These SNPs were located mainly on chromosomes 1, 10, and 21. Among them, three SNPs were associated with multiple traits (Table 2): LOC1:5146711 and LOC20:7178403 were associated with BW and CL, with SNP effects ranging from −58.05 to 78.12, and LOC21:8126695 was associated with BW and TL, with SNP effects of 118.46 and 2.77, respectively. Although only 1 SNP significant is related to CL was found, 40 suggestive SNPs related to BW, TL, BL, and AL also had certain reference value; a QQ plot of the p values is shown in Figure S8.

To further explore the relationship between morphological traits and genetics, we selected SNPs associated with body weight (chr 21, T -> C) and body length (chr 10, G -> T) in the hybrid puffer for further analysis. The results showed that these SNPs were homozygous in the obscure puffer and were not associated with growth traits. However, in the hybrid puffer, most of them were heterozygous and exhibited fast growth, while only a few individuals were homozygous and exhibited slow growth. Combined with morphological trait analysis, the results showed that only homozygous individuals had morphological traits similar to the obscure puffer, and the morphological trait analysis results of some individuals were overlapping with the obscure puffer. This result is consistent with the analysis of population structure, suggesting that the morphological traits of hybrid puffer are influenced by their parents, and that genetic variants play a regulatory role in morphological trait development.

In conclusion, the morphological traits were significantly correlated with genetics, and the morphological traits of the hybrid puffer were influenced by the obscure puffer. In addition, although the number of GWAS groups involved was limited and the false-positive rate was elevated, we comprehensively scanned all genes within a 300 kb range upstream and downstream of the significant SNPs (Table S11). Subsequently, candidate genes associated with growth and metabolism were selected, thereby providing a valuable gene library for subsequent association analysis. Simultaneously, it is important to direct our attention to chromosomes harboring multiple significant SNPs, such as chromosomes 1, 10, and 22.

3.9. Combining Selection Signatures and Association Analysis

We integrated selection signatures and GWAS approaches to identify the candidate genes associated with growth-related traits in pufferfish. In the selection signatures, the genes involved in processes or pathways related to growth and metabolism were subjected to analysis using the F_{st} and π ratio as described above. In total, 106 candidate genes were related to growth and metabolism (Table S13), including 56 genes that were independently identified using F_{st}, while the other 50 genes were identified using the π ratio, and 10 genes intersected. For example, *wnt2b*, *ccnd2*, *map2k4*, *ndst2*, *stat5b*, *atp6v1b2*, and *cdk1* were identified using F_{st}; *ngfr*, *smad4*, *rnf31*, and *star* were identified using the π ratio; and *itgav*, *ighv3-43*, *ighm*, *dsp*, *dsg2*, *dsc2*, and *atp6v1b2* were identified via intersection. Most of these genes were associated with GO terms or KEGG pathways. Notably, *wnt2b*, *ngfr*, *itgav*, and *atp6v1b2* have been associated with cellular growth and energy metabolism in the published literature. In the GWAS, a total of 254 candidate genes related growth-related traits were identified, including *rnf213*, *ngfr*, *smad4*, *wnt3*, *wnt5b*, and *dsp*, which are related to metabolism, growth, and immunity (Table S12).

Ultimately, through intersection analysis, a total of 13 growth-related regions and 13 significant potential candidate genes (*itgav*, *ighv3-43*, *ighm*, *atp6v1b2*, *pld1*, *xmrk*, *inhba*, *dsp*, *dsg2*, and *dsc2*, Table 3, Figure S9) associated with pufferfish growth were identified.

Table 3. Growth-related genes identified using both selection signatures and GWAS.

Chr	Gene	Symbol	Description	Selected Region	F_{st}	π Ratio	SNPs Number	Traits	p Value
1	*LOC101079631*	*itgav*	integrin alpha-V	28,970,001–29,070,000	0.71	1.82	-	-	-
5	*LOC101067887*	*ighv3-43*	immunoglobulin epsilon heavy chain-like	6,090,001–6,220,000	0.86	1.78	-	-	-
5	*LOC101075124*	*ighm*	Ig mu chain C region membrane-bound form	6,120,001–6,200,000	0.86	1.78	-	-	-
-	*LOC101072524*	*atp6v1b2*	V-type proton ATPase subunit B, brain isoform	390,001–490,000	0.89	4.45	-	-	-
10	*LOC101069674*	*pld1*	phospholipase D1-like	11,240,001–11,340,000	0.35	1.07	6	BL	8.48×10^{-5}
10	*LOC101073748*	*xmrk*	melanoma receptor tyrosine-protein kinase-like	11,880,001–12,000,000	0.30	1.23	1	TL	6.99×10^{-5}
10	*LOC101073973*	*inhba*	inhibin beta A chain-like	11,870,001–12,000,000	0.29	1.24	1	TL	6.99×10^{-5}
22	*LOC115247960*	*dsp*	desmoplakin-like, partial	1,290,001–1,490,000	0.58	1.98	2	TL	5.35×10^{-6}
22	*LOC101063843*	*dsp*	desmoplakin-like, partial	1,380,001–1,490,000	0.64	1.09	2	TL	5.35×10^{-6}
22	*LOC115247930*	*dsg2*	desmoglein-2-like, partial	1,340,001–1,49,0000	0.66	1.47	2	TL	5.35×10^{-6}
22	*LOC101064287*	*dsg2*	desmoglein-2-like, partial	1,250,001–1,450,000	0.57	2.03	2	TL	5.35×10^{-6}
22	*LOC101064516*	*dsc2*	desmocollin-2-like	1,240,001–1,440,000	0.57	2.04	2	TL	5.35×10^{-6}
22	*LOC115247939*	*dsc2*	desmocollin-2-like	1,330,001–1,490,000	0.68	1.74	2	TL	5.35×10^{-6}

BL, body length; TL, total length.

4. Discussion

4.1. The Growth Rate and Model

Growth is one of the most significant traits for the aquaculture industry, as it is directly related to economic efficiency [5]. At the same time, the growth rate is a crucial factor that optimizes feeding rates and breeding density. Hybrid puffer has a significant growth advantage over obscure puffer, but the growth rate and model were until unknown. Constructing a growth model to estimate or predict fish weight is essential for achieving better control of the output of an aquaculture system [6], and growth is influenced by several factors. Therefore, we controlled for factors that influence growth, such as environmental conditions, feeding practices, and density. The results show that there is a significant difference in growth and morphological traits between the hybrid and obscure puffers, which might be affected by hybrid dominance. Similar heterosis results have been reported for other aquatic species, such as snow trout [13] and groupers [11].

The growth model is a valuable reference for the aquaculture industry [27]. However, most studies on pufferfish have focused on early-stage and short-term growth models. For instance, previous research has revealed that the early developmental model exhibits negative allometric growth for *T. rubripes* ($b < 3$) [27], while the juvenile developmental model demonstrates uniform growth for the obscure puffer ($b \approx 3$) [52], indicating a lack of investigation into long-term growth models. Notably, this study shows that the obscure puffer has a negative allometric growth model and the hybrid puffer has a uniform growth model. These novel growth models imply regulatory mechanisms biased towards patrilineal inheritance mechanisms; similar results were found regarding growth rates among catfish (*Silurus lanzhouensis*) and *Pelteobagrus fulvidraco* (♀) × *Leiocassis longirostris* (♂) hybrids [12,53].

4.2. The Growth-Related Candidate Genes and SNPs

Growth is a complex quantitative trait regulated by multiple genes [54]. The analysis of selection signatures is a genomic analysis method that was used in this study to identify the specific regions of higher genetic differentiation between hybrid and obscure puffers, and thereby to elucidate the molecular mechanisms underlying hybridization. Similarly, GWAS constitutes a pivotal approach in aquaculture research, especially in SNP trait association research, such as growth, disease resistance, color differentiation, and hypoxia tolerance [23,55–57]. However, both methods have limitations; therefore, the objective of this study was to combine the use of selection signatures and GWAS to investigate the growth-related selection regions and genes in pufferfish.

In the selection signatures, the results demonstrated a close association between the candidate genes derived from the identified selection features, several crucial growth- and metabolism-related GO terms, and KEGG pathways (Figures 8 and 9). Remarkably, the regulation of muscle contraction (GO:0006937) inhibits epidermal growth factor receptor (EGFR)–extracellular signal-regulated kinase (ERK)1/2 signaling under EGF stimulation, which can also enhance growth ability [58], and cellular senescence (ko04218) regulates cellular aging and damage processes [59]. All these pathways are implicated in cellular growth, development, and programmed cell death through indirect mechanisms, highlighting the intricate biological complexity underlying growth traits. In the GWAS, several SNPs were associated with multiple traits. Similar results were found for mandarin fish and Pacific abalone [23], indicating that coordinated regulatory mechanisms may be involved. We hypothesized that growth-related traits in pufferfish are influenced by the coordinated regulation of multiple genes across different chromosomes. Similar observations have been reported in other fish species, such as the Asian sea bass, wherein *Lca371* on LG2 was found to be involved in the regulation of TL, BW, and BL [60]. In carp, SNP0626 on LG19 was significantly associated with BW and BL [61].

We identified several genes associated with growth-related traits according to the selection signatures and GWAS, including *itgav*, *dsp*, *dsg2*, *dsc2*, *rnf213*, *smad4*, *atp6v1b2*, *map2k4*, *stat5b*, and *inhba*. Several of these genes were associated with fish growth in previous studies [21,23]. The *itgav* gene functions as a major player in regulating cell proliferation, apoptosis, migration, maintaining the integrity and permeability of vascular wall, and other biological functions, and it is related to the growth of osteosarcoma tissue [62]. In zebrafish, the knockout of *rnf213* was associated with abnormal vascular development of the head [63], and it was demonstrated that it is associated with fat formation and highly expressed in fat samples from obese populations [64]. Meanwhile, these genes regulate cellular proliferation and apoptosis through complex regulatory networks, such as the Wnt, MAPK, and TGF-β signaling pathways and other synergistic regulatory mechanisms, affecting the growth traits of individuals. For example, *dsp* has been shown to be a downstream regulator of the arrhythmogenic right ventricular cardiomyopathy (ARVC) pathway (ko05412) and to carry out inhibitory regulation of the Wnt pathway [65], in turn affecting its downstream regulation of the cell-cycle protein (*ccnd2*), thereby affecting cellular value creation [66]. In the Wnt signaling pathway, *wnt3* is upstream of the regulatory genes TCF/LEF, which also regulate TCF/LEF and cellular value-added through *ccnd2* [67]. Further, in the TGF-β signaling pathway, *smad4* can not only directly regulate downstream gonadal development and apoptosis [68], but has been shown to have a regulatory effect on the cell growth cycle through the Wnt signaling pathway [69,70]. Similarly, *ngfr* and *map2k4*, as key genes in the MAPK signaling pathway, regulate the cell cycle, growth, and death [71]. Studies have shown that *map2k4* regulates cell growth by activating cell cycle-associated proteins and EMT signals downstream of the PI3K–Akt signaling pathway [72]. Taken together, these candidate genes do not act individually to regulate growth but function through multiple pathways for synergistic network regulation. Although the experimental group used in this study was not a traditional group, such as a full-sib family, but rather a comparison between the hybrid puffer and the parental obscure puffer, these were used because the use of different individuals for the analyses is still of interest for how to select the parental individuals in breeding research work, which provides a theoretical basis for the subsequent study of fish growth and development.

In this study, we identified valuable SNPs and growth-related candidate genes that could contribute to the development of pufferfish growth models. Our findings may be applicable to pufferfish farming and may significantly accelerate the breeding process. However, other relevant physiological regulations in pufferfish breeding and genetics still need to be studied.

5. Conclusions

Our research indicates that the growth of hybrid puffers is considerably superior to that of obscure puffers, and there are notable disparities in their morphological traits. Furthermore, the growth and morphological traits of hybrid puffers are concurrently influenced by the genetic effects of the parental obscure puffers. Growth models of hybrid and obscure puffers in the culture cycle were constructed in this study based on their morphological and growth traits, revealing that there are different growth models between the obscure and hybrid puffers. The obscure puffer exhibits a negative allometric growth model, while the hybrid puffer exhibits a uniform growth model. After integrating selection signatures and GWAS analysis, a total of 13 growth-related regions and 13 significant potential candidate genes (*itgav*, *ighv3-43*, *ighm*, *atp6v1b2*, *pld1*, *xmrk*, *inhba*, *dsp*, *dsg2*, and *dsc2*) associated with pufferfish growth were identified.

Supplementary Materials: The following supporting information can be downloaded at https://www.mdpi.com/article/10.3390/fishes9100404/s1, Figure S1: The growth trends in the BW and TL of the hybrid and obscure puffer. A: The BW growth trend. B: The TL growth trend; Figure S2: The different morphological traits correlations and kinship relationships between populations. A: The correlation for the obscure puffer. B: The correlation for the hybrid puffer. C: The kinship for the obscure puffer. D: The kinship for the hybrid puffer; Figure S3: The cluster tree between hybrid and obscure puffer based on unweighted pair-group method with arithmetic mean (UPGMA). Obscure puffer (red), hybrid puffer (blue); Figure S4: The experimental population structure constructed using SNPs. A: The population structure for obscure and hybrid puffers. B: The population structure for the obscure puffer. C: The population structure for the hybrid puffer. D: The cross-validation error analysis for the obscure and hybrid puffer population structure. E: The cross-validation error analysis for the obscure puffer population structure. F: The cross-validation error analysis for the hybrid puffer population structure; Figure S5: The GO enrichment analysis and gene classification conducted based on the overlapping F_{st} and π ratio; Figure S6: The KEGG enrichment analysis enrichment pathway classification. A: The KEGG pathway classification of genes from F_{st} selection regions. B: KEGG pathway classification of genes from π ratio selection regions. C: KEGG pathway classification of genes in the intersection region; Figure S7: Manhattan plots of 5 growth-related traits in the hybrid and obscure puffer groups. The down bars represent marker density on each chromosome. A, B, C, D, and E: Manhattan plots of 5 growth-related traits in the obscure puffer. F, G, H, I, and J: Manhattan plots of 5 growth-related traits in the hybrid puffer; Figure S8: QQ plots of 5 Manhattan plots in the obscure and hybrid puffer groups. A, B, C, D, and E: QQ plots of 5 Manhattan plots for the obscure puffer. F, G, H, I, and J: QQ plots of 5 Manhattan plots for the hybrid puffer; Figure S9: The intersection analysis of selection signatures and GWAS; Table S1: The growth data on the parental strain of the hybrid puffer; Table S2: The growth data regarding the genotyping by sequencing samples; Table S3: The growth data regarding the hybrid and obscure puffer groups; Table S4: The morphological traits of the hybrid and obscure puffer groups; Table S5: Spearman correlation analysis of the hybrid and obscure puffer groups; Table S6: The classification of SNPs in genotyping by sequencing results; Table S7: The classification ratio of SNPs in the genotyping by sequencing results; Table S8: The candidate genes screened using the F_{st} and π ratio; Table S9: The GO enrichment analysis of candidate genes; Table S10: The KEGG enrichment analysis of candidate genes; Table S11: The significant SNP information from the GWAS; Table S12: All candidate genes related to growth and metabolism in the GWAS analysis; Table S13: All candidate genes related to growth and metabolism in the selection signatures analysis.

Author Contributions: Conceptualization: Y.S. and Z.Z.; methodology: C.W. and Y.G.; software: C.W., Y.G., S.S. and M.W.; formal analysis: C.W., S.S. and Y.Y.; investigation: C.W. and Y.S.; validation: C.W., Z.S. and Y.W.; writing—original draft preparation: C.W.; writing—review and editing: Y.S. and Z.Z.; supervision: Y.S. and Z.Z.; Funding acquisition: Y.S. and Z.Z.; Visualization: Y.S. and C.W. All authors have read and agreed to the published version of the manuscript.

Funding: This work was supported by the Project for Seed Industry Vitalization of Jiangsu Province (JBGS [2021] 133) and the National Natural Science Foundation of China (32002424).

Institutional Review Board Statement: The animal study protocol was approved by the Ethics Committee of the College of Oceanography, Hohai University (approval code: hhuhy-22-01).

Data Availability Statement: The data are available from the corresponding author on reasonable request.

Conflicts of Interest: Zhenlong Sun and Yoahui Wang are employed at Zhogyang Seed Industry (Jiangsu) Co., Ltd. The authors declare that this employment did not influence the results of the study. The other authors declare no conflicts of interest.

References

1. Wang, D.; Wu, F.X. *China Fishery Statistical Yearbook*; China Agriculture Press Co., Ltd.: Beijing, China, 2023; pp. 25–26.
2. Gao, Y.; Zhang, H.; Gao, Q.; Wang, L.; Zhang, F.; Siva, V.S.; Zhou, Z.; Song, L.; Zhang, S. Transcriptome analysis of artificial hybrid pufferfish Jiyan-1 and its parental species: Implications for pufferfish heterosis. *PLoS ONE* **2013**, *8*, e58453. [CrossRef]
3. Shi, Y.; Zhang, G.; Zhu, Y.; Liu, J. Effects of photoperiod, temperature, and salinity on growth and survival of obscure puffer *Takifugu obscurus* larvae. *Aquaculture* **2010**, *309*, 103–108. [CrossRef]
4. Wei, Y.; Wang, J.; Zhang, X.; Duan, M.; Jia, L.; Xu, H.; Liang, M.; Liu, J. Fish protein hydrolysate supplementation in plant protein based diets for tiger puffer (*Takifugu rubripes*) is an effective strategy of fish meal sparing. *Aquac. Rep.* **2021**, *20*, 100720. [CrossRef]
5. Wang, T.; Wu, X.; Song, L.; Yang, Y.; Gong, S.; Zeng, L.; Tao, Y.; Zhong, C.; Meng, Z.; Liu, X. Identification of candidate growth-related SNPs and genes using GWAS and transcriptome analyses in leopard coral grouper (*Plectropomus leopardus*). *Aquaculture* **2023**, *574*, 739677. [CrossRef]
6. Aljehani, F.; N'Doye, I.; Laleg-Kirati, T.-M. Extended Kalman filter for fish weight estimation using augmented fish population growth model. *IFAC-PapersOnLine* **2023**, *56*, 9855–9861. [CrossRef]
7. Gui, J.-F.; Zhou, L.; Li, X.-Y. Rethinking fish biology and biotechnologies in the challenge era for burgeoning genome resources and strengthening food security. *Water Biol. Secur.* **2022**, *1*, 100002. [CrossRef]

8. Bai, J.; Li, C.; Tang, Z.; Wu, C.; Wei, Z. Comparative study of carbohydrate levels on growth, oxidative stress and glucolipid metabolism of hybrid fish between *Megalobrama amblycephala* (♀) × *Culter alburnus* (♂) and *Culter alburnus*. *Reprod. Breed.* **2023**, *3*, 131–142. [CrossRef]
9. Wang, S.; Tang, C.; Tao, M.; Qin, Q.; Zhang, C.; Luo, K.; Zhao, R.; Wang, J.; Ren, L.; Xiao, J. Establishment and application of distant hybridization technology in fish. *Sci. China Life Sci.* **2019**, *62*, 22–45. [CrossRef]
10. Fan, C.; Zhang, X.; Tang, L.; Zhang, X.; Li, J.; Li, Y.; Li, Q.; Wang, Z. Study on the effect of mass selection and hybridization on growth performance of Chinese pearl oyster *Pinctada martensii*. *Front. Mar. Sci.* **2022**, *9*, 851142. [CrossRef]
11. Gu, H.; Wang, H.; Deng, S.; Dai, X.; He, X.; Wang, Z. Hybridization with Percocypris pingi male can improve the cultivation characteristics of snow trout (*Schizothorax wangchiachii*), especially the resistance to white spot disease (*Ichthyophthirius multifiliis*). *Aquaculture* **2023**, *562*, 738805. [CrossRef]
12. Wang, H.; Lin, G.; Zhou, J.; Zong, Y.; Ning, X.; Wang, T.; Yin, S.; Zhang, K.; Ji, J. The hybrid *Pelteobagrus fulvidraco* (♀) × *Leiocassis longirostris* (♂) exhibits improved trait on hypoxia-tolerance. *Aquaculture* **2023**, *562*, 738859. [CrossRef]
13. Chen, S.; Tian, Y.; Li, Z.; Liu, Y.; Li, Z.; Duan, P.; Li, L.; Wang, X.; Wang, L.; He, X. Heterosis in growth and low temperature tolerance in Jinhu grouper (*Epinephelus fuscoguttatus* ♀× *Epinephelus tukula* ♂). *Aquaculture* **2023**, *562*, 738751. [CrossRef]
14. Hu, Y.; Hua, Y. The Development of Embryo and Fry of Hybrid Puffer (*Fugu Ocellatus*♀× *Fugu Obscurus*). *J. Nanjing Norm. Univ. (Nat. Sci. Ed.)* **1996**, *19*, 59–63.
15. Fan, W.; Liu, H.; Zhao, W.; Zhang, F. Embryonic, Larval and Juvenile Development of Hybrid between Tawny Puffer *Takifugu flavidus* (♀) × Redfin Puffer *Takifugu rubripes* (♂). *J. Fish. China* **2011**, *35*, 1065–1071. [CrossRef]
16. Zhang, N.; Li, J.; Jiang, J.; Pan, G.; Zhou, W. Effects of Salinity on Fertilized Egg Hatching and Larval Growth of Hybrid Puffer (*Takifugu flavidus* ♀× *Takifugu obscures* ♂). *Chin. Agric. Sci. Bull.* **2015**, *31*, 31–34.
17. Jiang, H.-B.; Bao, J.; Han, Y. Mitochondrial DNA sequence of the hybrid of *Takifugu flavidus* (♀)× *Takifugu rubripes* (♂). *Mitochondrial DNA Part A* **2016**, *27*, 2117–2118. [CrossRef] [PubMed]
18. Jin, X.; Zhu, H.; Shi, Y.; Chen, Z.; Wang, Y.; Gui, J.-F.; Zhao, Z. Host Hybridization Dominates over Cohabitation in Affecting Gut Microbiota of Intrageneric Hybrid *Takifugu* Pufferfish. *Msystems* **2023**, *8*, e01181-22. [CrossRef] [PubMed]
19. Park, I.-S.; Lee, S.; Yoo, G.-Y. Impact of dietary protein levels on growth, feed utilization, body composition, and hematological characteristics of juvenile hybrid pufferfish (*Takifugu obscurus* × *T. rubripes*). *Aquac. Rep.* **2022**, *22*, 100994. [CrossRef]
20. Tian, Y.; Wang, C.; Wang, Y.; Xiong, Y.; Liu, Y.; Yan, H.; Wu, A.; Gao, R.; Li, M.; Wang, L. Gill transcriptomes analysis of *Takifugu obscurus*, *Takifugu rubripes* and their hybrid offspring in freshwater and seawater. *Aquac. Rep.* **2024**, *37*, 102208. [CrossRef]
21. Omeka, W.; Liyanage, D.; Lee, S.; Lim, C.; Yang, H.; Sandamalika, W.G.; Udayantha, H.; Kim, G.; Ganeshalingam, S.; Jeong, T. Genome-wide association study (GWAS) of growth traits in olive flounder (*Paralichthys olivaceus*). *Aquaculture* **2022**, *555*, 738257. [CrossRef]
22. Baesjou, J.-P.; Wellenreuther, M. Genomic signatures of domestication selection in the Australasian snapper (*Chrysophrys auratus*). *Genes* **2021**, *12*, 1737. [CrossRef] [PubMed]
23. Zhou, Y.; Fu, H.-C.; Wang, Y.-Y.; Huang, H.-Z. Genome-wide association study reveals growth-related SNPs and candidate genes in mandarin fish (*Siniperca chuatsi*). *Aquaculture* **2022**, *550*, 737879. [CrossRef]
24. Zohar, Y.; Mylonas, C.C. Endocrine manipulations of spawning in cultured fish: From hormones to genes. *Aquaculture* **2001**, *197*, 99–136. [CrossRef]
25. Budi, D.S.; Puspitasari, S.; Febrianti, R.P.N.; Bodur, T.; Mukti, A.T. A novel technique for mass induction of propagation in small fish species: Hormone immersion. *Anim. Reprod. Sci.* **2023**, *255*, 107280. [CrossRef]
26. Zadmajid, V.; Bashiri, S.; Sharafi, N.; Butts, I.A.E. Effect of hCG and Ovaprim™ on reproductive characteristics of male Levantine scraper, *Capoeta damascina* (Valenciennes, 1842). *Theriogenology* **2018**, *115*, 45–56. [CrossRef] [PubMed]
27. Gao, X.; Cao, S.; Zhang, X.; Zhu, Z.; Hai-Bin, C.; Rui, X.; Zhao, K.-F.; Zhang, C.-X.; Liu, B.-L. Growth patterns and feeding characteristics in early developmental stages of *Takifugu rubripes* cultured in a recirculating aquaculture system. *Aquaculture* **2023**, *577*, 739981. [CrossRef]
28. Li, W.; Xiong, Y.; Wang, Z.; Zhang, Q.; Shen, X.; Liu, Q.; Yan, H.; Gao, R.; Liu, Y.; Pang, H. Effects of different photoperiod conditions on survival, growth, and gonadal development of *Takifugu rubripes* adults. *Aquaculture* **2023**, *564*, 739048. [CrossRef]
29. Sun, X.; Wang, Y. Growth models in aquaculture for hybrid and natural groupers based on early development stage. *Aquaculture* **2024**, *578*, 740026. [CrossRef]
30. Shi, X.; Chen, S.; Gan, C.; Liu, X.; Chen, F.; Ma, X.; Wu, L.; Zhang, Y.; Waiho, K.; Tian, X. Comparison of growth performance, feed utilization and morphology between homologous and heterologous juvenile Qihe crucian carp *Carassius auratus*. *Aquaculture* **2022**, *561*, 738634. [CrossRef]
31. Gao, F.-X.; Shi, Y.; Duan, W.; Lu, W.-J.; Huang, W.; Zhang, X.-J.; Zhao, Z.; Zhou, L.; Gui, J.-F. A rapid and reliable method for identifying genetic sex in obscure pufferfish (*Takifugu obscurus*). *Aquaculture* **2020**, *519*, 734749. [CrossRef]
32. Zhou, H.X.; Duan, G.Q.; Hu, Y.T.; Wang, H.; Jiang, H.; Ling, J.; Pan, T.S.; Chen, X.L. Analysis of Morphological Differences between Differential Body Colors Groups of Hybrid Yellow Catfish. *Fish. Sci.* **2021**, *40*, 726–732.
33. Haddon, M. *Modelling and Quantitative Methods in Fisheries*; Chapman and Hall/CRC: Boca Raton, FL, USA, 2011.
34. Dureuil, M.; Aeberhard, W.H.; Dowd, M.; Pardo, S.A.; Whoriskey, F.G.; Worm, B. Reliable growth estimation from mark–recapture tagging data in elasmobranchs. *Fish. Res.* **2022**, *256*, 106488. [CrossRef]

35. Chen, S.; Zhou, Y.; Chen, Y.; Gu, J. fastp: An ultra-fast all-in-one FASTQ preprocessor. *Bioinformatics* **2018**, *34*, i884–i890. [CrossRef] [PubMed]
36. Li, H.; Durbin, R. Fast and accurate short read alignment with Burrows–Wheeler transform. *Bioinformatics* **2009**, *25*, 1754–1760. [CrossRef] [PubMed]
37. Van der Auwera, G.A.; Carneiro, M.O.; Hartl, C.; Poplin, R.; Del Angel, G.; Levy-Moonshine, A.; Jordan, T.; Shakir, K.; Roazen, D.; Thibault, J. From FastQ data to high-confidence variant calls: The genome analysis toolkit best practices pipeline. *Curr. Protoc. Bioinform.* **2013**, *43*, 11.10.11–11.10.33. [CrossRef]
38. Wang, K.; Li, M.; Hakonarson, H. ANNOVAR: Functional annotation of genetic variants from high-throughput sequencing data. *Nucleic Acids Res.* **2010**, *38*, e164. [CrossRef] [PubMed]
39. Chang, C.C.; Chow, C.C.; Tellier, L.C.; Vattikuti, S.; Purcell, S.M.; Lee, J.J. Second-generation PLINK: Rising to the challenge of larger and richer datasets. *Gigascience* **2015**, *4*, s13742-015-0047-8. [CrossRef]
40. Wang, M.; Li, L.; Lin, H.; Zhou, Z.; Liu, B.; Zhong, J.; Pu, F.; Shi, Y.; Zhou, T.; Xu, P. Genome-wide association study identifies genomic loci of sex determination, gonadal weight and gonadosomatic index traits in *Takifugu bimaculatus*. *Aquaculture* **2022**, *546*, 737389. [CrossRef]
41. Geng, X.; Sha, J.; Liu, S.; Bao, L.; Zhang, J.; Wang, R.; Yao, J.; Li, C.; Feng, J.; Sun, F. A genome-wide association study in catfish reveals the presence of functional hubs of related genes within QTLs for columnaris disease resistance. *BMC Genom.* **2015**, *16*, 196. [CrossRef]
42. Yang, J.; Lee, S.H.; Goddard, M.E.; Visscher, P.M. GCTA: A tool for genome-wide complex trait analysis. *Am. J. Hum. Genet.* **2011**, *88*, 76–82. [CrossRef]
43. Kumar, S.; Stecher, G.; Li, M.; Knyaz, C.; Tamura, K. MEGA X: Molecular evolutionary genetics analysis across computing platforms. *Mol. Biol. Evol.* **2018**, *35*, 1547. [CrossRef] [PubMed]
44. Alexander, D.H.; Novembre, J.; Lange, K. Fast model-based estimation of ancestry in unrelated individuals. *Genome Res.* **2009**, *19*, 1655–1664. [CrossRef] [PubMed]
45. Pfeifer, B.; Wittelsbürger, U.; Ramos-Onsins, S.E.; Lercher, M.J. PopGenome: An efficient Swiss army knife for population genomic analyses in R. *Mol. Biol. Evol.* **2014**, *31*, 1929–1936. [CrossRef] [PubMed]
46. Gallone, B.; Steensels, J.; Prahl, T.; Soriaga, L.; Saels, V.; Herrera-Malaver, B.; Merlevede, A.; Roncoroni, M.; Voordeckers, K.; Miraglia, L. Domestication and divergence of Saccharomyces cerevisiae beer yeasts. *Cell* **2016**, *166*, 1397–1410. [CrossRef]
47. Nei, M.; Li, W.-H. Mathematical model for studying genetic variation in terms of restriction endonucleases. *Proc. Natl. Acad. Sci. USA* **1979**, *76*, 5269–5273. [CrossRef]
48. Lin, T.; Zhu, G.; Zhang, J.; Xu, X.; Yu, Q.; Zheng, Z.; Zhang, Z.; Lun, Y.; Li, S.; Wang, X. Genomic analyses provide insights into the history of tomato breeding. *Nat. Genet.* **2014**, *46*, 1220–1226. [CrossRef] [PubMed]
49. Gao, X. Multiple testing corrections for imputed SNPs. *Genet. Epidemiol.* **2011**, *35*, 154–158. [CrossRef]
50. Yin, L.; Zhang, H.; Tang, Z.; Xu, J.; Yin, D.; Zhang, Z.; Yuan, X.; Zhu, M.; Zhao, S.; Li, X. rMVP: A memory-efficient, visualization-enhanced, and parallel-accelerated tool for genome-wide association study. *Genom. Proteom. Bioinform.* **2021**, *19*, 619–628. [CrossRef]
51. Zhou, X.; Stephens, M. Genome-wide efficient mixed-model analysis for association studies. *Nat. Genet.* **2012**, *44*, 821–824. [CrossRef] [PubMed]
52. Qiao, Y.; Shi, Y.; Lu, G.; Cao, X.; Li, X.; Wang, W.; Fan, W. Growth performance of *Takifugu obscurus* cultured in different modes. *Fish. Sci. Technol. Inf.* **2021**, *48*, 132–136. [CrossRef]
53. Niu, J.-S.; Wang, T.; Li, Z.; Wang, Z.-W.; Ding, M.; Wang, M.-T.; Lian, Z.-Q.; Mei, J.; Wang, Y.; Zhou, L. Efficient breeding and growth advantage of all-male population in Lanzhou catfish (*Silurus lanzhouensis*). *Aquaculture* **2024**, *578*, 740023. [CrossRef]
54. Löhr, H.; Hess, S.; Pereira, M.M.; Reinoß, P.; Leibold, S.; Schenkel, C.; Wunderlich, C.M.; Kloppenburg, P.; Brüning, J.C.; Hammerschmidt, M. Diet-induced growth is regulated via acquired leptin resistance and engages a Pomc-somatostatin-growth hormone circuit. *Cell Rep.* **2018**, *23*, 1728–1741. [CrossRef]
55. Yang, M.; Wang, Q.; Chen, J.; Wang, Y.; Zhang, Y.; Qin, Q. Identification of candidate SNPs and genes associated with anti-RGNNV using GWAS in the red-spotted grouper, *Epinephelus akaara*. *Fish Shellfish Immunol.* **2021**, *112*, 31–37. [CrossRef] [PubMed]
56. Chen, D.; Zhang, Q.; Tang, W.; Huang, Z.; Wang, G.; Wang, Y.; Shi, J.; Xu, H.; Lin, L.; Li, Z. The evolutionary origin and domestication history of goldfish (*Carassius auratus*). *Proc. Natl. Acad. Sci. USA* **2020**, *117*, 29775–29785. [CrossRef]
57. Ding, J.; Gao, Z.; Wang, J.; Zhang, Y.; Wang, X.; Wu, X.; Zhu, J.; Shen, W. Genome-wide association and transcriptome analysis provide the SNPs and molecular insights into the hypoxia tolerance in large yellow croaker (*Larimichthys crocea*). *Aquaculture* **2023**, *573*, 739547. [CrossRef]
58. Sheng, Z.; Cao, X.; Deng, Y.-n.; Zhao, X.; Liang, S. SUMOylation of AnxA6 facilitates EGFR-PKCα complex formation to suppress epithelial cancer growth. *Cell Commun. Signal.* **2023**, *21*, 189. [CrossRef]
59. Muñoz-Espín, D.; Cañamero, M.; Maraver, A.; Gómez-López, G.; Contreras, J.; Murillo-Cuesta, S.; Rodríguez-Baeza, A.; Varela-Nieto, I.; Ruberte, J.; Collado, M. Programmed cell senescence during mammalian embryonic development. *Cell* **2013**, *155*, 1104–1118. [CrossRef]
60. Wang, C.M.; Bai, Z.Y.; He, X.P.; Lin, G.; Xia, J.H.; Sun, F.; Lo, L.C.; Feng, F.; Zhu, Z.Y.; Yue, G.H. A high-resolution linkage map for comparative genome analysis and QTL fine mapping in Asian seabass, *Lates calcarifer*. *BMC Genom.* **2011**, *12*, 174. [CrossRef]

61. Laghari, M.; Lashari, P.; Zhang, X.; Xu, P.; Xin, B.; Zhang, Y.; Narejo, N.; Sun, X. Mapping quantitative trait loci (QTL) for body weight, length and condition factor traits in backcross (BC1) family of Common carp (*Cyprinus carpio* L.). *Mol. Biol. Rep.* **2014**, *41*, 721–731. [CrossRef] [PubMed]
62. Li, Z.; Deng, L.; Li, Y.; Wang, Y. MiR-139 inhibits proliferation, migration and invasion of osteosarcoma cell line MG63 via down-regulating integrin subunit alpha V (ITGAV). *Tissue Cell* **2022**, *75*, 101720. [CrossRef]
63. Liu, W.; Morito, D.; Takashima, S.; Mineharu, Y.; Kobayashi, H.; Hitomi, T.; Hashikata, H.; Matsuura, N.; Yamazaki, S.; Toyoda, A. Identification of *RNF213* as a susceptibility gene for moyamoya disease and its possible role in vascular development. *PLoS ONE* **2011**, *6*, e22542. [CrossRef] [PubMed]
64. Sarkar, P.; Thirumurugan, K. New insights into TNFα/PTP1B and PPARγ pathway through RNF213-a link between inflammation, obesity, insulin resistance, and Moyamoya disease. *Gene* **2021**, *771*, 145340. [CrossRef] [PubMed]
65. Celeghin, R.; Risato, G.; Beffagna, G.; Cason, M.; Bueno Marinas, M.; Della Barbera, M.; Facchinello, N.; Giuliodori, A.; Brañas Casas, R.; Caichiolo, M. A novel DSP zebrafish model reveals training-and drug-induced modulation of arrhythmogenic cardiomyopathy phenotypes. *Cell Death Discov.* **2023**, *9*, 441. [CrossRef]
66. Jiang, P.; Mao, L.; Lei, X.; Luo, C.; Zhang, Y.; Zhong, X.; Yin, Z.; Xu, X.; Li, D.; Zheng, Q. miR-1297 inhibits osteosarcoma cell proliferation and growth by targeting CCND2. *Am. J. Cancer Res.* **2022**, *12*, 3464. [PubMed]
67. Rulifson, I.C.; Karnik, S.K.; Heiser, P.W.; Ten Berge, D.; Chen, H.; Gu, X.; Taketo, M.M.; Nusse, R.; Hebrok, M.; Kim, S.K. Wnt signaling regulates pancreatic β cell proliferation. *Proc. Natl. Acad. Sci. USA* **2007**, *104*, 6247–6252. [CrossRef]
68. Zhao, M.; Mishra, L.; Deng, C.-X. The role of TGF-β/SMAD4 signaling in cancer. *Int. J. Biol. Sci.* **2018**, *14*, 111. [CrossRef]
69. Du, X.; Li, Q.; Yang, L.; Liu, L.; Cao, Q.; Li, Q. SMAD4 activates Wnt signaling pathway to inhibit granulosa cell apoptosis. *Cell Death Dis.* **2020**, *11*, 373. [CrossRef]
70. Du, X.; Pan, Z.; Li, Q.; Liu, H.; Li, Q. SMAD4 feedback regulates the canonical TGF-β signaling pathway to control granulosa cell apoptosis. *Cell Death Dis.* **2018**, *9*, 151. [CrossRef] [PubMed]
71. Wei, J.; Huang, J.; Kuang, Y.; Li, Y.; Zhong, D.; Song, J. Metformin inhibits proliferation of oral squamous cell carcinoma cells by suppressing proteolysis of nerve growth factor receptor. *Arch. Oral Biol.* **2021**, *121*, 104971. [CrossRef] [PubMed]
72. Liu, S.; Huang, J.; Zhang, Y.; Liu, Y.; Zuo, S.; Li, R. MAP2K4 interacts with Vimentin to activate the PI3K/AKT pathway and promotes breast cancer pathogenesis. *Aging (Albany NY)* **2019**, *11*, 10697. [CrossRef] [PubMed]

fishes

MDPI

Article

A New Mutagenesis Tool for Songpu Mirror Carp (*Cyprinus carpio* L.) for Selective Breeding: Atmospheric-Pressure Room-Temperature Plasma Mutagenesis Technology

Xiaona Jiang [1,2], Chitao Li [1,2], Mei Shang [1,2], Xuesong Hu [1,2], Yanlong Ge [1,2] and Zhiying Jia [1,2,*]

[1] Heilongjiang River Fisheries Research Institute, Chinese Academy of Fishery Sciences, Harbin 150076, China
[2] Key Laboratory of Freshwater Aquatic Biotechnology and Breeding, Ministry of Agriculture and Rural Affairs, Harbin 150076, China
* Correspondence: jiazhiying@hrfri.ac.cn

Abstract: As a new, safe, and efficient method, Atmospheric-Pressure Room-Temperature Plasma (ARTP) mutagenesis has been widely applied in the field of microbial breeding and industrial applications, but it is rarely used in fish. In this study, ARTP mutagenesis technology was applied for the first time to a common carp strain, Songpu mirror carp (*Cyprinus carpio* L.), to increase genetic variation in this species. The appropriate experimental conditions were determined to include a radio frequency output power of 160 W and the processing of fertilized eggs for 360 s. The ARTP treatment group had a lower survival rate than the control group. The CV of morphological characters in the ARTP treatment group was significantly higher than that in the control group, and the CV of body weight was the highest ($p < 0.05$). In addition, the deformity rate in the ARTP treatment group was significantly higher than in the control group ($p < 0.05$). Individuals with high weight and no deformities were screened within the selection pressure of 1:15 of ARTP treatment group and fed in the same pool with the control group of the same age. The measurement of serum indices showed that, in the ARTP treatment group, TP, ALP, ALB, T-CHO, LDL levels were significantly higher than those in the control group ($p < 0.05$). Furthermore, the relative expressions of *SOD*, growth-related genes *GH*, *IGF-I*, protein synthesis-related genes *TOR* and *4EBP1* were significantly higher in the ARTP treatment group than in the control group ($p < 0.05$). In summary, Songpu mirror carp subjected to ARTP treatment showed a higher growth potential and antioxidant capacity.

Keywords: ARTP; Songpu mirror carp; CV; serum indicators; relative expression

Key Contribution: This research optimized the mutagenesis conditions of ARTP mutation breeding in fertilized Songpu mirror carp eggs for the first time.

1. Introduction

The purpose of selective breeding in aquaculture fish is to produce varieties with excellent economic traits, such as rapid growth, disease resistance and high meat quality. Improving the genetic variability of breeding materials is important in fish breeding, which is the main purpose of the application of mutagenesis technology. Mutations are the basis of genetic variation, and naturally occurring mutations play an important role in evolution. At present, due to the low incidence of natural mutations, artificial mutagenesis is commonly performed, for example, by applying physical radiation or chemical mutagens. Chemical mutagens mainly cause DNA alkylation damage or form base adducts [1,2]. ENU is a widely used chemical mutagen in fish that mainly induces single-base substitution [3,4]. ENU induces a relatively high mutation frequency. In addition, physical radiation, such as γ rays, X-rays, UV rays or particle radiation, can cause the replacement, reorganization and recombination of biomolecules or atoms. These mutagenesis methods can accelerate mutations in fish at the genome level and generate genetic diversity to improve targeted traits through targeted screening, but these mutagenesis methods are characterized by safety risks, and limited controllability of mutagenesis efficiency. ARTP mutagenesis technology is another safe and efficient mutagenesis method that can be applied after ion beam implantation and APDBD mutagenesis technology [5–9]. Compared with the application of UV and chemical mutagens, ARTP can cause more DNA damage, resulting in a higher mutation rate. ARTP irradiation is a fast and effective method for generating a library of mutants with sufficient diversity to improve the phenotype of microorganisms [8,10]. Radio-frequency atmospheric pressure glow discharge is utilized by ARTP to drive plasma working gas through the discharge region between two electrodes under the induction of the externally applied radio-frequency electric field, resulting in the formation of a room temperature plasma jet downstream of the plasma torch nozzle exit [8,10,11]. The main component of ARTP is the high density of active chemical species that penetrate cell walls and membranes, thereby damaging DNA molecules, causing mutations [6,7,12]. The ARTP biological mutation breeding systems instrument does not require a vacuum system and is low cost, presents uniform discharge, is stable and controllable, shows a high degree of safety, has been gradually applied in scientific research and industrial fields, and has achieved good results [13,14]. In addition, the plasma generation conditions are mild (atmospheric-pressure, room-temperature range), and are characterized by a high level of a large number of active particles, a broad mutation spectrum, and a high mutation rate [7,8]. The conditions for regarding ARTP working gas source type, flow rate, discharge power, processing time and other conditions can be controlled. By changing the operating conditions of the instrument, combined with the screening pressure and the application of high-throughput screening technology, the resulting mutation strength and mutation library can be greatly improved. ARTP is expected to become a new method for the efficient inducing of mutation for selective breeding. To date, ARTP has been successfully used in the breeding of more than 100 microorganisms [15], including species of both fungi [16] and bacteria [17]. In addition, it has been reported that the ARTP mutagenesis technique has been successfully applied in Japanese flounder (*Paralichthys olivaceus*) and blunt snout bream (*Megalobrama amblycephala*), in which the ARTP treatment time for fertilized eggs or sperm was optimized, and ARTP mutagenesis methods were established [10,18]. There are few reports of the application of ARTP in fish mutation breeding, and further research is needed.

Common carp (*Cyprinus carpio*) is the fourth most important freshwater fish in the world, with global aquaculture production reaching more than 4.23 million tons in 2020, accounting for nearly 8.6% of the world's total freshwater aquaculture production [19]. Songpu mirror carp has been bred on a large scale in most areas of China due to its high survival rate and strong resistance to diseases and cold temperature [20]. Therefore, Songpu mirror carp is the main species of freshwater fish employed for selective breeding research. The use of selective breeding methods to improve the economic traits of Songpu mirror carp is of great importance. Currently, mutagenesis breeding achieves a high mutation rate, and mutagenesis can be combined with other breeding methods, such as hybridization, selection and gynogenesis; thus, this approach may have significant practical value. In this study, ARTP technology was used for the first time in Songpu mirror carp to establish the optimal ARTP mutagenesis parameters for fertilized eggs and to evaluate the mutagenic effects.

2. Materials and Methods

2.1. Animal Experiment and Sample Collection

The Songpu mirror carp used in the study were similar in weight (3 ± 0.1 kg) and came from the Kuandian Fisheries Experiment Station of Heilongjiang River Fisheries Research Institute, Chinese Academy of Fishery Sciences. Mature female and male Songpu mirror carp were artificially induced, and eggs and sperm were manually collected from the mature female (n = 5) and male carp (n = 5) by gently pressing the abdomen, and then stored in a 4 °C refrigerator before use.

2.2. Mutation by ARTP Treatment

The sperm were added to the eggs and mixed in 21 °C water to obtain fertilized eggs. The fertilized eggs were distributed onto a filter screen to ensure that they did not overlap with each other, and an ARTP mutation instrument (ARTP-A, TMAXTREE Biotechnology Co., Ltd., Luoyang, China) was used for processing. Approximately 3000 eggs at the metaphase of first mitosis were placed into multiple glass petri dishes containing 1 mL of water. The glass petri dishes were then exposed to plasma for ARTP mutation system parameter selection, including input power, helium gas flow, treatment distance, and treatment time. A hatching rate of approximately 50% (the highest mutagenesis rate) was selected for ARTP in Songpu mirror carp [10], in which the input powers were divided into 0 w, 120 w, 160 w, 200 w, helium gas flow rate of 15 L/min treatment distance of 2 mm, and treatment time of 0 s, 120 s, 240 s or 360 s, respectively. Finally, the fertilized eggs subjected to ARTP treatment were transferred to a 21 °C water bath until fertilization.

2.3. Fertilization Rate and Hatchability Rate

The fertilization rates and hatchability rates for the ARTP treatment and the control groups were calculated [21]. The egg/sperm ratio was 1:2 × 10^7 (egg/spermatozoa). Here, the fertilization rate refers to the ratio of the number of fertilized embryos to the total number of embryos. The hatching rate indicates the ratio of the number of surviving embryos to the number of fertilized embryos. The calculation of the fertilization rate was generally carried out after 8 h of egg incubation, whereas the hatchability rate was evaluated shortly after the fry emerged [22]. At least three replicates per experimental group were used.

2.4. Morphological Characteristics of Individuals Subjected to ARTP Treatment

The fertilized eggs for the ARTP treatment and control groups were cultured in the same environment. After 5-month culture, morphological traits, such as the weight (W), standard length (SL), body height (H), body width (BW) and head length (HL) were measured. The types and numbers of deformities were calculated. Individuals with high W and no deformities were screened with a selection pressure of 1:15 and electronically marked. After 14 months, morphological traits, such as the W, SL, H, BW and HL were measured.

2.5. Sample Collection and Indicator Determination

Three fish were randomly selected for tail sample collection in the ARTP treatment and the control groups at the 14th month, and Anle fish (MS-222, 100 mg/L, Beijing Green Hengxing Biological Technology Co., Beijing, China) was used to anaesthetize the experimental fish. Blood was collected from the tail vein and stored in a premade heparin anticoagulant tube at 4 °C for 1–2 h, followed by centrifugation (Multifuge X1R, Thermo Fisher Scientific, Waltham, MA, USA) at 1342 *g* for 10 min. The upper serum was drawn off, dispensed into centrifuge tubes and placed at −20 °C for use in the determination of serum biochemical indicators. The liver, intestine, and dorsal muscles (at the same position) were collected from the fish, mixed with samples, and placed in a −80 °C freezer for the determination of corresponding indicators. The evaluated serum biochemical indicators were determined by the immunoturbidimetry method, including TP, ALB, ALT, AST, ALP, T-CHO, TG, HDL, LDL, UA and TBA, and all indicators were measured using a biochemical analyzer (BS350E, Mindary, Shenzhen, China).

2.6. RNA Extraction and RTq–PCR

RNA extraction was performed on the tissues (liver, intestine, and dorsal muscles) collected from all experimental fish. According to the manufacturer's instructions, total RNA was extracted from common carp tissues using the RNeasy Mini Kit (Qiagen, Hilden, Germany). According to the instructions, each cDNA was synthesized from 1 µg of total RNA using the PrimeScript™ RT reagent Kit with gDNA Eraser (TaKaRa, Beijing, China). Specific primers were obtained using Primer Premier 5.0. RTq–PCR was– performed according to the instructions of TB Green™ Premix Ex Taq™ II (TaKaRa, Beijing, China) using an ABI7500 system (Life Technologies, Carlsbad, CA, USA). Real-time PCR amplification was performed using a total volume of 20 μL that contained 10 μL of 2× TB Green™ Premix Ex Taq™ II, 0.8 μL of forward primer, 0.8 μL reverse primer, 0.4 μL ROX Reference Dye, 2 μL of template cDNA, and 6 μL sterilized water. The reaction conditions were as follows: pre-denaturation, 95 °C, 30 s; PCR reaction, 95 °C for 5 s, according to the annealing temperature of different specific primer reactions for 30 s; number of cycles, 40 times. The final step was a temperature of 95 °C, 15 s, then 60 °C, 60 s, then 95 °C, 15 s, and held at 4 °C. Primer specificity was confirmed via dissociation curve analysis. *β-actin* was used as an internal reference gene. Double-distilled water was included in place of the template as the negative control. The relative gene expression levels of *GH*, *IGF-I*, *S6K*, *TOR*, *4EBP1*, *SOD*, and *CAT* genes were determined using the $2^{(-\Delta\Delta Ct)}$ method [23]. The primers used in this study are shown in Table 1.

Table 1. Primers used in this experiment.

Gene	Primers	Sequence 5′-3′
GH	F R	TCAAGGGATGTCTCGATGGT CTACAGGGTGCAGTTGGAAT
IGF-I	F R	GGGCCTAGTTCAAGACGG AGTGGCTTTGTCCAGGTAA
S6K	F R	TGGAGGAGGTAATGGACG ACATAAAGCAGCCTGACG
TOR	F R	CCACAACGCAGCCAACAA GCCACAGAATAGCAACCCT
4EBP1	F R	GCTACCTCACGACTATTGC TTCTTGCTTGTCACTCCTG
SOD	F R	TGTGGGGTTCTGCCTCTTG TGGGAACATAGTGAGGGAGA
CAT	F R	TGGTGGATAATAACAGTTGGG ACACGATACAACACTGCTGC
β-actin	F R	GGCAGGTCATCACCATCGG TTGGCATACAGGTCTTTACGG

2.7. Data and Statistical Analysis

The CV of different morphological characteristics and the GRw, AGRw and IGRw of each stage of W in each stage were calculated according to the following formulas:

$$CV(\%) = SD/mean \times 100;$$

$$GW(g/g) = (Wx + 1 - Wx)/Wx;$$

$$AGRw(g/d) = (Wx + 1 - Wx)/(tx + 1 - tx);$$

$$IGRw\ (\%/d) = (\ln Wx + 1 - \ln Wx) \times 100/(tx + 1 - tx).$$

where SD is the standard deviation, and the mean is the average value of a morphological trait. Wx + 1 and Wx are the Ws at time tx + 1 and tx, respectively.

Four morphological parameters were compared, including W/SL, H/SL, BW/SL and HL/SL. In this study, the significance of the differences in all data analyses were evaluated by using the Bonferroni *t* test in SAS 9.1 (SAS Institute Inc, Cary, NC, USA). The data are shown as the mean ± SD of at least three replications. All of the data were checked for normal distribution by one-sample Kolmogorov–Smirnov test and homogeneity of variances by Levene's test. All the experiments were performed at least three times.

3. Results

3.1. Determination of the Experimental Conditions of ARTP Treatment of Fertilized Eggs

To determine a suitable ARTP treatment time and input power, the fertilization rates and hatchability rates of Songpu mirror carp were calculated under different input power levels and correspondingly different treatment time periods. Compared with the control group (0 w), the fertilization rates gradually decreased with increasing input power and treatment time periods (Figure 1a). In the treatment groups, the fertilization rates were highest (84.8 ± 2.5%) when the treatment time was 120 s and the input power was 120 W and lowest (50.7 ± 2.9%) under a treatment time of 360 s and an input power of 200 W. There was a significant difference between the fertilization rates of the treatment and the control groups ($p < 0.05$), except when the treatment time was 120 s with an input power of 120 w. In addition, the fry survival rates gradually decreased with increasing input power and treatment time, similar to the trend for fertilization rates (Figure 1b). In the treatment groups, when the input power was 120 W and the treatment time was 120 s, the hatchability rate was highest (72.8 ± 3.1%) while, when the input power was 200 W and the treatment time was 360 s, the hatchability rate was lowest (26.8 ± 5.0%). The survival rates of fry in the treatment group were significantly different from those in the control group except when the input power was 120 w and the treatment time was 120 s ($p < 0.05$). To ensure a high possibility of mutation, but also certain fertilization rates and hatchability rates, a final input power of 160 W and a treatment time of 360 s were selected for Songpu mirror carp mutagenesis. Types of abnormal larvae that hatched from ARTP-treated eggs were mainly divided into short tail trunk and large cardio-coelom, tail folding and shortened trunk, in Songpu mirror carp.

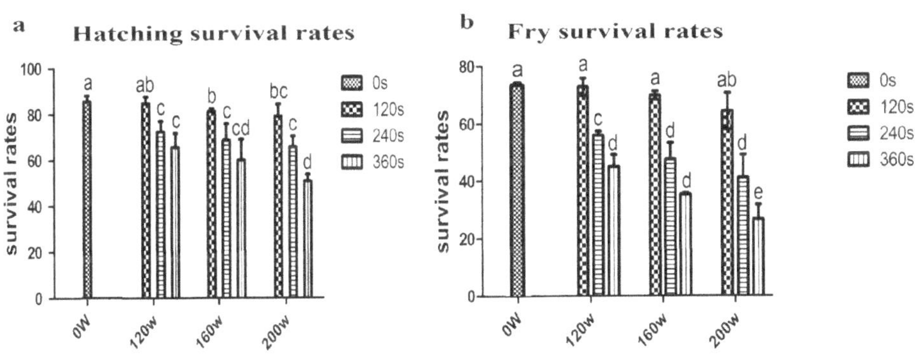

Figure 1. Hatching rates and fry survival rates for Songpu mirror carp under different conditions. (**a**) Survival rates of fertilized Songpu mirror carp eggs under different treatment times and output power levels. (**b**) Hatchability rates of fertilized Songpu mirror carp eggs under different treatment times and output powers. The treatment time was 0 s and the output power was 0 W in the control group, and the other groups constituted the ARTP treatment group. Lowercase letters in the column chart indicate significant differences determined by using the Bonferroni *t* test in SAS 9.1 ($p < 0.05$).

3.2. Morphological Characteristics of Songpu Mirror Carp

Fertilized Songpu mirror carp eggs were mutagenized with an input power of 160 w and an ARTP treatment time of 360 s. After 5 months of culture, the W, SL, H, BW and HL of Songpu mirror carp in the ARTP treatment and the control groups were measured. There were significant differences in each trait between the ARTP treatment groups and the control groups ($p < 0.05$) (Table 2) (Figure 2). The results also showed that the CV of each trait was higher in the control group at 5 months after fertilization. In addition, the difference values in the SLs and BWs of the ARTP treatment groups were greater than those of the control group. The deformity rate of the ARTP treatment group was calculated as 19.2%, mainly representing deformities of the mouth (10.4%), lake of operculum (4.6%) and fin (0.9%) skull (0.5%), and back height (2.8%) in Songpu mirror carp. The deformity rate of the control group was 1.1%, mainly corresponding to the appearance of the operculum (Supplementary Table S1) (Figure 3). The above results showed that the ARTP treatment group had more morphological differences between individuals than the control group at 5 months after fertilization.

Table 2. Morphological traits of Songpu mirror carp ARTP treatment group and control group.

Time After Fertilization	Item	Group	Trait				
			W (g)	SL (mm)	H (mm)	W (mm)	HL (mm)
5 months	Mean + SD	ARTP	80.33 ± 34.54 [b]	134.26 ± 22.07 [b]	51.33 ± 7.89 [b]	26.35 ± 4.87 [b]	44.43 ± 7.21 [b]
		Control	127.37 ± 45.82 [a]	156.06 ± 19.35 [a]	63.83 ± 8.25 [a]	30.4 ± 4.49 [a]	52.79 ± 7.45 [a]
	CV(%)	ARTP	43.00	16.44	15.37	18.48	16.23
		Control	35.97	12.40	12.93	14.78	14.10
	Difference	ARTP	180.50	145.59	47.85	44.21	40.38
		Control	298.90	118.28	57.83	33.89	60.49

Notes: The significance of the difference in each column is indicated by different lowercase letters ($p < 0.05$).

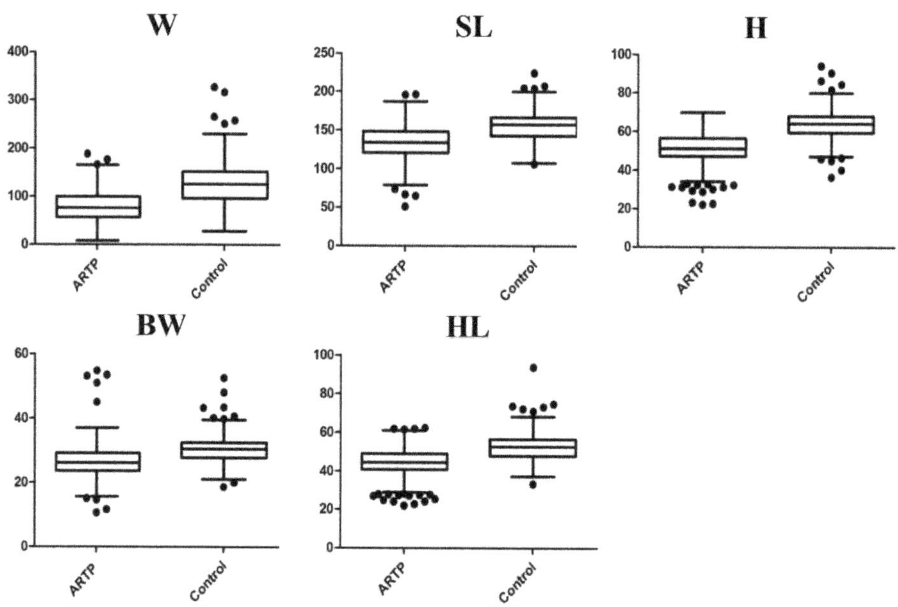

Figure 2. Box plots of W and phenotypic traits in Songpu mirror carp. Box plots of the W, SL, H, BW and HL of Sonpu mirror carp in three periods in the ARTP treatment and control groups initiated at 5 months after fertilization. The black dots in the figure indicate values greater than 1.5 times the interquartile range.

Figure 3. Malformation types in Songpu mirror carp, arranged from large to small according to total length. Types of deformities include lack of fin (**A**,**B**,**E–G**), deformity of mouth and skull (**C**), lack of operculum (**D**,**H**,**I**), scales on the back (**C**,**J**), and high back (**F**,**I–K**). Bar indicates 50 mm.

3.3. Comparative Analysis of Morphological Parameters

In order to screen out the common carp line with high body mass and fast growth rate, only W was used as the screening basis, and the selection pressure ratio was 1:15 when selecting larger individuals at the 5th month after fertilization. The ratios of several morphological parameters of Songpu mirror carp included the W/SL, H/SL, BW/SL and HL/SL ratios. The W/SL and H/SL of the ARTP treatment group were significantly lower than those of the control group at 5 months after fertilization ($p < 0.01$) (Figure 4). At 14 months after fertilization, the W/SL, H/SL and BW/SL of the ARTP treatment group were significantly lower than those of the control group ($p < 0.01$). There were no significant differences in HL/SL between the ARTP treatment and the control groups. In addition, the results showed that the W/BL, H/BL, and HL/BL of the ARTP treatment group were significantly lower than those of the control group ($p < 0.01$), while the coefficient of variation was higher than that in the control group (Table 3). The results showed that there were certain morphological differences between the Songpu mirror carp in the ARTP treatment group and the control group.

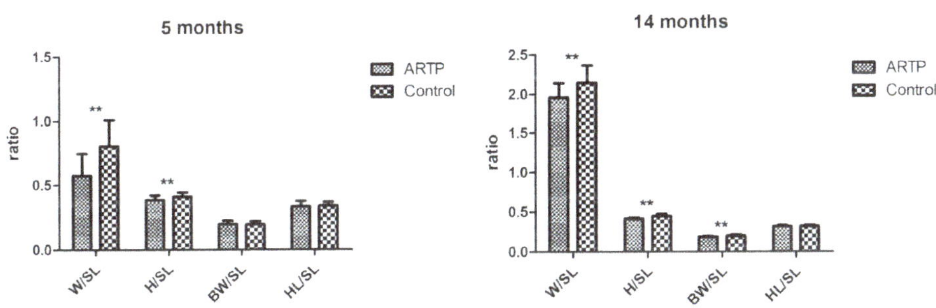

Figure 4. Morphological parameter ratios of different stages in Songpu mirror carp. The ratios of morphological parameters of Songpu mirror carp at 5 months and 14 months after fertilization, including W/SL, BH/SL, BW/SL and HL/SL Statistically significant differences were defined at $p < 0.05$ (** $p < 0.01$).

Table 3. The morphological parameter ratios of Songpu mirror carp ARTP treatment group and control group.

Time After Fertilization	Item	Group	Trait			
			W/SL (g/mm)	H/SL	BW/SL	HL/SL
5 months	Mean ± SD	ARTP	0.58 ± 0.17^{b}	0.38 ± 0.03^{b}	0.19 ± 0.04	0.33 ± 0.03^{b}
		Control	0.80 ± 0.21^{a}	0.41 ± 0.03^{a}	0.20 ± 0.02	0.34 ± 0.03^{a}
	CV(%)	ARTP	29.73	8.55	20.24	8.70
		Control	25.60	8.13	11.66	8.69

Note: The significance of the difference in each vertical line is indicated by lowercase letters ($p < 0.05$).

3.4. Rate of W Gain in Songpu Mirror Carp

To better compare the difference in Songpu mirror carp W between the ARTP treatment and control groups, the GRw, AGRw and IGRw of W were calculated in different periods. The W of the fish in the ARTP treatment group in the three periods was significantly lower than that in the control group ($p < 0.05$) (Table 4). At 5 months–14 months after fertilization, the GRw, the AGRw and the IGRw of the ARTP treatment group were higher than those of the control group. The results showed that the growth rate of Songpu mirror carp in the ARTP treatment group was faster than that in the control group.

Table 4. Comparison of growth for three time periods (Means + SD).

Group	W (g)		5 Months–14 Months		
	5 Months	14 Months	GRw (%)	AGRw (g/d)	IGRw (%/d)
ARTP	81.87 ± 16.40^{d}	439.60 ± 58.60^{b}	4.37 ± 2.57	1.32 ± 0.16	0.62 ± 0.47
Control	127.37 ± 45.82^{c}	478.10 ± 61.65^{a}	2.75 ± 0.35	1.30 ± 0.06	0.49 ± 0.11

Notes: Data are presented as mean ± SD of three replicates. Different superscript lowercase letters indicate significant differences ($p < 0.05$).

3.5. Serum Biochemical Parameters of Songpu Mirror Carp

Eleven serum indicators were measured in the ARTP treatment and control groups (Table 5). The contents of TP, ALP, ALB, T-CHO and LDL were significantly higher in the ARTP treatment group than in the control groups ($p < 0.05$). In the control group, AST/ALT and AST contents were significantly higher than those in the ARTP treatment group ($p < 0.05$). The above results showed that there were differences in serum physiological and biochemical indicators between the ARTP treatment group and the control group.

Table 5. Serum biochemical parameters in ARTP treatment group and control group.

Group	TP (g/L)	ALT (U/L)	AST/ALT	AST (U/L)	ALP (U/L)	ALB (g/L)	T-CHO (mmol/L)	TG (mmol/L)	HDL (mmol/L)	LDL (mmol/L)	UA (mmol/L)	TBA (μmol/L)
ARTP	40.17 ± 1.72 a	137.10 ± 8.66	2.67 ± 0.29 b	318.10 ± 2.71 b	73.43 ± 13.54 a	15.67 ± 0.58 a	4.02 ± 0.21 a	0.46 ± 0.13	2.26 ± 0.14	1.67 ± 0.15 a	63.57 ± 28.73	0.47 ± 0.15
Control	32.27 ± 3.27 b	136.10 ± 14.00	3.93 ± 0.58 a	535.70 ± 59.28 a	8.20 ± 2.26 b	10.53 ± 1.17 b	3.04 ± 0.29 b	0.59 ± 0.07	2.09 ± 0.20	1.25 ± 0.15 b	69.73 ± 18.14	0.50 ± 0.00

Notes: Data are presented as mean ± SD of three replicates. Along the same row, different letters indicate significant differences between groups based on one-way analysis of variance (ANOVA) ($p < 0.05$).

3.6. Expression of Genes Related to Protein Synthesis in Songpu Mirror Carp

The relative expression levels of *S6K*, *TOR* and *4EBP1* in the dorsal muscles and intestinal tissues were determined in the ARTP treatment and control groups (Figure 5). In the dorsal muscles, the relative expression levels of *TOR* and *4EBP1* in the ARTP treatment group were significantly higher than those in the control group ($p < 0.01$). In intestinal tissue, the relative expression levels of *S6K*, *TOR* and *4EBP1* were significantly higher in the ARTP treatment group than in the control group ($p < 0.05$). There were significant differences in the relative expression of genes related to protein synthesis in the dorsal muscles and intestinal tissue between the ARTP treatment and control groups.

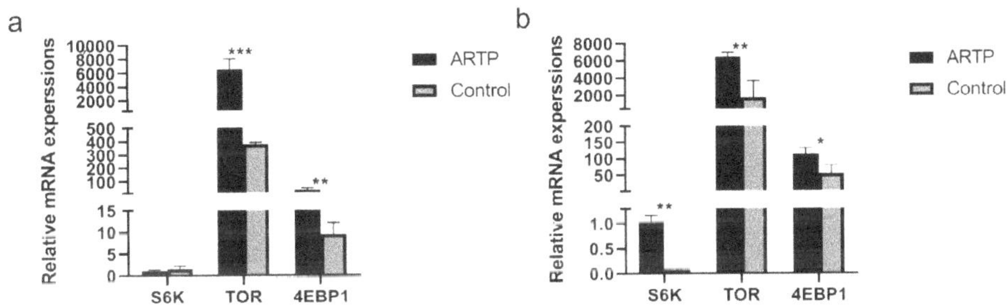

Figure 5. Relative expression levels of protein synthesis-related genes in the dorsal muscles (**a**) and intestine (**b**) in the ARTP treatment and control groups. Relative mRNA expression of *S6K* compared to that in the ARTP treatment group. Statistically significant differences were defined at $p < 0.05$ (* $p < 0.05$; ** $p < 0.01$; *** $p < 0.001$).

3.7. Expression of Antioxidant-Related and GH/IGF-1 Axis Genes in Songpu Mirror Carp

The relative expression levels of *SOD, CAT, GH* and *IGF-I* were determined in the ARTP treatment and control groups (Figure 6). The relative expression of *SOD* in the liver was significantly higher in the ARTP treatment group than in the control group ($p < 0.01$). In the dorsal muscles, the relative expression levels of *GH* and *IGF-I* in the ARTP treatment group were significantly higher than those in the control group ($p < 0.05$). There were significant differences in the relative expression of growth-related genes and antioxidant-related genes in the liver and dorsal muscles between the ARTP treatment and control groups.

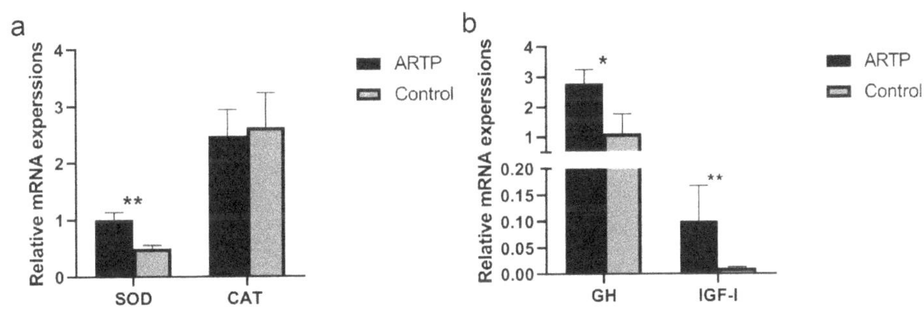

Figure 6. Relative expression levels of antioxidant-related genes in the liver (**a**) and growth-related genes in dorsal muscles (**b**) in the ARTP treatment and control groups, respectively. Relative mRNA expression levels of *SOD* compared to those in the ARTP treatment group and *GH* compared to those in the control group. Statistically significant differences were defined at $p < 0.05$ (* $p < 0.05$; ** $p < 0.01$).

4. Discussion

ARTP mutagenesis represents a new stable, efficient, safe and reliable technology with the advantages of a high mutation rate and a large mutation pool [8]. At present, ARTP mutagenesis technology has been widely used in microbial breeding, but it is less commonly used in fish mutation breeding. In this study, ARTP technology was used for the first time in the mutagenesis breeding of fertilized carp eggs, and the mutation conditions were optimized. When ARTP mutagenesis is applied in microorganism breeding, lethality and a positive mutation rate are considered as appropriate mutation conditions [24]. However, this study used fertilized eggs of Songpu mirror carp as the experimental objects and the survival rate of breeding and emergence as the basis for screening ARTP mutation conditions. The fertilized eggs of Songpu mirror carp are semitransparent and spherical, with a size of 1.75–1.89 mm (average 1.82 ± 0.06 mm). The egg diameter of Japanese flounder eggs is approximately 1 mm [18]; however, the fertilized eggs of Songpu mirror carp had higher radio frequency output power and helium flow rate than those of Japanese flounder, but shorter processing time in this study. In mutation breeding, the CV represents the degree of morphological differences between individuals [25]. Previous research has shown that, in ARTP mutant–Japanese flounder, mutant genes are related to growth and immune pathways [18]. In this study, it was found that the CV of morphological characteristics of the ARTP group was higher than that of the control group, indicating that many morphological differences between individuals appeared under ARTP treatment and might indicate the presence of mutations in relevant genes in Songpu mirror carp. In addition, this study found that the CV in W was highest in the ARTP treatment group, reaching 43%. We suggest that the potential for W selection increased in Songpu mirror carp after ARTP treatment. In this study the W of the fish in the mutagenized groups was significantly lower than that in the control group, and the ratios of W, H, and BW to SL were also significantly lower than those in the control group ($p < 0.05$), while the HL did not change significantly, indicating that the mutagenized group showed the morphological character of a larger head. In addition, previous studies have shown embryo deformity rates of 16.1%9–1.3% in grass carp after the sperm were mutagenized with different ENU densities [26]. After fertilized redtail notho (*Nothobranchius guentheri*) eggs were mutagenized by exposure to fast neutrons and X-rays, the malformation rate of the resulting larvae was 16.8% [27]. In this study, the malformation rate of Songpu mirror carp in the ARTP treatment group was 19.2% at 5 months after fertilization. The results showed that the ARTP mutagenesis technique could improve the deformity rate in the mutagenesis groups compared to other mutagenesis methods, which might allow an increased gene mutation rate to be obtained in Songpu mirror carp.

The purpose of applying mutation breeding in fish is to obtain excellent traits and more genetic diversity. Although the W of the fish in the ARTP treatment group was lower than that in the control group, it was found that the GRw, AGRw and IGRw of W in the ARTP treatment group were greater than those in the control group. We suggest that the growth rate of Songpu mirror carp was in the later stage was accelerated by ARTP treatment, and these fish could be used as later-breeding parents. Based on the differences in morphological characteristics between the ARTP treatment and control groups, ARTP mutagenesis technology can be applied as a useful method in Songpu mirror carp breeding to speed up the selection of important economic traits.

Biochemical indicators in fish serum reflect fish growth health [28]. The results of this study showed that the ALP, ALB and T-CHO levels in the ARTP treatment group were significantly higher than those in the control group ($p < 0.05$), and the AST/ALT ratio and AST levels were significantly lower than those in the control group ($p < 0.05$). Elevated levels of ALP indicate a state of stress that may lead to tissue damage in fish, especially in the liver [29]. Similarly, an increase in ALP levels was previously found in common carp after xenobiotic exposure [30]. ALB TP and T-CHO in serum can be used as indicators of fish health related to conditions such as liver disease and immune dysfunction [31,32]. AST and ALT are considered biomarkers for assessing liver function and cell membrane permeability, and elevated levels of these markers can lead to liver dysfunction. ARTP treatment might have beneficial effects on the liver function of Songpu mirror carp. In addition, LDL is known to be associated with atherosclerosis and other vascular diseases, such as thrombosis [33]. In this study, it was found that the LDL level in the serum of fish in the ARTP treatment group was significantly higher than that in the control group ($p < 0.05$), indicating that ARTP treatment might have harmful effects on the formation, transport and metabolism of cholesterol in Songpu mirror carp. The liver is the main organ reflecting stress responses, such as antioxidant capacity, and can secrete SOD and CAT to alleviate the effects of lipid peroxidation and cell dysfunction on cell function [34,35]. The relative expression of SOD in the liver of the fish in the ARTP treatment group was significantly higher than that in the control group ($p < 0.05$), indicating that the ARTP treatment group fish showed a higher antioxidant capacity. These results were similar to the serum indicator results obtained in this study.

The growth of fish is regulated by multiple factors, among which protein deposition and the expression of *GH* and *IGF-I* in muscle are important influencing factors, and protein deposition is the result of the balance between protein synthesis and degradation [36,37]. The muscle growth process is linked to protein deposition [38,39], and protein deposition appears to be a major determinant of fish weight gain [40,41]. In fish, *GH* induces muscle growth by regulating the expression of genes encoding products such as myostatin and myogenic regulatory factor, while *IGF-I* can stimulate the proliferation and differentiation of myoblasts to promote muscle growth [42–45]. In addition, studies have shown that ARTP mutagenized mutants under systematic genome analysis have more mutation sites in genes related to protein synthesis and metabolism [18,46,47]. *S6K*, *TOR* and *4EBP1* have long been employed as markers of protein deposition and protein synthesis in tissues [48,49]. The results of this study showed that the relative expression levels of *TOR* and *4EBP1* in dorsal muscles and intestinal tissue and *GH* and *IGF-I* in dorsal muscles were significantly higher in the ARTP treatment groups than those in the control group, indicating that the muscle growth rate in the ARTP treatment group was faster than that in untreated Songpu mirror carp. Possible reasons for the greater GRw, AGRw and IGRw in the ARTP treatment group than in the control group were further verified based on tissue gene expression levels.

5. Conclusions

This study optimized the mutagenesis conditions of ARTP mutation breeding in fertilized Songpu mirror carp eggs for the first time and provided meaningful theoretical guidance for subsequent research on fertilized Songpu mirror carp eggs.

Supplementary Materials: The following supporting information can be downloaded at: https://www.mdpi.com/article/10.3390/fishes9110448/s1, Table S1: The deformity rate of Songpu mirror carp at 5 months after hatching in the ARTP treatment and control groups, respectively.

Author Contributions: Z.J. and X.J. conceived the project and designed the experiments. X.J. wrote the manuscript. C.L. and M.S. performed the experiments. Y.G. and X.H. analyzed the data. The manuscript was revised and approved by Z.J. All authors have read and agreed to the published version of the manuscript.

Funding: This work was supported by the National Key R&D Program of China (2022YFD2400102), China Agriculture Research System (grant number CARS-45-07), the Central Public-interest Scientific Institution Basal Research Fund, the Chinese Academy of Fishery Sciences (CAFS) (NO. 2023TD35) and the National Key Research and Development Plan (2023YFD2400204-4).

Institutional Review Board Statement: All animal procedures in this study were conducted according to the guidelines for the care and use of laboratory animals of the Heilongjiang River Fisheries Research Institute, Chinese Academy of Fishery Sciences (CAFS). The studies involving animals were reviewed and approved by the Committee for the Welfare and Ethics of Laboratory Animals of the Heilongjiang River Fisheries Research Institute, CAFS (approval number: 20210910-001 (approved on 5 September 2021)).

Data Availability Statement: The datasets used and/or analyzed during the current study are available from the corresponding author on reasonable request.

Conflicts of Interest: None of the authors have conflicts of interest that may influence the content of this manuscript.

Abbreviations

Absolute growth rate, AGRw; Albumin, ALB; Alanine aminotransferase, ALT; Alkaline phosphatase, ALP; Atmospheric pressure dielectric barrier discharge plasm, APDBD; Atmospheric-pressure room-temperature plasma, ARTP; Aspartate aminotransferase, AST; Beta-actin, β-actin; Catalase, CAT; Coefficients of variation, CV; Factor 4E binding protein 1, 4EBP1; Growth hormone, GH; High-density lipoprotein, HDL; Insulin-like growth factor

1, IGF-I; Instantaneous growth gain, IGRw. Low-density lipoprotein, LDL; N–N-ethyl-N-nitrosourea, ENU; Relative growth rate, GRw; S6 kinase, S6K; Superoxide dismutase, SOD; Target of rapamycin, TOR; Total bile acid, TBA; Total cholesterol, T-CHO; Triglyceride, TG; Total protein, TP; Uric acid, UA; Ultraviolet, UV.

References

1. Miao, Z.H.; Rao, V.A.; Agama, K.; Antony, S.; Kohn, K.W.; Pommier, Y. 4-nitroquinoline-1-oxide induces the formation of cellular topoisomerase I-DNA cleavage complexes. *Cancer Res.* **2006**, *66*, 6540–6545. [CrossRef] [PubMed]
2. Slamenová, D.; Gábelová, A.; Ruzeková, L.; Chalupa, I.; Horváthová, E.; Farkasová, T.; Bozsakyová, E.; Štětina, R. Detection of MNNG-induced DNA lesions in mammalian cells; validation of comet assay against DNA unwinding technique, alkaline elution of DNA and chromosomal aberrations. *Mutat. Res.* **1997**, *383*, 243–252. [CrossRef] [PubMed]
3. Knapik, E.W. ENU mutagenesis in zebrafish—From genes to complex diseases. *Mamm. Genome* **2000**, *11*, 511–519. [CrossRef]
4. Mullins, M.C.; Hammerschmidt, M.; Haffter, P.; Nüsslein-Volhard, C. Large-scale mutagenesis in the zebrafish: In search of genes controlling development in a vertebrate. *Curr. Biol.* **1994**, *4*, 189–202. [CrossRef]
5. Fang, M.; Jin, L.; Zhang, C.; Tan, Y.; Jiang, P.; Ge, N.; Li, H.; Xing, X. Rapid mutation of Spirulina platensis by a new mutagenesis system of atmospheric and room temperature plasmas (ARTP) and generation of a mutant library with diverse phenotypes. *PLoS ONE* **2013**, *8*, e77046. [CrossRef]
6. Guo, L.; Li, H.; Wang, L.; Wang, S.; Zhao, H.; Sun, W.; Xing, X.; Bao, C. Genetic effects of radio-frequency, atmospheric-pressure glow discharges with helium. *Appl. Phys. Lett.* **2008**, *92*, 221504. [CrossRef]
7. Wang, L.; Huang, Z.; Li, G.; Zhao, H.; Xing, X.; Sun, W.; Li, H.; Gou, Z.; Bao, C. Novel mutation breeding method for Streptomyces avermitilis using an atmospheric pressure glow discharge plasma. *J. Appl. Microbiol.* **2010**, *108*, 851–858. [CrossRef]
8. Zhang, X.; Zhang, X.; Li, H.; Wang, L.; Zhang, C.; Xing, X.; Bao, C. Atmospheric and room temperature plasma (ARTP) as a new powerful mutagenesis tool. *Appl. Microbiol. Biotechnol.* **2014**, *98*, 5387–5396. [CrossRef] [PubMed]
9. Zhang, X.; Zhang, C.; Zhou, Q.; Zhang, X.; Wang, L.; Chang, H.; Li, H.; Oda, Y.; Xing, X. Quantitative evaluation of DNA damage and mutation rate by atmospheric and room-temperature plasma (ARTP) and conventional mutagenesis. *Appl. Microbiol. Biotechnol.* **2015**, *99*, 5639–5646. [CrossRef]
10. Su, X.; Zhao, S.; Xu, W.; Shuang, L.; Zheng, G.; Zou, S. Efficiently whole-genomic mutagenesis approach by ARTP in blunt snout bream (*Megalobrama amblycephala*). *Aquaculture* **2022**, *555*, 738241. [CrossRef]
11. Zhang, X.; Wu, Y.; Ma, F.; Wang, L.; Zhang, C.; Li, H.; Xing, X. Application of ARTP mutagenesis in breeding of microbial biocatalysts for food and feed processing industry. *Biotechnol Bus.* **2019**, *3*, 13–24. (In Chinese) [CrossRef]
12. Li, H.; Wang, Z.; Ge, N.; Le, P.; Wu, H.; Lu, Y.; Wang, L.; Chong, Z.; Bao, C.; Xing, X. Studies on the Physical Characteristics of the Radio-Frequency Atmospheric-Pressure Glow Discharge Plasmas for the Genome Mutation of *Methylosinus trichosporium*. *Plasmas IEEE Trans. Plasma Sci.* **2012**, *40*, 2853–2860. [CrossRef]
13. Li, J.; Guo, S.; Hua, Q.; Hu, F. Improved AP-3 production through combined ARTP mutagenesis, fermentation optimization, and subsequent genome shuffling. *Biotechnol. Lett.* **2021**, *43*, 1143–1154. [CrossRef]
14. Ye, L.; Ye, R.; Hu, F.; Wang, G. Combination of atmospheric and room temperature plasma (ARTP) mutagenesis, genome shuffling and dimethyl sulfoxide (DMSO) feeding to improve FK506 production in *Streptomyces tsukubaensis*. *Biotechnol. Lett.* **2021**, *43*, 1809–1820. [CrossRef] [PubMed]
15. Ottenheim, C.; Nawrath, M.; Wu, J. Microbial mutagenesis by atmospheric and room-temperature plasma (ARTP): The latest development. *Bioresour. Bioprocess.* **2018**, *5*, 12. [CrossRef]
16. Shi, F.; Tan, J.; Chu, J.; Wang, Y.; Zhuang, Y.; Zhang, S. A qualitative and quantitative high-throughput assay for screening of gluconate high-yield strains by *Aspergillus niger*. *J. Microbiol. Methods* **2015**, *109*, 134–139. [CrossRef]
17. Yuan, L.; Wang, L.; Ma, K.; Guo, L.; Xing, X. Characteristics of Hydrogen Production of an Enterobacter aerogenes Mutant Generated by a New Atmospheric and Room Temperature Plasma (ARTP). *Biochem. Eng. J.* **2011**, *55*, 17–22.
18. Hou, J.; Zhang, X.; Wang, G.; Sun, Z. Novel breeding approach for Japanese flounder using atmosphere and room temperature plasma mutagenesis tool. *BMC Genom.* **2019**, *20*, 323. [CrossRef]
19. FAO. *The State of World Fisheries and Aquaculture 2022: Towards Blue Transformation*; FAO: Rome, Italy, 2022. [CrossRef]
20. Hu, X.; Li, C.; Shang, M.; Ge, Y.; Jia, Z.; Wang, S.; Shi, L. Inheritance of growth traits in Songpu mirror carp (*Cyprinus carpio* L.) cultured in Northeast China. *Aquaculture* **2017**, *477*, 1–5. [CrossRef]
21. Hou, J.; Wang, G.; Zhang, X.; Sun, Z.; Liu, H. Cold-shock induced androgenesis without egg irradiation and subsequent production of doubled haploids and a clonal line in Japanese flounder, *Paralichthys olivaceus*. *Aquaculture* **2016**, *464*, 642–646. [CrossRef]
22. Galo, J.M.; Ribeiro, R.P.; Streit-Junior, D.P.; Albuquerque, D.M.; Fornari, D.C.; Roma, C.F.; Guerreiro, L.R. Oocyte quality of tambaqui (*Colossoma macropomum*) during the reproductive season. *Braz. J. Biol.* **2015**, *75*, 279–284. [CrossRef] [PubMed]
23. Livak, K.J.; Schmittgen, T.D. Analysis of relative gene expression data using real-time quantitative PCR and the 2(-Delta Delta C(T)) Method. *Methods* **2001**, *25*, 402–408. [CrossRef] [PubMed]

24. Li, H.; Luo, W.; Wang, Q.; Yu, X. Direct fermentation of gelatinized cassava starch to acetone, butanol, and ethanol using *Clostridium acetobutylicum* mutant obtained by atmospheric and room temperature plasma. *Appl. Biochem. Biotechnol.* **2014**, *172*, 3330–3341. [CrossRef]
25. Bu, H.Y.; Ou, X.M.; Ma, J.K. Determination of Chlorantraniliprole Residue in Environmental Aquatic Samples by High Performance Liquid Chromatography. *Chin. J. Spectrosc. Lab.* **2008**, *25*, 1230–1234.
26. Jiang, X.; Sun, C.; Zhang, Q.; Zou, S. ENU-induced mutagenesis in grass carp (*Ctenopharyngodon idellus*) by treating mature sperm. *PLoS ONE* **2011**, *6*, e26475. [CrossRef]
27. Zhang, Y.; Li, X.; Yuan, R. Effect of Promoting Growth of Fry and Hatchability by Irradiating Carp Embryos with the Ra-Be(n,γ) Neutron Source and -60Coγ(n,γ)-ray source. *Acta Sci. Nat. Univ. Pekin.* **1984**, *1*, 24–30.
28. Seibel, H.; Baßmann, B.; Rebl, A. Blood Will Tell: What Hematological Analyses Can Reveal About Fish Welfare. *Front. Vet. Sci.* **2021**, *8*, 616955. [CrossRef]
29. Latif, A.; Khalid, M.; Ali, M. An assessment of naphthalene stress on renal and hepatic functional integrity in *Labeo rohita*. *Int. J. Curr. Eng. Technol.* **2014**, *4*, 319–324.
30. Rao, J.V. Toxic effects of novel organophosphorus insecticide (RPR-V) on certain biochemical parameters of euryhaline fish, *Oreochromis mossambicus*. *Pestic. Biochem. Physiol.* **2006**, *86*, 78–84. [CrossRef]
31. Banace, M. Mirvagefei AR. Effect of sub-lethal Diazinon Concentrations on Blood Plasma Biochemistry. *Int. J. Environ. Res.* **2008**, *2*, 189–198.
32. Yang, J.; Wang, T.; Lin, G.; Li, M.; Zhu, R.; Yiannikouris, A.; Zhang, Y.; Mai, K. The Assessment of Diet Contaminated with Aflatoxin B1 in Juvenile Turbot (*Scophthalmus maximus*) and the Evaluation of the Efficacy of Mitigation of a Yeast Cell Wall Extract. *Toxins* **2020**, *12*, 597. [CrossRef] [PubMed]
33. Huang, H.; Weng, B.; Hsuuw, Y.; Lee, Y.; Chen, K. Dietary Supplementation of Two-Stage Fermented Feather-Soybean Meal Product on Growth Performance and Immunity in Finishing Pigs. *Animals* **2021**, *11*, 1527. [CrossRef] [PubMed]
34. Dawood, M.A.O.; Ali, M.F.; Amer, A.A.; Gewaily, M.S.; Mahmoud, M.M.; Alkafafy, M.; Assar, D.H.; Soliman, A.A.; Van Doan, H. The influence of coconut oil on the growth, immune, and antioxidative responses and the intestinal digestive enzymes and histomorphometry features of Nile tilapia (*Oreochromis niloticus*). *Fish. Physiol. Biochem.* **2021**, *47*, 869–880. [CrossRef]
35. Smith, C.M.; Ryan, P.J.; Hosken, I.T.; Ma, S.; Gundlach, A.L. Relaxin-3 systems in the brain—The first 10 years. *J. Chem. Neuroanat.* **2011**, *2*, 262–275. [CrossRef]
36. Salto, R.; Vílchez, J.D.; Cabrera, E.; Guinovart, J.J.; Girón, M.D. Activation of ERK by sodium tungstate induces protein synthesis and prevents protein degradation in rat L6 myotubes. *FEBS Lett.* **2014**, *588*, 2246–2254. [CrossRef] [PubMed]
37. Suryawan, A.; Davis, T.A. Regulation of protein degradation pathways by amino acids and insulin in skeletal muscle of neonatal pigs. *J. Anim. Sci. Biotechnol.* **2014**, *5*, 8. [CrossRef]
38. Adams, G.R.; Haddad, F. The relationships among IGF-1, DNA content, and protein accumulation during skeletal muscle hypertrophy. *J. Appl. Physiol.* **1996**, *81*, 2509–2516. [CrossRef]
39. Young, V.R. Regulation of protein synthesis and skeletal muscle growth. *J. Anim. Sci.* **1974**, *38*, 1054–1070. [CrossRef]
40. Dumas, A.; De Lange, C.F.; France, J.; Bureau, D.P. Quantitative description of body composition and rates of nutrient deposition in rainbow trout (*Oncorhynchus mykiss*). *Aquaculture* **2007**, *273*, 165–181. [CrossRef]
41. Yadav, A.K.; Mandal, S.C.; Patel, A.B.; Maurya, P.K. Evaluation of dietary protein requirement for the growth performance of minor carp, *Cirrhinus reba* (Hamilton, 1822) fingerlings. *Aquac. Res.* **2019**, *50*, 3343–3349. [CrossRef]
42. Coolican, S.A.; Samuel, D.S.; Ewton, D.Z.; McWade, F.J.; Florini, J.R. The mitogenic and myogenic actions of insulin-like growth factors utilize distinct signaling pathways. *J. Biol. Chem.* **1997**, *272*, 6653–6662. [CrossRef]
43. Fuentes, E.N.; Valdés, J.A.; Molina, A.; Björnsson, B.T. Regulation of skeletal muscle growth in fish by the growth hormon—Insulin-like growth factor system. *Gen. Comp. Endocrinol.* **2013**, *192*, 136–148. [CrossRef]
44. Montserrat, N.; Capilla, E.; Navarro, I.; Gutiérrez, J. Metabolic Effects of Insulin and IGFs on Gilthead Sea Bream (*Sparus aurata*) Muscle Cells. *Front. Endocrinol.* **2012**, *3*, 55. [CrossRef]
45. Rius-Francino, M.; Acerete, L.; Jiménez-Amilburu, V.; Capilla, E.; Navarro, I.; Gutiérrez, J. Differential effects on proliferation of GH and IGFs in sea bream (*Sparus aurata*) cultured myocytes. *Gen. Comp. Endocrinol.* **2011**, *172*, 44–49. [CrossRef]
46. Gong, M.; Zhang, H.; Wu, D.; Zhang, Z.; Zhang, J.; Bao, D.; Yang, Y. Key metabolism pathways and regulatory mechanisms of high polysaccharide yielding in *Hericium erinaceus*. *BMC Genom.* **2021**, *22*, 160. [CrossRef]
47. Zhu, Z.; Chen, W.; Zhou, H.; Cheng, H.; Luo, S.; Zhou, K.; Zhou, P.; Xia, L.; Ding, X. ARTP and NTG compound mutations improved Cry protein production and virulence of *Bacillus thuringiensis* X023. *Appl. Microbiol. Biotechnol.* **2022**, *106*, 4211–4221. [CrossRef]
48. Fan, Z.; Wu, D.; Li, J.; Zhang, Y.; Xu, Q.; Wang, L. Dietary protein requirement for large-size Songpu mirror carp (*Cyprinus carpio* Songpu). *Aquac. Nutr.* **2020**, *26*, 1748–1759. [CrossRef]
49. Ruvinsky, I.; Meyuhas, O. Ribosomal protein S6 phosphorylation: From protein synthesis to cell size. *Trends Biochem. Sci.* **2006**, *31*, 342–348. [CrossRef]

fishes

MDPI

Article

Heritability Estimates for Growth Traits and Correlation Analysis between Weight and Metamorphosis Rate in the Bullfrog *Rana* (*Aquarana*) *catesbeiana*

Wencheng Xu [1,2], Yanzhe Wang [1,2], Guodong Wang [1,2,*], Lili Zhang [1,2], Guiling Zhang [1,2], Zhipeng Huo [1,2] and Hui Ge [3]

1 Key Laboratory of Healthy Mariculture for the East China Sea, Ministry of Agriculture and Rural Affairs, Fisheries College, Jimei University, Xiamen 361021, China; wenchengxv@163.com (W.X.); wangyanzhe7@163.com (Y.W.); llzhang@jmu.edu.cn (L.Z.); lingzi1995@163.com (G.Z.); a18759273202@163.com (Z.H.)

2 State Key Laboratory of Mariculture Breeding, Fisheries College, Jimei University, Xiamen 361021, China

3 Fisheries Research Institute of Fujian, 7 Shanhai Road, Huli, Xiamen 361000, China; gehuizlj@163.com

* Correspondence: gdongwang@163.com

Abstract: Metamorphosis is a crucial process in the life cycle of *Rana* (*Aquarana*) *catesbeiana*. *R. catesbeiana* tadpoles, in their short larval period, possess a high survival rate and also a highly competitive ability in the amphibious stage. In actual seed production, the economic traits of larval period and metamorphosis rate are used as quantifiable indicators of quality for individuals and populations, respectively. However, studies of these economic traits in larval cultivation and production are still lacking. In this study, we constructed 40 full-sib families of *R. catesbeiana* and measured the weight and metamorphosis rate of tadpoles at different developmental stages. Subsequently, we calculated the phenotypic and genetic association between weight and metamorphosis rate in tadpoles and assessed the heritability of these two traits. The heritabilities of weight at three developmental stages were all higher than 0.40 and decreased with advancement of the developmental stage; the heritability of the metamorphosis rate was 0.18 ± 0.20, a moderate level. Correlation analysis of weight and metamorphosis rate at each developmental stage in each tadpole family showed that weight at stages 25-I, 25-II, and 25-III was significantly correlated at the phenotypic level but non-significantly at the genetic level. The metamorphosis rate was only moderately associated with stage 25-III weight (0.38, $p < 0.05$). The results of this study confirm the importance and transportability of tadpole weight in actual seed production and provide basic data and a potentially optimized direction for the selective breeding of high-metamorphosis-rate bullfrogs.

Keywords: *Rana* (*Aquarana*) *catesbeiana*; weight; metamorphosis rate; heritability; phenotypic and genetic correlation

Key Contribution: This is the first study on the phenotypic and genetic association between weight and metamorphosis rate in bullfrog tadpoles. The metamorphosis rate was moderately associated with weight at stage 25-III, providing basic data for the selective breeding of high-metamorphosis-rate bullfrogs.

1. Introduction

The bullfrog *Rana* (*Aquarana*) *catesbeiana* was introduced into China in the 1960s and cultivated in Guangdong, Guangxi, Fujian, Jiangxi, Zhejiang, and Anhui provinces in the 1990s. By 2022, the national cultivated output of *R. catesbeiana* reached 700,000 tons, an increase of about 10% over 2021, and the output value of the whole industrial chain reached nearly CNY 100 billion. Compared to fish and shrimp culture, bullfrogs offer advantages of rapid growth, being a rich protein source produced efficiently with regard to land, feed, and water, and being well suited to industrial production. However, with the continuous expansion of bullfrog production, there are many problems, such as a lack of policy and financial support, low investment in science and technology, a lack of approved drugs, and environmental pollution. Moreover, the most important problem is the lack of excellent bullfrog lines, which is constraining the development of bullfrog production. Therefore, breeding improved lines is a pressing issue for the sustainable development of bullfrog farming.

In production, it generally takes 150–200 days for a fertilized egg to develop into a commercial frog of 250 g, in which the tadpole stage is 60–120 days, accounting for about one-half of the whole growth cycle. Development is a continuous, hierarchical, and irreversible process. The metamorphosis of *R. catesbeiana* is a crucial stage in the transition of tadpoles from aquatic larvae to amphibious juveniles, during which there is a significant change in physiological structure and lifestyle. Tadpole metamorphosis includes cell proliferation, cell differentiation, cell death, and tissue specificity, which is mainly regulated by thyroid hormones produced from the hypothalamic–pituitary–thyroid axis [1,2]. It is generally believed that individuals with a larger size or shorter larval period could have a high-level ability to resist predation and avoid injury in the amphibious stage compared to other individuals [3,4]. This view has been verified in the pool frog (*Pelophylax lessonae*) and edible frog (*Pelophylax esculentus*) [4]. Individuals with a short larval period have stronger resistance to stress and disease resistance [5,6] and could develop into larger frogs [7,8] possessing higher fecundity [6,9–12]. As one of the most important farmed frogs, stronger resistance to stress and disease resistance in bullfrogs would be conducive to production. Tadpoles with a short larval period are favored for production. Hence, the genetics of metamorphosis in the bullfrog, *R. catesbeiana*, could have great significance in production.

The metamorphosis rate of tadpoles refers to the proportion of individuals completing metamorphosis relative to the total number of tadpoles in the population, a metric which can be used to compare the metamorphosis performance between populations. It is impossible to quantify a tadpole's metamorphosed performance in terms of metamorphosis rate; instead, it is typically determined by the amount of time that passes between fertilization and the end of metamorphosis. In production, it is hard to observe the metamorphosis time in a large number of individuals at the same time.

Owing to the varying degrees of association among distinct traits in organisms, attributes can be measured indirectly in addition to directly [13]. Studies of the wood frog, *Rana sylvatica* [14], and the African clawed frog, *Xenopus laevis* [15], showed that the duration of metamorphosis correlates with individual size at the beginning of metamorphosis. Studies of metamorphosis in *R. catesbeiana* populations show that large individual tadpoles usually undergo a shorter larval period as a tadpole [16]. Thus, we suggest that the weight of the larval stage at a given period could indicate the larval period in tadpoles.

In the process of breeding, bullfrogs have a breeding habit in which a single father and single dam produce offspring. This habit allows families to spontaneously form and makes it easy to identify the parents and offspring in a small area. In this study, we measured the weight and metamorphosis rate of tadpoles at different developmental stages, evaluated their heritabilities, and estimated the breeding values for the weight and metamorphosis rate of tadpoles. The goal was to find the correlation between the weight and metamorphosis rate of tadpoles and to provide basic data for *R. catesbeiana* production.

2. Materials and Methods

2.1. Ethics Statement

Rare or protected animals were not included in the experiments of this study. This study was carried out according to the guidelines of the Declaration of Helsinki and was approved by the Animal Care Advisory Committee of Jimei University (Approval No. 2019-0906-003, 6 September 2019).

2.2. Family Construction

The parental *R. catesbeiana* were from the Yifan Biotechnology Company Limited (Xiamen, China). Parents of bullfrogs were separated and held temporarily according sex in a greenhouse pool of 400 cm × 600 cm × 100 cm in flowing water and with a water depth of 10 cm. The cultivation density was approximately 90 animals per square meter, and the feed was equivalent to 2% of the weight of the bullfrogs. The feed was bullfrog compound feed from Xiamen Jiakang Feed Company Limited (Xiamen, China). Following sexual maturity, sires and dams weighing 350–400 g and 400–500 g, respectively, were chosen and incubated in 90 cm × 90 cm × 90 cm boxes. In 10-day short-term cultivation, these bullfrogs were combined in pairs to construct 120 full-sib families labeled 23MYFM1–23MYFM120. Afterwards, non-spawning and low-hatching-rate families were excluded, and 40 families with more spawning were selected as experimental materials.

2.3. Tadpole Cultivation

The tadpoles were incubated in glass tanks of 65 cm × 45 cm × 45 cm with a water depth of 15 cm. Each family had 600 tadpoles, which were grown for 120 days after being split into three groups of 200 tadpoles each, with the production density adjusted according to actual production. After 30 days, the density was reduced to 100 individuals per tank and to 20 individuals per tank after 65 days. During the cultivation period, the total daily feeding of each group was consistent, and the ration was equally distributed among feedings. Tadpoles were fed twice a day (8:00 and 20:00) at a rate of 7% of body weight, with adjustment of the ration based on the weight of the tadpoles. One-third of the culture water was exchanged every day. Throughout the experimental period, the average air temperature was 26–30 °C, the average water temperature was 25–28 °C, the dissolved oxygen was 6.0 mg/L, and the pH was 7.0–7.6.

2.4. Data Measurement

2.4.1. Weight of Tadpoles in Each Period

According to Gosner stages (GSs) [17], tadpole stages 0–20 are called the embryonic stage, stages 21–24 are the outer gill stage, and stage 25 is the tadpole stage, where they can swim freely for feeding. In stage 26, the hind legs begin to develop until the tail disappears completely at stage 46, marking the completion of metamorphosis. Stage 25 of *R. catesbeiana* is 45–110 days long, accounting for four fifths of the entire larval stage. High inter-individual variation exists in the deviation of developmental stages. In this study, the developmental stage at which a tadpole's BMW (body maximum width) was greater than 2.5 mm was defined as stage 25-I, corresponding to the staging system frequently employed in actual production, which occurs roughly 10–29 days after incubation. Stage 25-II, or roughly 30–59 days after incubation, was described as the point at which the tadpole's BMW was greater than 4.0 mm. Stage 25-III, roughly 60–80 days after incubation, was defined as the point at which the tadpole's BMW was greater than 9.0 mm. These definitions indicated the various stages of stage 25 growth, and the weight of each tadpole in each family was recorded at each stage.

2.4.2. Metamorphosis Rate

The number of juvenile frogs that had completed metamorphosis after 120 days was recorded, and then, the metamorphosis rate of each family was calculated. The metamorphosis rate was defined as the proportion of juvenile frogs that completed metamorphosis as a fraction of the total number of juvenile frogs and tadpoles in captivity for each family.

2.4.3. Statistical Analysis

The heritability of traits was evaluated by the restricted maximum likelihood method using the R package ASReml-R 4.2 [18]. The variance of the traits was calculated using the animal model as follows:

$$y_{ijk} = \mu + a_{ijk} + f_{ij} + e_{ijk} \quad (1)$$

where y_{ijk} represents the observation from the parent i and parent j for individual k; μ represents the population means; a_{ijk} represents the additive genetic effect value, namely, the breeding value; f_{ij} represents the common random effect of all all-sib families; and e_{ijk} denotes the random residuals. The formula for heritability evaluation is as follows:

$$h^2 = \sigma_a^2 / \sigma_a^2 + \sigma_f^2 + \sigma_e^2 \quad (2)$$

where σ_a^2 is the additive genetic variance, σ_f^2 is the random effect variance, and σ_e^2 is the residual variance. Weight heritability was calculated using the animal model; the residual variance was calculated using ASReml-R default parameters. A heritability value less than 0.10 is regarded as "low" heritability, 0.10–0.40 is "moderate", and greater than 0.4 is "high" heritability [19]. The significance level of heritability was then tested using the method of Liu et al. [20]. The formula is as follows:

$$t = h^2 / \sigma \quad (3)$$

The mean of individual weight breeding values was used as the weight-related breeding value of families. Correlation was analyzed using SPSS 22.0, after which the genetic correlation between breeding values of weight and metamorphosis rate was assessed using the Pearson coefficient.

3. Results

By accounting for the weight and relative proportion of different developmental stages in each family at stage 25 (Figure 1a,b), family number 23MYFM11 had the maximum average weight of 0.05 g (Figure 1c), which was significantly higher than those in the other families. Moreover, the minimum average weight of the family was only 0.01 g. At stage 25-II, the average weight of the 40 families was 0.32 g (Figure 1d). Individuals in 50% of families in stage 25-III weighed between 1.40 and 1.90 g, and several individuals in family 23MYFM48 weighed as much as 3.87 g (Figure 1e). As shown in Figure 1, there was no correlation between weight at stage 25-I and stage 25-III. For instance, family 23MYFM48 presented the lowest mean of weight at stage 25-I but presented the opposite result at stage 25-III.

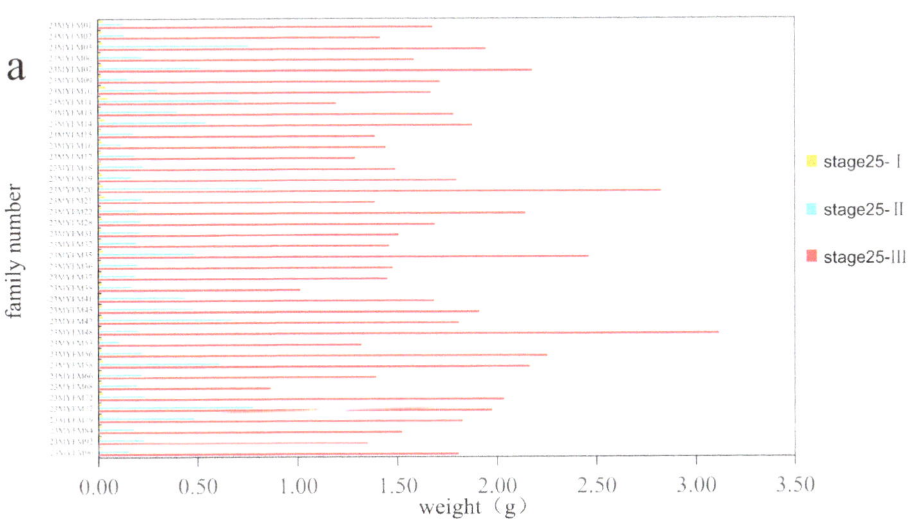

Figure 1. *Cont.*

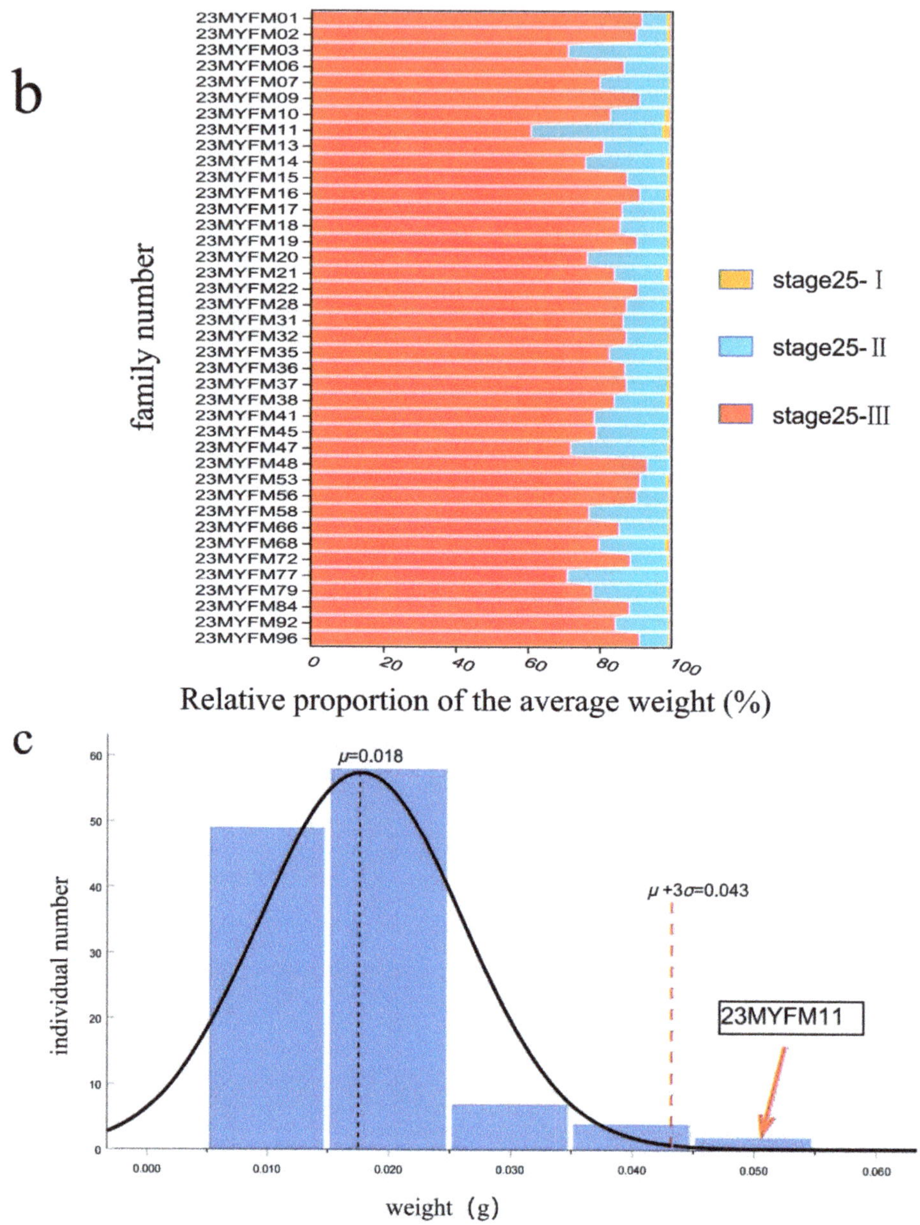

Figure 1. *Cont.*

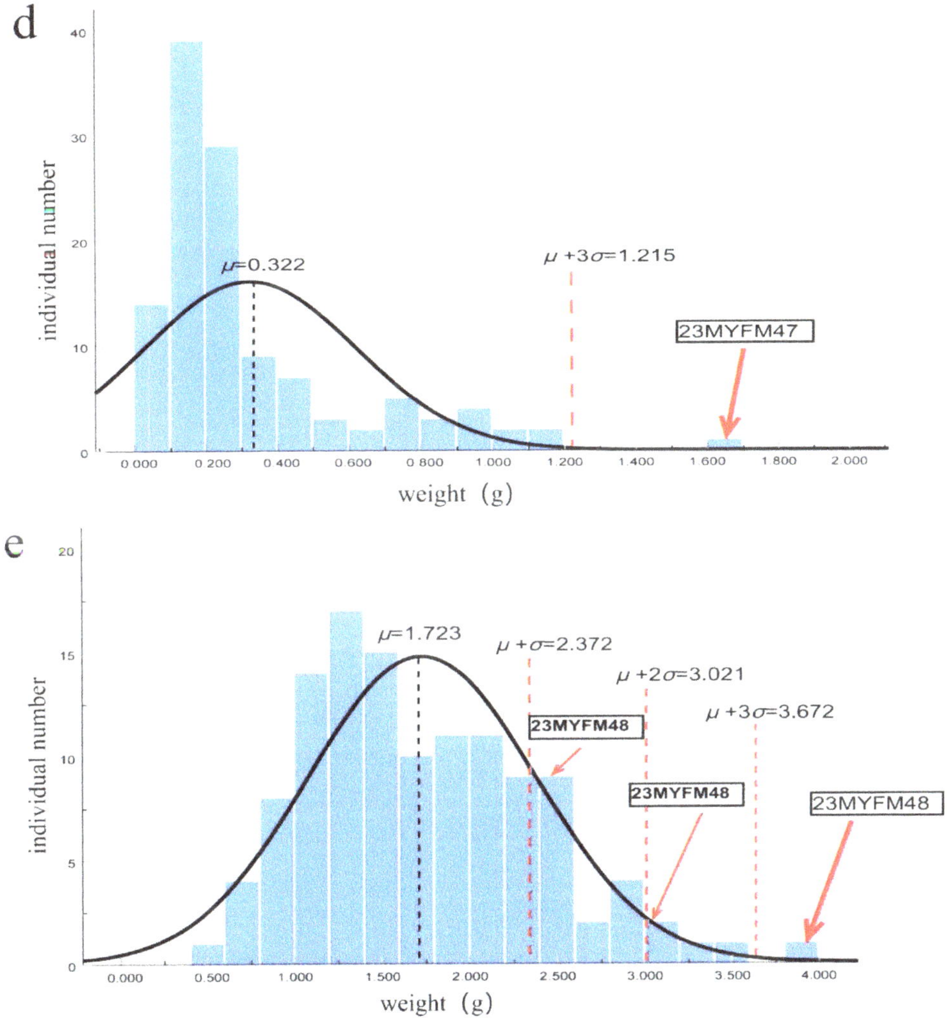

Figure 1. The weight distribution of each family. (**a**) Average weight of each family at each growth stage. (**b**) Relative proportion of the average weight at each developmental stage in each family. (**c–e**) Normal distributions of individual weight in stages 25-I, 25-II, and 25-III, respectively. μ represents the mean, σ represents the standard deviation, and the weight value beyond the discrete interval (beyond $\mu \pm 3\sigma$) represents the weight value of an individual which is significantly higher than those of others.

After 120 days of cultivation (Figure 2a), the metamorphosis rate was concentrated between 6–16%, and the number of families with a metamorphosis rate of 8–10% accounted for one-fourth of the total families (Figure 2b). In actual production, the metamorphosis rate of tadpoles is about 10–20%, and results of our study align with actual production. The highest metamorphosis rate was 28.3% (families 23MYFM20, 23MYFY77), which was significantly higher than other families and higher than most groups in actual production. The lowest metamorphosis rate was only 1.7% (families 23MYFM2, 23MYFM36, 23MYFM47, 23MYFM84, 23MYFM92, and 23MYFM96). The data showed that the earliest individual had undergone metamorphosis after 78 days, and some individuals did not complete metamorphosis for 150 days. Therefore, there is a large variation in the larval period among families, as well as between individuals in the same family.

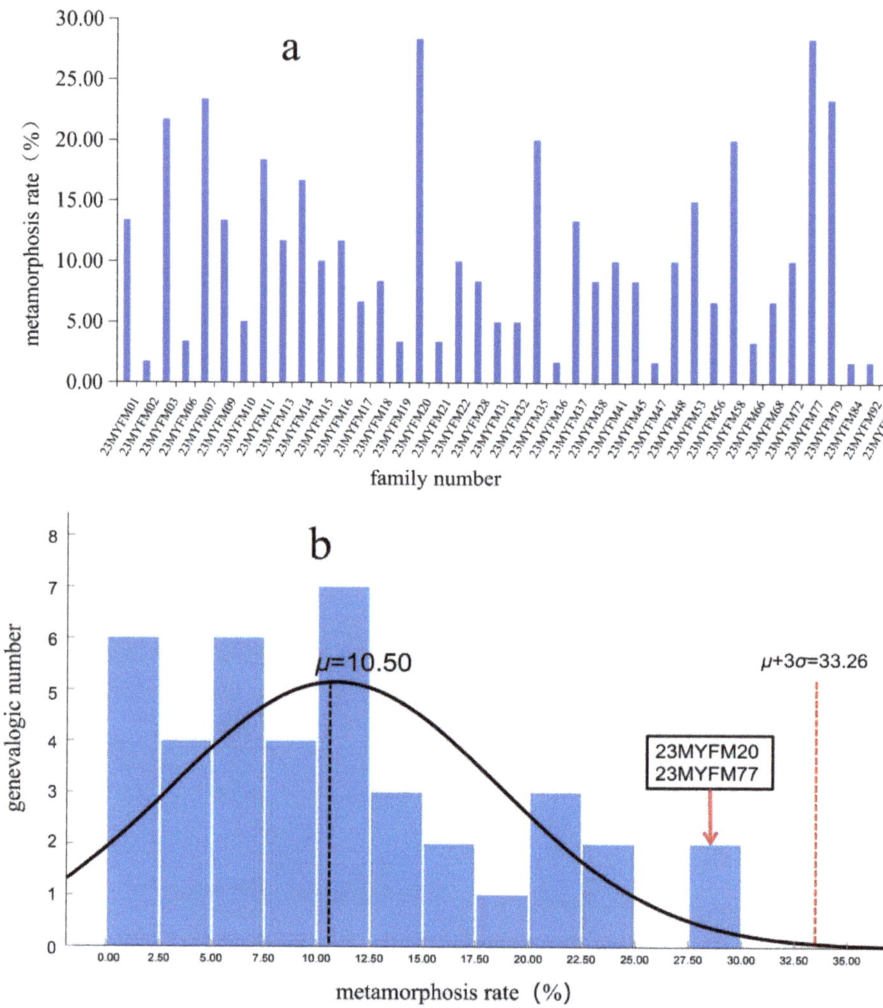

Figure 2. The metamorphosis rate of each family. (**a**) Metamorphosis rate of each family. (**b**) Normal distribution of metamorphosis rate. μ represents the mean and σ represents the standard deviation.

As shown in Table 1, a relatively low heritability was estimated for the metamorphosis rate (0.18 ± 0.20), while the weight at stages 25-I, II, and III showed a high heritability. Moreover, stages 25-I and II showed a strong correlation with the metamorphosis rate (0.57 and 0.38, $p < 0.05$), and the correlation coefficient with stage 25-I was not significant (0.03, $p > 0.05$). The metamorphosis rate was significantly correlated only with 25-III's weight at the genetic level (0.38, $p < 0.05$), but not growth (0.26, $p > 0.05$) or weight at stages 25-I (0.21, $p > 0.05$) and 25-II (−0.03, $p > 0.05$). Given the number of comparisons undertaken, Table 1's significance level was Bonferroni-adjusted [21]. Only the metamorphosis rate was significantly associated with the phenotypic correlation of the stage 25-II weight (0.57, $p < 0.005$).

Table 1. Heritability of metamorphosis rate, stage 25-I weight, stage 25-II weight, stage 25-III weight, growth, with phenotypic (above diagonal) and genetic (below diagonal) correlations between these traits.

	h^2	Growth	Stage 25-I Weight	Stage 25-II Weight	Stage 25-III Weight	Metamorphosis Rate
growth	0.13 ± 0.06		-	-	-	-
stage 25-I weight	0.99 ± 0.00	0.20		0.40 *	−0.03	0.21
stage 25-II weight	0.46 ± 0.12	0.60 *	0.22		0.42 *	0.57 **
stage 25-III weight	0.44 ± 0.20	0.66 *	−0.03	−0.09		0.38 *
metamorphosis rate	0.18 ± 0.20	0.26	0.21	−0.03	0.38 *	

Significance: * $p < 0.05$; ** $p < 0.005$.

4. Discussion

4.1. Metamorphosis of Tadpoles

Tadpoles undergo metamorphosis to develop into amphibious juveniles, and the time of actual production is concentrated at 85–95 d. In contrast, tadpoles of the rice field frog, *Quasipaa spinosa,* take about three years to complete metamorphosis [22], and there is huge variation between individuals. Hence, we regard *R. catesbeiana* as more suitable for practical production due to its relatively short larval period and its advantage of high economic value.

To adapt to amphibious life, the morphology of tadpoles needs to change, including limb differentiation, gill and tail disappearance, alterations in the liver, and remodeling of the skin, gut, and cranial cartilage [23]. The distribution and vulnerability to predation of *R. catesbeiana* increase when it transitions from a tadpole in aquatic life to a young amphibious frog, and it strengthens its defense against predation [24]. Tadpole transformation has a direct impact on the financial advantages that farmers receive from selling young frogs during production. In conclusion, research on the metamorphosis of tadpoles has scientific significance as well as significance for guiding practical production. The occurrence of metamorphosis can be used as an evaluation index for production.

4.2. Weight and Metamorphosis Rate of Tadpoles

We found that the metamorphosis rate of tadpoles was moderately related to their weights at stages 25-II and 25-III. Tadpole metamorphosis is affected by hormone levels and nutrient accumulation. Body size gradually increases during growth but decreases after metamorphosis, which includes the disappearance of the tail and the growth of limbs. Some studies have shown that thyroid hormone content in tadpoles is kept at an extremely low level before metamorphosis, gradually increases with the process of metamorphosis, reaches the highest level at the climax of metamorphosis, and then gradually decreases [25,26]. Similarly, from the beginning of stage 21, the level of lipids begins to rise and reaches its highest level before the climax period of metamorphosis, then decreases in the late stage of metamorphosis [27]. The liver lipids of tadpoles of the Omei brown frog, *Rana omeimontis,* play a role in promoting and regulating metamorphosis [28]. The lipid accumulation of tadpoles is mostly concentrated in the tail and viscera, and the external manifestations are tail length and body broadening [26]. Tadpole weight is correlated with metamorphosis. Tadpoles with a longer tail and a greater body width were shown to weigh more and have a shorter larval duration and larger bodies after metamorphosis.

From the perspective of predation, the individual size of amphibians is positively correlated with their ability for rapid movement, and larger individuals have a higher survival rate when avoiding natural enemies in the wild [29]. Larger individuals have an advantage in dealing with adverse factors such as the environment [30] as well as interspecies [25,26] and intraspecific competition [27]. According to certain research [4], individuals undergoing shorter larval periods could have a larger body size, earlier feeding capacity, and lower mortality rate, whereas those with longer larval periods could have poor feeding ability and greater mortality rate. In production, smaller individuals with long larval periods could be preyed upon by young frogs with shorter larval periods and larger individual size. Further, sudden changes in their living environment during transformation are likely to cause mortality of almost 80% of smaller tadpoles. We also found that tadpoles died during the period of metamorphosis and subsequent development, with a mortality rate of about 10–20%, and the majority consisted of individuals with a long larval period. The major phenotypes of these individuals were lacking forelegs, although their tails had nearly or completely vanished, but their front legs were still covered in skin and could not stretch. Therefore, large tadpoles will have a stronger ability to complete metamorphosis and have a high survival rate.

Selective breeding is important to aquaculture production. Some selection methods, such as index selection methods, BLUP breeding value estimation, and other derived BLUP methods, have been widely used to breed high-performance families for certain traits in salmon and trout [31,32]. Metamorphosis rate, as a population-level trait, reflects the metamorphosis of the families in a population. In production, various factors affect the rate of metamorphosis in a family. The larval periods of some individuals may be too long, and individuals may die in large numbers. In this study, the breeding environment was basically the same. The larval periods of same-family individuals varied greatly. Individual mortality was higher in some families. These result in a low rate of metamorphosis within the family. The heritability of metamorphosis rate was only 0.18 $\pm$ 0.20. This estimate may indicate that the metamorphosis rate is affected by a variety of other factors. Correspondingly, a family's metamorphosis rate may be higher if its tadpoles are heavier and there are more individuals with shorter larval periods. In the process of breeding, the combination of a solitary sire and a solitary dam to produce offspring can produce families naturally. The family selection method for *R. catesbeiana* has natural advantages. As a result, the weight in the larval stage can be used to select breeders for bullfrog species with high rates of metamorphosis.

4.3. Heritability and Correlation between Weight and Metamorphosis Rate

Berven [33] discovered varied heritability for growth (0.07–0.66) and moderate heritabilities for development (0.35) in wood frogs. The heritability of early growth and development traits differs greatly among different individuals in Acuracea [33]. Couch's spadefoot, *Scaphiopus couchii,* has no substantial heritability for growth, although having high heritabilities of development (0.87) [34]. Tadpoles of the grass frog *Rana temporaria* exhibit low growth heritability (0–0.1) and significant development heritability (0.1–0.3) [35,36]. Spring peeper *Pseudacris crucifer* does not seem to have any appreciable heredity (less than 0.1) for either growth or development [37,38]. Lastly, the moderate heritability of growth rate for *Rana temporaria* tadpoles was 0.17 $\pm$ 0.11 [35]. Thus, the heritability of life-history traits can be quite variable among anurans. Similarly, the estimated heritability of weight and metamorphosis rate in this study was 0.13–0.99, with a wide range, which is similar to results of the above studies.

In this study, the heritability of weight at stage 25-I was at an extremely high level (0.99 $\pm$ 0.00). Kaplan [39] suggested that weight at this stage is greatly affected by the maternal effect. At this stage, the tadpole relies mainly on the nutrients stored in the body for nutrient supply. In production, a tadpole during this period does not feed much, there is a high probability of deformity and mechanical damage, and weak individuals are more likely to die during the process. This was also the case in our study, and the remaining individuals had small differences in weight. We think that individuals that survived showed relatively high levels of genetic adaptation under the environmental conditions of this study. At stages 25-I (48 d) and 25-III (100 d), the heritability of weight gradually reduced to 0.46 $\pm$ 0.12 and 0.44 $\pm$ 0.20, respectively. In production, the tadpoles at stage 25-II are about to enter the stage of development where their diets alter and their internal organs start to develop into those of the adult stage [40,41]. During this stage, most tadpoles show signs such as abdominal enlargement and the abdomen appears red. Similarly, individuals with such indications have a mortality rate of up to 80%. In comparison to previous stages, the tadpole's growth rate in stage 25-II was faster. In this same period, tadpole sizes began to differ from one another. After Bonferroni correction was performed to minimize errors in the interpretation of interpretation of results, only the rate of metamorphosis was significantly associated with the phenotypic correlation of stage 25-II weight. In stage 25-III, the body-size difference between individuals was more obvious. With the growth and development of tadpoles, the difference in weight between individuals also increased. Some studies compared the different growth rates in different living environments of the same species [42], and concluded that the differences in the growth and development of bronze frog *Rana clamitans* are induced by the environment. For example, both feeding level and diet density during production have significant effects on tadpole size [43,44]. The development of the same group in the same location will also change with the environment [45]. Weight was quite different among individual tadpoles in the same family in the same breeding environment in our study. Different phenotypes are affected by environmental selection in different ways. Individual differences increase in response to specific stresses as environmental effects increase in the latter stages of tadpole development. Rose [46] wrote that this is due to the genetic differences in response to the living environment and interspecific competition. Thus, the differences are directly manifested as feeding and further reflected by the variation in weight. During the production process, weaker-feeding individuals will grow too small, have weaker disease resistance, or may die, whereas stronger-feeding individuals will grow larger, faster, and have stronger disease resistance. The present study found a gradual decrease in weight heritability with the advance of tadpole growth. Introducing time as a fixed effect, and combining the weight heritability of data from each developmental period, the estimated growth heritability (0.13 ± 0.06) was significantly lower than in a single developmental period. Tadpole size significantly affects both fitness and natural selection [35]. In conclusion, we think that the variations in tadpoles' larval weight are caused by their distinctive genetic makeup. These results might indicate that compared with other stages, the development of stage 25-II is more related to subsequent growth, and thereby affect metamorphosis.

We found a significant phenotypic correlation between metamorphosis rate and both weights at stages 25-II (0.57, $p < 0.05$) and 25-III (0.38, $p < 0.05$), while there was no correlation with stage 25-I (-0.03, $p > 0.05$). There was a significant phenotypic correlation between weights in adjacent stages. In its genetic correlations, the metamorphosis rate was only moderately significantly associated with stage 25-III (0.38, $p < 0.05$), but not stages 25-I (0.21, $p > 0.05$) and 25-II (-0.03, $p > 0.05$). The genetic correlation between weight at the three stages was also not significant. As a study of the Pacific treefrog (*Pseudacris regilla*) showed [19], while there was no genetic association between individuals of different stages, there was a significant correlation between them in phenotype. From the genetic correlation between metamorphosis rate and weight at each stage, there was a significant correlation with metamorphosis in stage 25-III. This is similar to the barking treefrog *Hyla gratiosa* [47], which represents a similar size at metamorphosis. The advantage of weight mentioned above will make individual tadpoles more suitable for the environment, and continuously affect their viability after metamorphosis. Thus, we believe that the significant correlation of metamorphosis rate with weight in stage 25 is consistent with biological characteristics.

In this study, the heritability of the metamorphosis rate in bullfrog tadpoles was 0.18 ± 0.20, which is moderate and has selective breeding potential. Therefore, the proportion of genetic control in phenotypic effects was increased by positive assortative mating between families with high metamorphosis rates. Groups with high metamorphosis rates were selected by measuring tadpole weight before metamorphosis to shorten the production time and improve production efficiency. The stage 25-III weight as a breeding index is comparable to the selection that occurs after metamorphosis since it is near the onset of metamorphosis. There was minimal heredity at stages 25-II and 25-III, and the transformation rate was not substantially correlated with stage 25-II. To increase production efficiency and identify the weight during developmental stages that have a strong genetic correlation with the metamorphosis rate, it is necessary to weigh tadpoles between stages 25-II and 25-III. This practice will increase the accuracy and efficiency of subsequent breeding efforts to produce high-metamorphosis-rate lines.

5. Conclusions

The metamorphosis of tadpoles is crucial in the whole growth and development of *R. catesbeiana* and is important for the transition from aquatic tadpoles to amphibious young frogs. At the same time, the short larval period also makes *R. catesbeiana* individuals have greater survival at the amphibious stage. A short larval period has a natural advantage, and it is feasible to use it evaluate the rate of metamorphosis. According to the correlation between the weight of tadpoles and their metamorphosis rate, the metamorphosis rate of families can be estimated by the weight of tadpoles at different developmental stages, as well as by weighing tadpoles between stages 25-II and 25-III and estimating their heritability as well as the genetic and phenotypic association. The results of this study confirm the importance and portability of tadpole weight in actual seed production and provide basic data and a potential optimized direction for the selective breeding of high-metamorphosis-rate bullfrogs.

Author Contributions: W.X.: validation, data analysis, writing—original draft. Y.W.: data analysis, writing—review and editing. G.W.: conceptualization, methodology, funding acquisition. L.Z.: conceptualization, project administration, resources, supervision. G.Z.: methodology. Z.H.: resources. H.G.: supervision. All authors have read and agreed to the published version of the manuscript.

Funding: This research was supported by the National Natural Science Foundation of China (grant number 31602172) and the Industry-Academia-Research Project of Fujian (grant number 2021N5012).

Institutional Review Board Statement: This study was carried out according to the guidelines of the Declaration of Helsinki and was approved by the Animal Care Advisory Committee of Jimei University (Approval No. 2019-0906-003, 6 September 2019).

Data Availability Statement: The data presented in this study are available on request from the corresponding author.

Acknowledgments: We are grateful to all those who contributed to this manuscript.

Conflicts of Interest: The authors declare that there are no conflicts of interest.

References

1. Kikuyama, S.; Kawamura, K.; Tanaka, S.; Yamamoto, K. Aspects of amphibian metamorphosis: Hormonal control. *Int. Rev. Cytol.* **1993**, *145*, 105–148.
2. Rose, C.S. Hormonal control in larval development and evolution—Amphibians. In *The Origin and Evolution of Larval Forms*; Elsevier: Amsterdam, The Netherlands, 1999; p. 167-VI.
3. Werner, E.E. Amphibian metamorphosis: Growth rate, predation risk, and the optimal size at transformation. *Am. Nat.* **1986**, *128*, 319–341. [CrossRef]
4. Altwegg, R.; Reyer, H.U. Patterns of natural selection on size at metamorphosis in water frogs. *Evolution* **2003**, *57*, 872–882.
5. Wilbur, H.M. Competition, predation, and the structure of the *Ambystoma-Rana sylvatica* community. *Ecology* **1972**, *53*, 3–21. [CrossRef]
6. Brockelman, W.Y. An analysis of density effects and predation in *Bufo americanus* tadpoles. *Ecology* **1969**, *50*, 632–644. [CrossRef]
7. Collins, J.P. A Comparative Study of the Life History Strategies in a Community of Frogs. Ph.D. Dissertation, University of Michigan, Ann Arbor, MI, USA, 1975.
8. Berven, K.A. The genetic basis of altitudinal variation in the wood frog *Rana sylvatica* II. An experimental analysis of larval development. *Oecologia* **1982**, *52*, 360–369. [CrossRef] [PubMed]
9. Smith-Gill, S.J.; Gill, D.E. Curvilinearities in the competition equations: An experiment with ranid tadpoles. *Am. Nat.* **1978**, *112*, 557–570. [CrossRef]
10. Berven, K.A. Mate choice in the wood frog, Rana sylvatica. *Evolution* **1981**, *35*, 707–722. [CrossRef]
11. Wilbur, H.M. Interactions of food level and population density in *Rana sylvatica*. *Ecology* **1977**, *58*, 206–209. [CrossRef]
12. McDiarmid, R.W.; Altig, R. *Tadpoles: The Biology of Anuran Larvae*; University of Chicago Press: Chicago, IL, USA, 1999.
13. Brooks, R.; Endler, J.A. Direct and indirect sexual selection and quantitative genetics of male traits in guppies (*Poecilia reticulata*). *Evolution* **2001**, *55*, 1002–1015. [CrossRef] [PubMed]

14. Wilbur, H.M.; Collins, J.P. Ecological aspects of amphibian metamorphosis: Nonnormal distributions of competitive ability reflect selection for facultative metamorphosis. *Science* **1973**, *182*, 1305–1314. [CrossRef] [PubMed]
15. McCoy, K.A.; McCoy, M.W.; Amick, A.; Guillette, L.J., Jr.; St. Mary, C.M. Tradeoffs between somatic and gonadal investments during development in the African clawed frog (*Xenopus laevis*). *J. Exp. Zool. Part A Ecol. Genet. Physiol.* **2007**, *307*, 637–646. [CrossRef] [PubMed]
16. Collins, J.P. Intrapopulation variation in the body size at metamorphosis and timing of metamorphosis in the bullfrog, *Rana catesbeiana*. *Ecology* **1979**, *60*, 738–749. [CrossRef]
17. Gosner, K.L. A simplified table for staging anuran embryos and larvae with notes on identification. *Herpetologica* **1960**, *16*, 183–190.
18. Gilmour, A.R.; Gogel, B.J.; Cullis, B.R.; Welham, S.; Thompson, R. *ASReml User Guide Release 1.0*; University of Hamburg: Hamburg, Germany, 2002.
19. Watkins, T.B. A quantitative genetic test of adaptive decoupling across metamorphosis for locomotor and life-history traits in the Pacific tree frog, *Hyla regilla*. *Evolution* **2001**, *55*, 1668–1677.
20. Liu, X.L.; Chang, Y.Q.; Xiang, J.H.; Cao, X.B. Estimates of genetic parameters for growth traits of the sea urchin, *Strongylocentrotus intermedius*. *Aquaculture* **2005**, *243*, 27–32. [CrossRef]
21. Rice, W.R. Analyzing tables of statistical tests. *Evolution* **1989**, *43*, 223–225. [CrossRef]
22. Traijitt, T.; Kitana, N.; Khonsue, W.; Kitana, J. Chronological changes in the somatic development of Hoplobatrachus rugulosus (Wiegmann, 1834) (Anura: Dicroglossidae). *Trop. Nat. Hist.* **2021**, *21*, 184–199.
23. Dodd, M.; Dodd, J. The biology of metamorphosis. *Physiol. Amphib.* **1976**, *3*, 467–599.
24. Loman, J. Food, feeding rates and prey size selection in juvenile and adult frogs, *Rana arvalis* Nilss. and *R. temporaria* L. *Ekol. Pol.* **1979**, *27*, 581–601.
25. Tollrian, R. Predator-induced morphological defenses: Costs, life history shifts, and maternal effects in *Daphnia pulex*. *Ecology* **1995**, *76*, 1691–1705. [CrossRef]
26. Segers, F.H.; Taborsky, B. Juvenile exposure to predator cues induces a larger egg size in fish. *Proc. R. Soc. B Biol. Sci.* **2012**, *279*, 1241–1248. [CrossRef]
27. Michimae, H.; Nishimura, K.; Tamori, Y.; Wakahara, M. Maternal effects on phenotypic plasticity in larvae of the salamander *Hynobius retardatus*. *Oecologia* **2009**, *160*, 601–608. [CrossRef] [PubMed]
28. Zhu, W.; Zhang, M.; Chang, L.; Zhu, W.; Li, C.; Xie, F.; Zhang, H.; Zhao, T.; Jiang, J. Characterizing the composition, metabolism and physiological functions of the fatty liver in *Rana omeimontis* tadpoles. *Front. Zool.* **2019**, *16*, 42. [CrossRef] [PubMed]
29. Kusano, T. Growth and survival rate of the larvae of *Hynobius nebulosus tokyoensis* Tago (Amphibia, Hynobiidae). *Popul. Ecol.* **1981**, *23*, 360–378. [CrossRef]
30. Pakkasmaa, S.; Merilä, J.; O'Hara, R. Genetic and maternal effect influences on viability of common frog tadpoles under different environmental conditions. *Heredity* **2003**, *91*, 117–124. [CrossRef] [PubMed]
31. Gjedrem, T.; Robinson, N.; Rye, M. The importance of selective breeding in aquaculture to meet future demands for animal protein: A review. *Aquaculture* **2012**, *350*, 117–129. [CrossRef]
32. Vandeputte, M.; Quillet, E.; Chevassus, B. Early development and survival in brown trout (*Salmo trutta fario* L.): Indirect effects of selection for growth rate and estimation of genetic parameters. *Aquaculture* **2002**, *204*, 435–445. [CrossRef]
33. Berven, K.A. The heritable basis of variation in larval developmental patterns within populations of the wood frog (*Rana sylvatica*). *Evolution* **1987**, *41*, 1088–1097.
34. Newman, R.A. Genetic variation for larval anuran (*Scaphiopus couchii*) development time in an uncertain environment. *Evolution* **1988**, *42*, 763–773.
35. Laurila, A.; Karttunen, S.; Merilä, J. Adaptive phenotypic plasticity and genetics of larval life histories in two *Rana temporaria* populations. *Evolution* **2002**, *56*, 617–627. [PubMed]
36. Sommer, S.; Pearman, P.B. Quantitative genetic analysis of larval life history traits in two alpine populations of *Rana temporaria*. *Genetica* **2003**, *118*, 1–10. [CrossRef]
37. Travis, J.; Emerson, S.B.; Blouin, M. A quantitative-genetic analysis of larval life-history traits in *Hyla crucifer*. *Evolution* **1987**, *41*, 145–156.
38. Emerson, S.B.; Travis, J.; Blouin, M. Evaluating a hypothesis about heterochrony: Larval life-history traits and juvenile hind-limb morphology in *Hyla crucifer*. *Evolution* **1988**, *42*, 68–78. [PubMed]
39. Kaplan, R.H. The implications of ovum size variability for offspring fitness and clutch size within several populations of salamanders (*Ambystoma*). *Evolution* **1980**, *34*, 51–64.
40. Atkinson, B.G.; Warkman, A.S.; Chen, Y. Thyroid hormone induces a reprogramming of gene expression in the liver of premetamorphic *Rana catesbeiana* tadpoles. *Wound Repair Regen.* **1998**, *6*, S323–S336. [CrossRef]
41. Rollins-Smith, L.A. Metamorphosis and the amphibian immune system. *Immunol. Rev.* **1998**, *166*, 221–230. [CrossRef] [PubMed]
42. Berven, K.A.; Gill, D.E.; Smith-Gill, S.J. Countergradient selection in the green frog, *Rana clamitans*. *Evolution* **1979**, *33*, 609–623. [CrossRef]
43. Murray, D.L. The effects of food and density on growth and metamorphosis in larval wood frogs (*Rana sylvatica*) from central Labrador. *Can. J. Zool.* **1990**, *68*, 1221–1226. [CrossRef]
44. Berven, K.A.; Chadra, B.G. The relationship among egg size, density and food level on larval development in the wood frog (*Rana sylvatica*). *Oecologia* **1988**, *75*, 67–72. [CrossRef]

45. Travis, J. Variation in development patterns of larval anurans in temporary ponds. I. Persistent variation within a *Hyla gratiosa* population. *Evolution* **1983**, *37*, 496–512. [CrossRef] [PubMed]
46. Rose, C.S. Integrating ecology and developmental biology to explain the timing of frog metamorphosis. *Trends Ecol. Evol.* **2005**, *20*, 129–135. [CrossRef] [PubMed]
47. Travis, J. Anuran size at metamorphosis: Experimental test of a model based on intraspecific competition. *Ecology* **1984**, *65*, 1155–1160. [CrossRef]

fishes

MDPI

Article

Identification and Characterization of microRNAs in Morphological Color Change of Polychromatic Midas Cichlids (*Amphilophus citrinellus*)

Guoqiang Wu [1,2], Xidong Mu [1], Yi Liu [1], Chao Liu [1], Xuejie Wang [1], Yexin Yang [1] and Hongmei Song [1,*]

[1] Key Laboratory of Tropical and Subtropical Fishery Resources Application and Cultivation, Ministry of Agriculture and Rural Affairs, Pearl River Fisheries Research Institute, Chinese Academy of Fishery Sciences, Guangzhou 510380, China; wgq361731661@163.com (G.W.); muxd@prfri.ac.cn (X.M.); liuyi@prfri.ac.cn (Y.L.); liuchao@prfri.ac.cn (C.L.); wxj@prfri.ac.cn (X.W.); yangyexin@prfri.ac.cn (Y.Y.)

[2] College of Fisheries and Life Science, Shanghai Ocean University, Shanghai 201306, China

* Correspondence: shm@prfri.ac.cn

Abstract: As a representative genetic and economic trait, pigmentation has a strong impact on speciation and adaptation. However, information and reports on microRNAs (miRNAs) associated with pigmentation remain limited. The Midas cichlid fish, with three typical distinct stages of body color pattern, "black-gray-gold", is an ideal model system for investigating pigmentation traits. In this study, miRNA libraries from scale tissues with the attached epidermis of Midas cichlids at three distinct stages of color transformation, black (B), transition (T), and gold (G), were sequenced using Illumina sequencing technology. In total, 53 (B vs. G), 88 (B vs. T), and 57 (T vs. G) miRNAs were differentially expressed between the respective groups. Target genes of the identified miRNAs were predicted, and the results showed that multiple target genes were related to pigmentation and pigment–cell differentiation. The miRNA–mRNA regulatory network suggests that miR-183-x and miR-133-x were predicted to be involved in regulating morphological color changes in Midas cichlids. The results advance our understanding of potential functions of miRNAs in skin pigment differentiation and early skin color fading of fishes.

Keywords: Midas cichlid; microRNAs; morphological color change; deep sequencing; target genes

Key Contribution: The Midas cichlid, which exhibits three distinct stages of body color pattern—black, gray, and gold—is an ideal model system for investigating pigmentation traits. This research found that the microRNAs miR-183-x and miR-133-x may be involved in regulating morphological color changes in Midas cichlids, potentially playing a role in skin pigment–cell differentiation and early skin color fading.

1. Introduction

Body color is a unique and important heritable economic trait of fish which plays an important role in mate selection, predation, camouflage, and behavioral processes, as well as in development [1,2]. In many fishes, a typical "body color fading" phenomenon occurs from the larval to the juvenile stage. After the larva fish hatches out of the membrane, melanocytes began to differentiate and form, and the number of melanovytes gradually increases [3]. The skin color of the fish is initially black, and then other pigment cells gradually differentiate and form, and the skin fades to different degrees, showing various colors or stripes in juveniles and adults [3,4]. In aquaculture production, the black of the blood hybrid parrot fish (*Vieja synspilum* ♀× *Amphilophus citrinellus* ♂) is completely faded in juveniles and adults, whereas koi carp (*Cyprinus carpio* var. koi) and *Astronotus ocellatus* are partially faded, and their striped body color is usually improved with use of pigmented feed to increase their commercial value. The complete degree of black body coloration fading in the early stages of fish development is the key to the effect of later color enhancement. However, it is common for the degree of fish body color not to meet the requirements of production in aquaculture [5]. At present, there are few studies on the biological characteristics of early body black coloration fading in fishes, so it is necessary to understand its molecular regulation in order to provide a theoretical basis for seeking an effective method to artificially improve fish body coloration.

The formation of fish body coloration is a complex biological process which is affected by many factors, such as heredity, nutrition, physiology, and environment [4,6]. It is not only related to pigment synthesis, but also depends on the type, number, and distribution of pigment cells and the interaction between pigment cells [7]. It has been reported that there are seven different kinds of pigment cells in fish, and different fishes have different types of pigment cells [8,9]. The differentiation, formation, and body color development of fish chromatophores are controlled by complex and well-balanced programs of gene activation and silencing [10,11]. Transcriptome analysis has been performed on non-model fishes with different skin colors, and numerous pigmentation-related genes have been identified and their functions studied, laying a foundation for the investigation of morphological in heredity in fishes [12–14].

MicroRNAs (miRNAs) are a class of endogenous single-stranded non-coding small RNAs 19–24 nt in length that negatively regulate gene expression at the post-transcriptional level by binding to target mRNAs in completely or incompletely complementary ways, thereby affecting numerous biological functions including body color formation [15]. MiRNAs play an important role in cell growth, tissue differentiation, and signal transduction, with temporal and tissue specificity [16–18]. The sequence evolution at miRNA binding sites could lead to rapid phenotypic evolution [19,20]. A series of miRNAs involved in regulating the formation of skin, body color, and hair color have been identified in koi carp (*Cyprinus carpio* L.), rainbow trout (*Oncorhynchus mykiss*), giant salamander (*Andrias davidianus*), Chinese soft-shelled turtle (*Pelodiscus sinensis*), and other animals [21–24]. Many genes and transcriptional regulators controlling skin and pigment cell formation in animals are regulated by miRNAs [25]. In a previous study, it was found that miR-508-3p can affect melanin production in alpacas by regulating the microphthalmia-associated transcription factor (*mitf*) [26]. Additionally, miR-206 plays a regulatory role in the skin color pigmentation by targeting the melanocortin 1 receptor (*mc1r*) in koi carp [27]. Overexpression of miR-137 can change skin color from black to brown in mice [28]. Overexpressing miR-148a-3p in alpaca melanocytes causes the expression of *mitf* pigmentation-associated protein tyrosinase (*try*) to be reduced, thereby affecting a decrease in the overall melanin content of the analyzed cells [29]. This evidence suggests that miRNAs could be involved in the formation of skin color patterns. However, few miRNAs associated with early black body color fading in fish have been reported.

The Midas cichlid (*Amphilophus citrinellus*) is a popular ornamental fish and the male parent of the hybrid blood parrot fish. The Midas cichlids have four types of pigment cells, including melanocytes, xanthophores, erythrophores, and iridocytes [30]. Body coloration is key ecological trait driving species formation [31]. At the early stage of development, the Midas cichlids show a typical body-color fading phenomenon (Figure 1) [3]. After fading, the whole body is uniform bright yellow without markings. It is an ideal biological model for studying the mechanism of early body-color determination in fish [32]. The gold/dark polymorphism is a Mendelian trait with the gold morph being the dominant form [11]. Kratochwil et al. reported that the 8.2 kb insertion within an intron of *goldentouch* determines the dark/gold polymorphism. However, the *goldentouch* expression does not differently change during ontogeny of gold Midas cichlids [33]. The molecular mechanism of morphological color change needs further study. Henning et al. [32] performed Illumina RNA-SEQ sequencing on the skin tissues of three typical body-color transition periods in Midas cichlids [33] and screened and identified several key genes related to pigment synthesis, such as dopachrome tautomerase (*dct*), solute carrier family24, member5 (*slc24a5*), tyrosinase (*tyr*), and tyrosinase-related protein 1 (*tyrp1*). However, whether there is a potential miRNA regulatory role on these genes related to body color remains unclear.

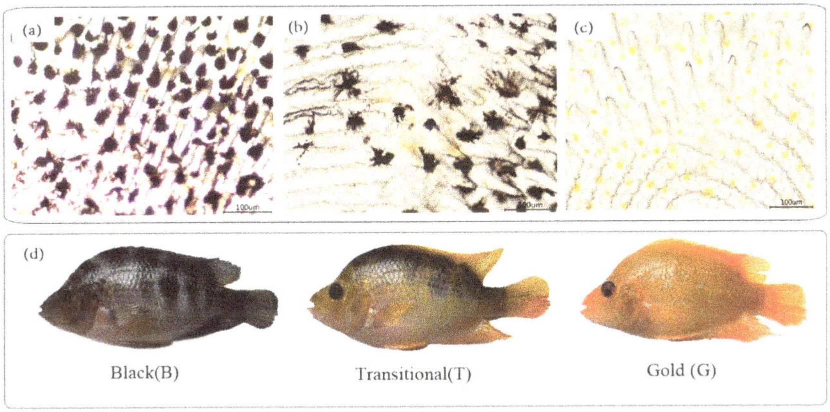

Figure 1. Morphological color change in the Midas cichlid. Viable melanophores and xanthophores in the dispersed and aggregated states are shown in (**a**). Melanophores of transitional fish showing a mixture of aggregated, dispersed, and dead melanophores are shown in (**b**). A scale of a gold fish completely lacking melanophores is shown in (**c**). A representative individual at three stages (B, T, and G) is shown in (**d**).

To explore the mechanism of miRNAs regulating body-color-related genes at the post-transcriptional level during the body-color fading stage of Midas cichlids, we used high-throughput sequencing analysis to identify miRNAs in fish of three typical phenotypes representing distinct stages of color transformation, which include of the black (B) stage (unfaded black period), transitional (T) stage (transition period into fading), and gold (G) stage (faded, completely golden period) (Figure 1). To understand the function of miRNAs in regulating body-color formation in fish, several differentially expressed miRNAs were verified using quantitative real-time PCR (qRT-PCR), and potential target genes of the miRNAs were predicted and analyzed using Gene Ontology (GO) enrichment and Kyoto Encyclopedia of Genes and Genomes (KEGG) pathway analysis.

2. Materials and Methods

2.1. Fish Farming and Sample Collection

Referring to the reports by Guoqiang Wu [34] and Henning [32], the body-color fading process of Midas cichlids is divided into three stages: B stage (black fish showing no signs of gold or yellow coloring), T stage (fish showing gray coloration with clear patterns of both gold and black throughout the body), and G stage (complete, or almost-complete gold coloring throughout the body) (Figure 1). A total of 100 black Midas cichlids (average body length 56.1 ± 0.5 mm) were collected from the Ornamental Fish Culture Base of the Pearl River Fisheries Research Institute, Chinese Academy of Fishery Sciences (Guangzhou, China) at 40 days after the larval fish hatched out of the membrane. All the fish were reared in a 200-L glass tank, and the water temperature was maintained at 26 ± 2 °C. Three fish were randomly selected on days 50, 65, and 85, corresponding to each of the three coloration periods. The fish were immersed in 100 mg/L MS-222 (Sigma-Aldrich, St. Louis, MO, USA) for anesthesia. Scale samples were quickly taken, placed in liquid nitrogen, and stored at −80 °C for long-term storage. All fish were cultured, and experiments were conducted in accordance with the Regulation for the Administration of Affairs Concerning Experimental Animals for the Science and Technology Bureau of China throughout the study.

2.2. Total RNA Extraction, Construction, and Sequencing of a Small RNA Library

Approximately 10–12 scales on both sides of the dorsal fin of each fish were collected and stored in RNA later solution (OMEGA, Tarzana, CA, USA) at 4 °C. Total RNA from the three groups of samples was extracted using TRizol reagent (Invitrogen, Carlsbad, CA, USA) according to the manufacturer's instructions. RNA integrity was confirmed with agarose gel electrophoresis, and the concentration was detected with an OD_{260} reading using an Agilent 2100 Bioanalyzer (Agilent Technologies, Santa Clara, CA, USA). The RNA samples were size-fractionated with 15% polyacrylamide gel electrophoresis, and the 18–30 nt fraction was enriched. Then, 3′ RNA and 5′ RNA adapters were ligated to the purified RNA pools (TruSeq® Small RNA Sample Preparation Kit, Illumina, San Diego, CA, USA) using T4 ligase (New England Biolabs, Ipswich, MA, USA), and small RNAs connected to both splices were reverse transcribed using PCR. PCR products of about 140 bp length were retained and purified to generate the sequencing libraries, and each library was sequenced using the Illumina HiSeq™ 2500 Genome Analyzer by Guangzhou Kidio Biotechnology Co., Ltd. (Guangzhou, China).

2.3. Basic Analysis of Sequencing Data

All sequencing reads obtained from the Illumina HiSeq™ 2500 were quality-controlled using FastQC software (http://www.bioinformatics.babraham.ac.uk/projects/fastqc/, accessed on 6 April 2022) to obtain clean labels by removing invalid sequences, including low-quality reads, 3′ adapter null reads, 5′ adapter contaminant reads, and reads shorter than 18 nt. All of the clean tags were mapped to small RNAs in the GeneBank (http://www.ncbi.nlm.nih.gov/genbank, accessed on 13 July 2022) and Rfam (http://rfam.sanger.ac.uk, accessed on 5 August 2022) databases, and the rRNA, scRNA, snoRNA, snRNA, and tRNA sequences in the sample were removed. Subsequently, all of the reads were aligned with a reference genome for *Amphilophus citrinellus* (https://www.ncbi.nlm.nih.gov/assembly/GCA_000751415.1, accessed on 26 September 2022) to remove repeat sequences and fragments from mRNA degradation identified by mapping to exons or introns. The remnant reads were mapped to the miRBase database (http://www.mirbase.org/, accessed on 29 September 2022) to identify conserved miRNAs. All of the unannotated tags were compared to the reference genome to identify the novel miRNA candidates according to their genomic positions and hairpin structures using the software Mireap-V0.2 with default parameters (http://sourceforge.net/projects/mireap, accessed on 12 October 2022). Conserved miRNAs with letters x and y indicated that miRNAs are processed from the 5′ arm and 3′ arm of the miRNA precursor, respectively.

2.4. Differential Expression Analysis of miRNAs

To investigate the differentially expressed miRNAs in the samples from the three periods, the expression level of identified miRNAs was calculated, and the formula was as follows:

$$TPM = \frac{T \times 10^6}{N}$$

where TPM is transcripts per million, T is the read number of miRNA, and N is the total number of miRNA reads.

Then, the analyses of differentially miRNA expression were carried out using edgeR packages, with $p < 0.05$, $|\log 2 \text{ ratio multiples}| \geq 2$ as the standard of miRNA differential expression [21].

2.5. Validation of miRNA Expression by Stem-Loop qRT-PCR

Ten miRNAs, including eight known miRNAs and two novel miRNAs, with differential expression, were selected to verify the accuracy of deep sequencing using stem-loop RT-qPCR. DNase-treated RNA (1 μg) was reverse transcribed into cDNA by the tail addition method, and U6 snRNA was used as an internal reference. Universal reverse primer, along with forward primers in the stem-loop, were used because they provide more specificity and sensitivity than linear primers [35]. Primer sequences are shown in Table 1. qRT-PCR was performed by using a QuantStudio6 Flex Real-Time PCR Detection System (Applied Biosystems, Foster City, CA, USA).

Table 1. Primers used in the present study for tailing reaction qRT-PCR.

miRNAs ID	Primer	Sequence (5′-3′)
miRNA RT-primer	RT	TTACCTAGCGTATCGTTGACAGCTTTTTTTT TTTTTTTTTTTTVN
miR-4492-y	Forward	CGGGGCTGGGCTCGCGCC
miR-183-x	Forward	TATGGCACTGGTAGAATTCACT
miR-451-x	Forward	AAACCGTTACCATTACTGAGTTT
miR-1335-y	Forward	CGGCTGAGGTGGGATCCC
miR-6937-x	Forward	TGGCTCTAAGGGCTGGGTC
miR-133-x	Forward	AGCTGGTAAAATGGAACCAAAT
novel-m0009-5p	Forward	GTGAATCCTTGGACCCATTGTCA
miR-551-y	Forward	GCGACCCATCCTTTGTTTCTG
miR-193-y	Forward	AACTGGCCTACAAAGTCCCAGT
novel-m0091-3p	Forward	GTTCAAATCCGGATGCCCCCT
U6	Forward	CTCGCTTCGGCAGCACA
Universal Primer	Reverse	TTACCTAGCGTATCGTTGAC

Total RNA was extracted from nine Midas cichlids scale samples (3 B, 3 T, and 3 G). DNase-treated RNA (1 μg) was reverse-transcribed into cDNA with the tail addition method using the FastQuantity RT Kit (KR106 TIANGE). Real-time qPCR was performed by using the QuantStudio6 Flex Real-Time PCR Detection System. A 20 μL PCR reaction that included 10 μL qPCR master mix, 0.6 μL miRNA primers (Table 1), 0.6 μL universal primer, 6 μL cDNA, and 2.8 μL ddH_2O was added in triplicate into 96-well plates. Amplification was performed with an initial denaturation at 95 °C for 15s, followed by 40 cycles of at 95 °C for 15 s and 65 °C for 30 s, and a final 70 cycles at 60 °C for 30 s. Melting curve analyses were performed following amplifications. All reactions were performed in triplicate for each sample. The relative expression of miRNAs was measured using the threshold Ct method and calculated using $2^{-\triangle\triangle Ct}$ and U6 snRNA normalized mRNA expression. All primers for stem-loop qRT-PCR are listed in Table 1.

2.6. Target Gene Prediction and Functional Enrichment Analysis

Target gene prediction of miRNAs in the present study was performed by using the RNAhybrid [22], Miranda (v3.3a) [36], and TargetScan (Version: 7.0) [37] methods. The intersection of target gene prediction results obtained using the three methods was taken as the result of miRNA target gene prediction, followed by GO function analysis and KEGG Pathway analysis for these target genes.

3. Results

3.1. Small RNA Sequence Analysis

Nine small RNA libraries from *A. citrinellus* skin in distinct stages of color transformation were sequenced by using the Illumina HiSeq^{TM} 2500. A total of 15,764,028, 16,435,309, and 19,766,435 raw reads were obtained from the B, T, and G stages, respectively. After filtering out the low-quality reads and removing adaptor sequences, 10,910,911 (70.68%), 15,463,079 (95.88%), and 17,184,005 (88.28%) clean reads were retrieved from the B, T, and G, respectively. The length distribution of sRNA read in Figure 2 shows the mode read length of 22 nucleotides. After aligning with small RNA sequence in the Rfam database, rRNA, tRNA, snRNA, and snoRNA were annotated and further analyzed in the subsequent step (Figure 3).

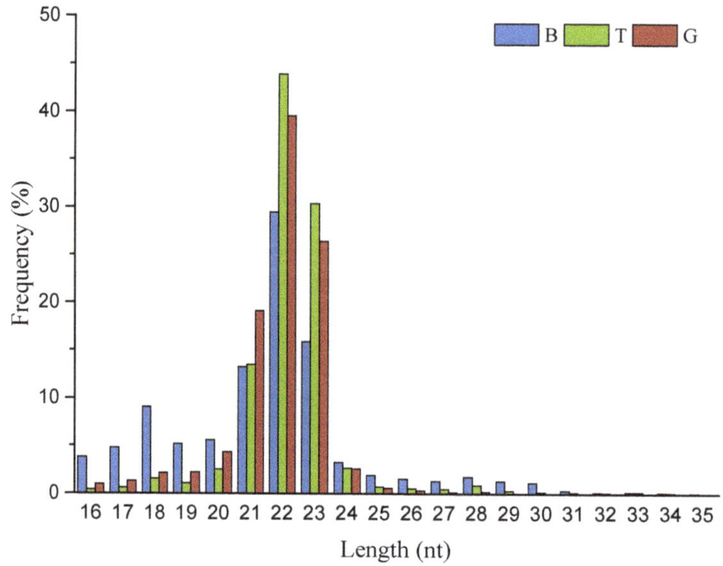

Figure 2. Length distribution and abundance of small RNAs in the three distinct stages of color transformation in *Amphilophus citrinellus* scale. B: the black phase, T: the transitional phase; G: the golden phase.

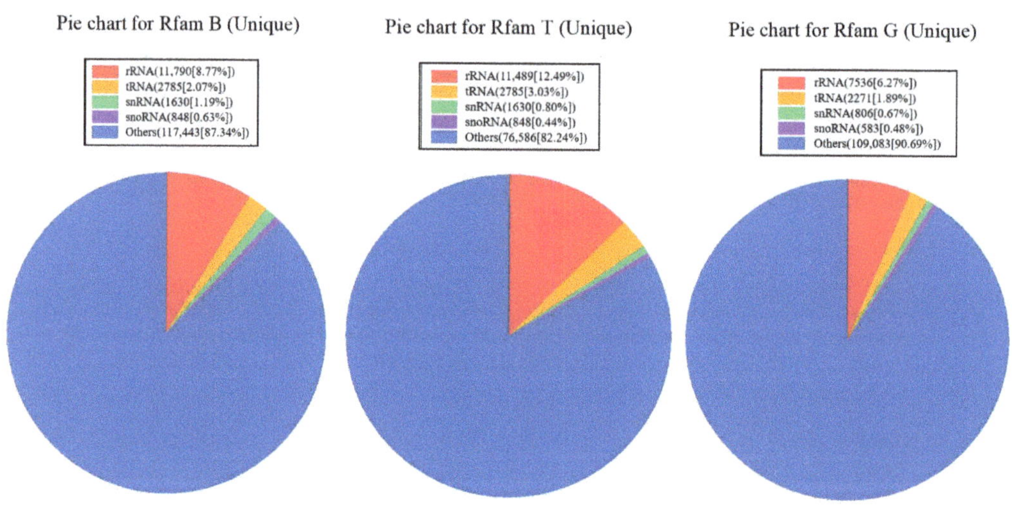

Figure 3. Unique Small RNA reads were BLAST searched against the Rfam database of non-coding RNAs to annotate rRNA, tRNA, snoRNA, snRNA, and other RNAs.

Potential rRNA, scRNA, snoRNA, snRNA, and tRNA in the samples were found and removed as much as possible by comparison with the Rfam database (Figure 3). The remaining sequences were compared with the miRBase database. Finally, 345, 281, and 362 known miRNAs were identified in the B, T, and G phases, respectively. Overall, 245 miRNAs were expressed in all three phases, while 55, 10, and 76 miRNAs were expressed specifically in the B, T, and G phases, respectively. These results suggest that some miRNAs expressed only at specific developmental stages play an important role in the regulation of post-transcriptional expression levels. The unannotated sequence revealed

279, 441, and 402 novel miRNAs using secondary structure prediction during the B, T, and G stages, respectively. Table 2 lists the 10 miRNAs with the highest expression levels in the three different periods.

Table 2. Highly expressed known miRNAs in the black (B), transitional (T), and golden (G) stages, three distinct stages of color transformation in *Amphilophus citrinellus* scale.

miRNA	B	miRNA	T	miRNA	G
miR-26-x	159,339	miR-199-x	130,621	miR-199-x	120,493
miR-199-x	127,301	miR-26-x	121,902	miR-100-x	85,094
miR-181-x	94,227	miR-181-x	83,262	miR-26-x	79,633
miR-205-x	82,376	miR-21-x	74,605	miR-181-x	61,932
miR-22-y	47,845	miR-10-x	68,839	let-7-x	57,923
let-7-x	46,787	miR-100-x	64,030	miR-205-x	55,605
miR-21-x	45,663	miR205-x	62,212	miR-21-x	55,018
miR-200-y	38,329	let-7-x	46,821	miR-146-x	49,708
miR-203-y	32,515	miR-146-x	43,975	miR-10-x	44,171
miR-199-y	30,214	miR-22-y	34,915	miR-200-y	39,404

3.2. Analysis of miRNA Expression Trends and Differences

Compared with group B, 53 miRNAs were significantly differentially expressed in group G, including 8 miRNAs up-regulated and 45 miRNAs down-regulated, while 88 miRNAs were significantly differentially expressed in group T, including 9 miRNAs up-regulated and 79 miRNAs down-regulated (Figure 4). Among them, miR-133-y and miR-133-x in groups T and G were significantly down-regulated compared with group B (>13-fold change), and the expression of novel-m0045-5p was significantly down-regulated, while the expression of novel-m0091-3p and miR-1582-y was significantly up-regulated only in group T (Table 2).

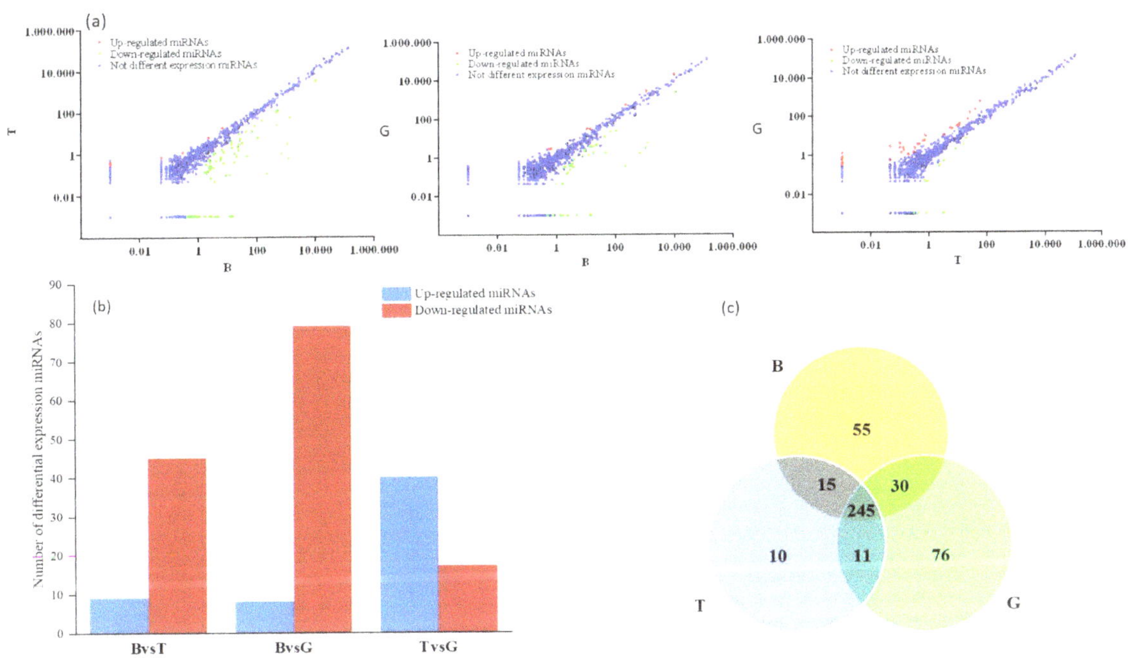

Figure 4. Differentially expressed miRNAs in three distinct stages of color transformation (B, T, and G). (**a**) Scatter plots of three pairwise comparisons. Red, green, and blue show the expression of

miRNAs up-regulated, down-regulated, and non-significant, respectively. (**b**) Red and blue indicate numbers of significantly up- and down-regulated miRNAs, respectively. (**c**) Venn diagram showing the distribution of known differentially expressed miRNAs. B: the black phase; T: the transitional phase; G: the golden phase.

Analysis of differential expression of miRNAs is an important means to study the spatial and temporal expression patterns of miRNAs related to their physiological significance. We randomly selected eight differentially expressed miRNAs that were already known (miR-4492-y, miR-183-x, miR-451-x, miR-1335-y, miR-6937-x, miR-133-x, miR-193-y, and miR-551-y) and two novel miRNAs with differential expression (novel-m0009-5P and novel-m0091-3p) for further study. Their expression patterns were verified using qRT-PCR, and the results showed that the qRT-PCR expression profile was consistent with those from RNA-sequencing (RNA-seq) analysis (Figures 5 and 6), indicating that the RNA-seq analysis results were accurate and reliable.

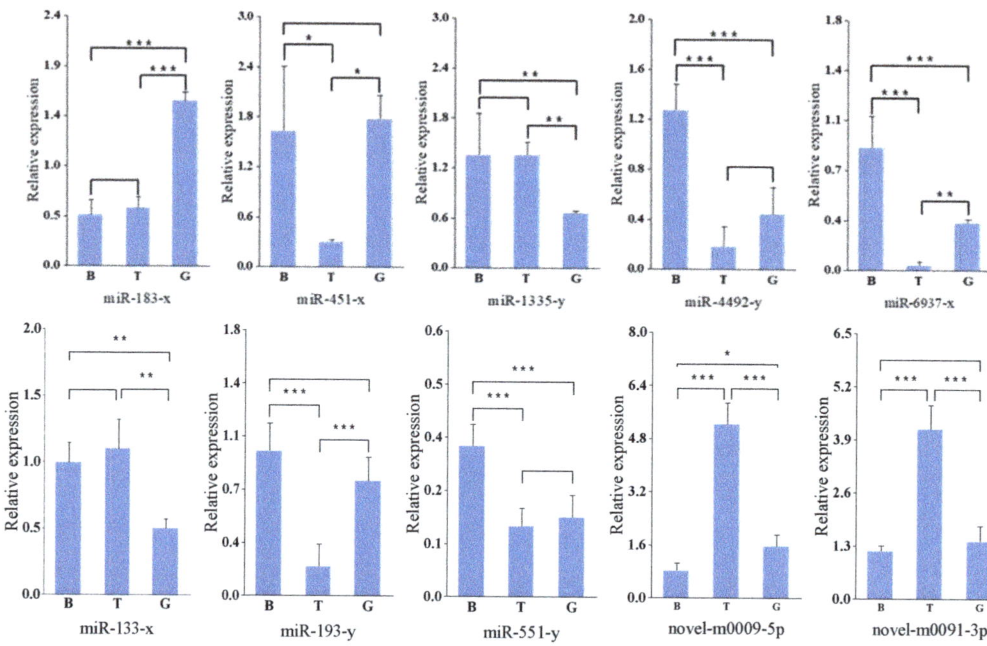

Figure 5. Relative expression of 10 miRNAs in three distinct stages of color transformation (B: the black phase; T: the transitional phase; G: the golden phase) using Stem-loop qRT-PCR. All samples were run in triplicate. Error bars represent standard deviation from the mean (* represents $p < 0.05$, ** represents $p < 0.01$, *** represents $p < 0.001$).

Figure 6. qRT-PCR validation of known and novel miRNAs with expression in three distinct stages of color transformation of *Amphilophus citrinellus* (B, T, and G). The expression of miRNAs was normalized to the abundance of U6. Each column represents the mean ± standard error (n = 3 each). B: the black phase; T: the transitional phase; G: the golden phase.

3.3. Target Prediction and Enrichment Analysis for miRNAs

RNAhybrid, miRanda, and Targetscan algorithms were used to predict target genes for the 1195 miRNAs identified, and 66,685 potential target genes were found (Figure 7). To predict the metabolic pathways of target genes of differentially expressed miRNAs in body-color fading of Midas cichlids, the predicted target genes were further annotated using GO enrichment and KEGG pathway analysis. GO enrichment analysis showed that 36,797 (B vs. G), 42,528 (B vs. T), and 38,133 (T vs. G) target genes were grouped into 20, 18, and 11 subclasses of biological processes, cellular components, and molecular functions, respectively (Figure 8).

KEGG pathway analysis showed annotations for 264, 279, and 278 pathways in B vs. G, B vs. T, and T vs. G groups, respectively, among which the melanin production pathway, mTOR signaling pathway, and Wnt (wingless-type MMTV integration site family) signaling pathway were related to body-color formation (Figure 9).

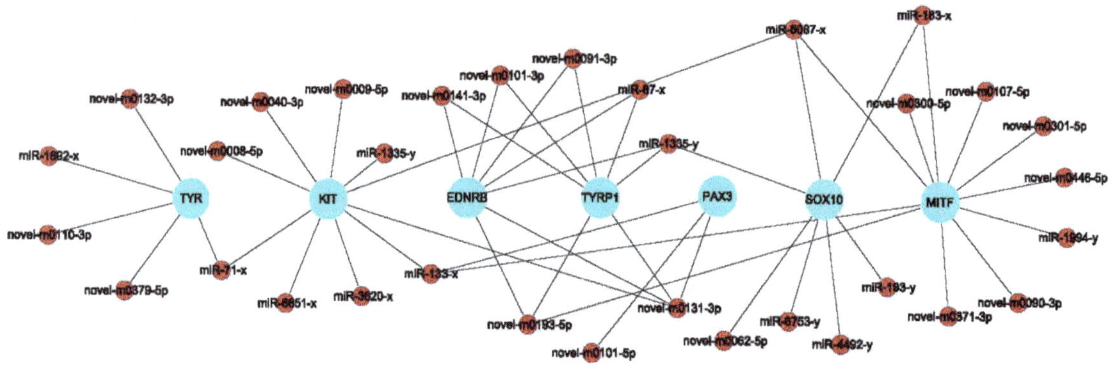

Figure 7. Network of putative interactions between miRNAs and their target genes related to the pigmentation process. The regulation network of miRNAs and their target genes was illustrated using Cytoscape. Red and blue circles represent miRNAs and target genes, respectively.

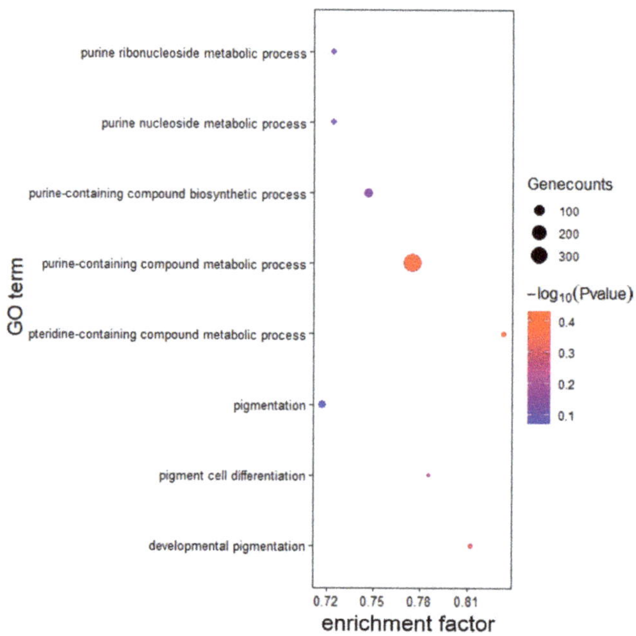

(**a**) B vs. T.

Figure 8. *Cont.*

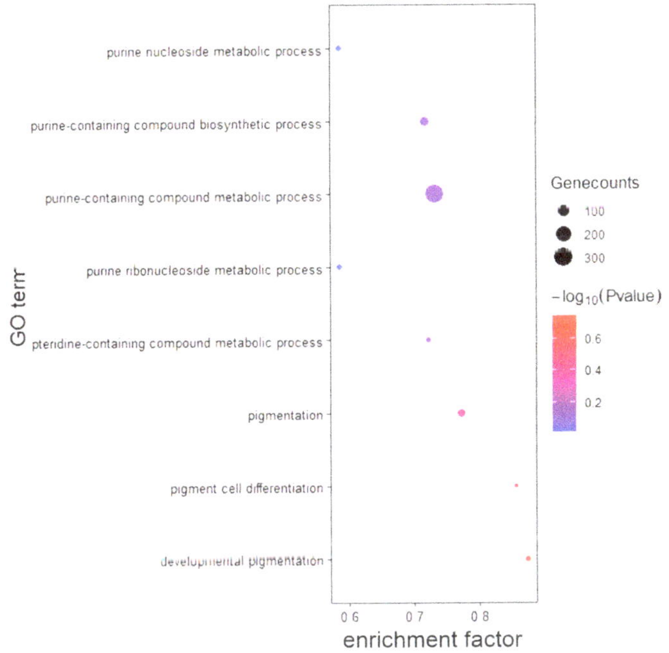

(**b**) B vs. G.

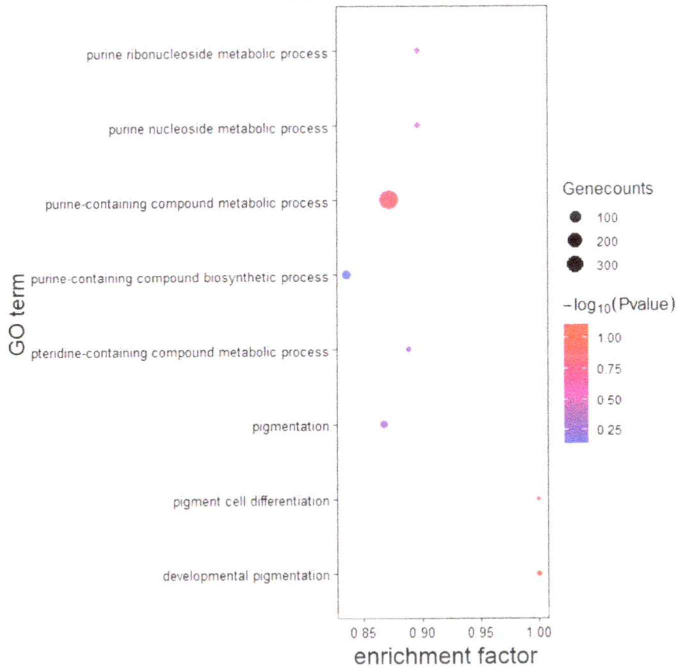

(**c**) T vs. G.

Figure 8. Gene Ontology (GO) classification of differential expression genes in B vs. G, B vs. T, and T vs. G groups. (**a**) B vs. T, (**b**) B vs. G, and (**c**) T vs. G.

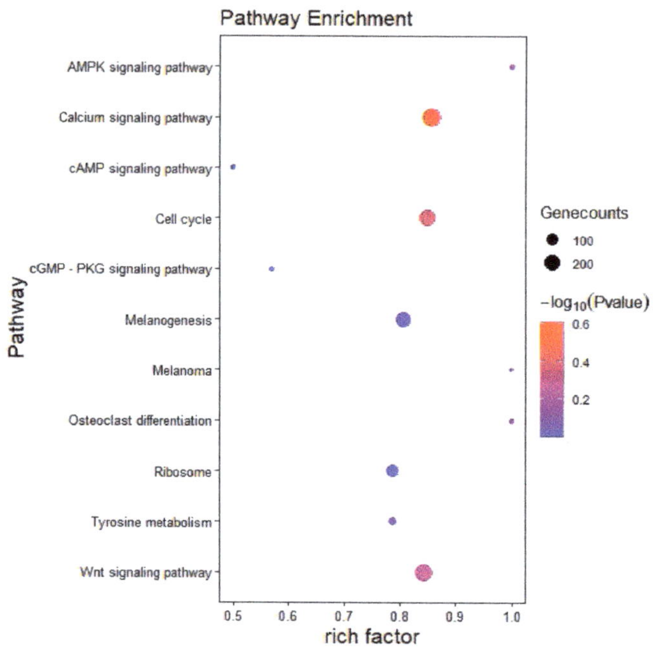

(**a**). B vs. T.

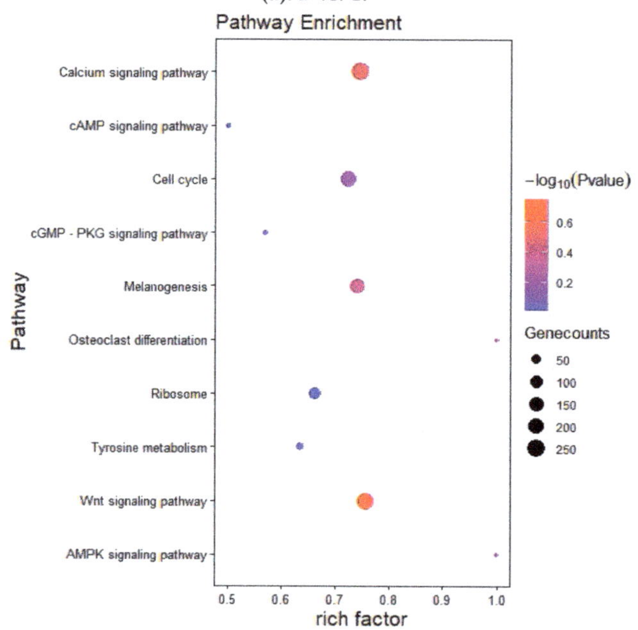

(**b**). B vs. G.

Figure 9. *Cont.*

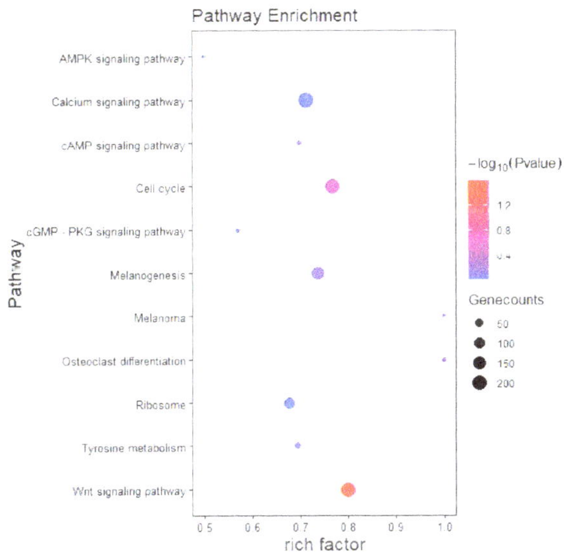

(**c**). T vs. G.

Figure 9. KEGG pathways-enriched analysis for target genes of differentially expressed miRNAs in B vs. T, B vs. G, and T vs. G groups. (**a**) B vs. T, (**b**) B vs. G, and (**c**) T vs. G.

4. Discussion

Studies on the expression and potential functions of miRNA in evolutionarily diverse aquatic species are limited. The Midas cichlid is a popular ornamental fish with a typical body color fading phenomenon at the early stage of development [19]. It is an ideal biological model for studying the mechanism of early body color determination in fishes [32]. We explored the miRNA transcriptome in Midas cichlids scales using deep sequencing technology, and provided a foundation for functional studies on the relationship between miRNA expression and fish skin pigment development.

Using this sequencing approach, we obtained a large number of clean reads. The model sized 22 nt was in line with classical size of Dicer-cleaved products in vertebrates, which was consistent with the majority and is similar to other teleosts, such as hybrid snakehead (*Channa maculate* ♀× *C. argus* ♂) [38], common carp (*C. carpio*) [39], and largemouth bass (*Micropterus salmoides*) [40].

The 10 miRNAs with the highest expression levels in the three different periods were listed in Table 1. Among these, MiR-199-x, miR-181, and miR-26a, which are related to immunity and disease regulation, showed high expression levels in all three periods without significant difference ($p > 0.05$). The miR-199a-5p can directly regulate the expression of Indian hedgehog (IHH) and reduce chondrocyte hypertrophy and matrix degradation via the IHH signal pathway in primary human chondrocytes [41], and is also involved in the regulation of melanoma cell metastasis-related genes [42]. The miR-181a/b inhibits expression of genes involved in synaptic transmission, neurite outgrowth, and mitochondrial respiration [43]. The miR-26a is regulated in various malignant tumors and may be involved in the genesis and development of tumors [44]. In addition, the high expression of miR-200, miR-199, and miR-25 in the three periods were all associated with body-color regulation in previous reports. Yi et al. [45] reported that the miR-200 family is expressed in the epidermis of mice, while the miR-199 family is greatly expressed in hair follicles, demonstrating the tissue expression specificity of miRNAs during skin development. MiR-25, ad a key regulatory miRNA of *mitf*, participates in the body-color differentiation of white and brown alpaca and plays a key role in melanocyte development, survival, and differentiation [46]. Due to the evolutionary conservation of miRNA, we infer that miR-200, miR-199, and miR-25 are also involved in regulating skin pigmentation during body-color fading in Midas cichlid.

Target gene prediction analysis revealed that the target genes corresponding to miR-183-x were SRY box 10 (*sox10*) and *mitf*, while the target genes corresponding to miR-133-x included *mitf*, *kit*, and paired-box 3 (*pax3*). Furthermore, *sox10* was found to be a common target gene of miR-183-x, miR-4492-y, and miR-193-y (Figure 7), whereas *mitf* was a common target gene of miR-133-x and miR-183-x (Figure 7), thus reflecting the complexity of the regulatory network between miRNAs and their target genes. The miR-1 has been reported to co-regulates the target gene histone deacetylase 4 (HDAC4) with miR-133a, which is involved in muscle development regulation [47]. MiR-183 plays a regulatory role in many human cancers, including colorectal, melanoma, prostate, and breast cancer, as well as osteoporosis [48–52]. However, in this study, many target genes corresponding to miR-183-x and miR-133, such as *mitf*, *tyr*, *tyrp1*, agouti-signaling protein (*asip*), melanocortin 1 receptor (*mc1r*), transcription factor *sox10*, and *pax3* are all directly or indirectly involved in the regulation of melanin synthesis [53–57]. Henning et al. [32] conducted a transcriptome analysis of Midas cichlids in three distinct stages of color transformation and identified the differentially expressed genes related to melanosome composition and differentiation, such as *tyr*, *tyrp1*, and *slc24a5*. These genes were down-regulated in the melanin synthesis pathway during body-color transformation. Previous studies have shown that the *mitf*, *mc1r*, and *tyr* showed significant expression during the morphological color change of *Amphilophus citrinellus* [3,5,56]. We suggest that miR-133-x and miR-183-x may be involved in the regulation of dark pigment cells during ontogeny. Previous studies have reported that miR-196a and miR-206 play a regulatory role in koi skin pigmentation by targeting the *mitf* and *mc1r* genes, respectively [27,58]. Yan et al. [59] found that miR-429 targets *foxd3* silencing in the common carp, thereby affecting the expression of *mitf* and its downstream genes including *tyr*, *tyrp1*, and *tyrp2* to regulate skin pigmentation. However, no similar targeting relationship was found in Midas cichlids compared to koi carp, suggesting potential species differences in the miRNAs regulating body color.

It has been reported that the gene *goldentouch* harboring the transposon insertion determines the gold polymorphism and the 8.2 kb insertion located within an intron of *goldentouch* determines the dark/gold polymorphism; however, *goldentouch* expression does not differently change during the ontogeny of gold Midas cichlids [33]. Therefore, it is necessary to investigate the relationship between miRNAs and the gene *goldentouch*. To screen for the target miRNAs, all identified miRNAs were mapped to the *goldentouch* 3′ UTR (we selected 2 kb sequence after the goldentouch coding sequence from genome as its 3′ UTR). A total of ten most like miRNAs were predicted; however, these miRNAs were not shown to be significantly differently expressed during ontogeny, which is consistent with *goldentouch* gene expression (Figure 5). We suggest that differently expressed miRNAs are mainly involved in melanin synthesis regulation during the ontogeny of gold Midas cichlids by regulating pigmentation-related target genes.

Although the target-gene function of B vs. G, B vs. T, and T vs. G groups have some divergence in the GO-enriched functional classification, the cellular processes subcategory has the largest number of genes in biological processes among the three groups. In cell components, the number of genes in cell and cell parts projected the largest number of genes, while in molecular functions, the largest number of genes was the subclass binding. In addition, there were eight Go terms related to body-color formation, including pigmentation, pigment cell differentiation, and developmental pigmentation process (Figure 8). KEGG pathway analysis can help to understand the interaction between target genes in specific biological functions and suggest the systematic behavior of organisms through genome or transcriptomic contents [60]. In KEGG enrichment analysis, the three groups of target genes were enriched in pathways related to body-color retention, such as the melanin production pathway, Wnt signaling pathway, etc. *pax3* and *sox10* in the Wnt signaling pathway regulate *tyr* by regulating the expression of *mitf*, thus affecting pigment formation [54]. In human melanocytes, *pax3* and *sox10* jointly induce melanocyte differentiation and melanin production [61]. These studies suggest that some differentially expressed miRNAs may be involved in skin pigmentation dynamic during the body color fading process of Midas cichlids by regulating pigmentation-related pathways through target genes.

5. Conclusions

In the present study, the miRNA libraries of Midas cichlid scales at three distinct stages of body-color transformation (black, transition, and gold periods, represented as B, T, and G) were constructed using high-throughput sequencing analysis. Overall, 345, 281, and 362 known miRNAs and 279, 441, and 402 novel miRNAs were obtained from the B, T, and G phase groups, respectively. In addition, differentially expressed miRNAs in the three groups (B vs. G, B vs. T, and T vs. G) were analyzed, and 10 miRNAs were verified using qRT-PCR, validating the reliability of sequencing results. The differential expression of miRNAs in different periods also preliminarily presented the time-specific expression pattern of miRNAs. Subsequently, prediction of target genes of differentially expressed miRNAs and enrichment analysis of GO and KEGG pathways further clarified that miRNAs may be involved in skin pigmentation regulation during morphological color change by regulating expression of target genes. For example, miR-133-x is likely to regulate several key genes related to body color, such as *mitf*, *kit*, and *pax3*, and miR-183-x may regulate *mitf* and *sox10*, which needs further verification. Overall, the result provide important information resources for the miRNA transcriptome of three distinct color stages of the Midas cichlids scale, which can help in further studies on the miRNA regulation mechanism of morphological color change.

Author Contributions: G.W.: investigation, visualization, writing—original draft; H.S.: supervision, conceptualization, methodology, writing—review and editing; Y.L.: writing—review and editing; X.M.: supervision; C.L.: supervision, X.W.; resources, Y.Y.; data curation. All authors have read and agreed to the published version of the manuscript.

Funding: This research was supported by the National Freshwater Genetic Resource Center (no. FGRC18537), the Guangzhou Science and Technology Planning Project (202002030047), the Basic and Applied Basic Research Foundation of Guangdong Province (2020A1515010304), and the National Natural Science Foundation of China (802037).

Institutional Review Board Statement: All fish experiments in the present study were approved by the Pearl River Fisheries Research Institute and the Chinese Academy of Fishery Sciences under contract, and the experimental process complied with protocols of international guidelines for the ethical use of animals in research (Approval Code: LAEC-PRFRI-2023-05-04; Approval Date: 22 May 2023).

Informed Consent Statement: Not applicable.

Data Availability Statement: All data generated or analyzed during this study are included in this published article. The datasets generated for this study can be found in the CNCB-NGDS Genome Sequence Archive (GSA) database under GSA accession number CRA006100 (https://bigd.big.ac.cn/gsa/browse/CRA006100, accessed on 11 November 2022). The reference genome for this study is the *Amphilophus citrinellus* genome (https://www.ncbi.nlm.nih.gov/assembly/GCA_000751415.1, accessed on 11 February 2023).

Conflicts of Interest: The authors declare that they have no known competing financial interests or personal relationships that could have influenced the work reported in this paper.

References

1. Zhang, G.; Tang, L.; Huang, J.; Wang, Y.; Wang, H.; Fan, Y.; Yuan, X.; Liu, W.; Peng, L.; Liu, J.; et al. Formation of asymmetric body color in the caudal fin of common carp (*Cyprinus carpio*). *Aquaculture* **2023**, *577*, 739970. [CrossRef]
2. Marcoli, R.; Jones, D.B.; Massault, C.; Marc, A.F.; Moran, M.; Harrison, P.J.; Cate, H.S.; Lopata, A.L.; Jerry, D.R. The skin structure in multiple color variants of barramundi (*Lates calcarifer*): A histological, immunohistochemical and ultrastructural overview. *Aquaculture* **2023**, *576*, 739859. [CrossRef]

3. Chen, X.; Wu, G.; Song, H.; Wang, X.; Mou, X.; Liu, Y.; Liu, C.; Hu, Y. Expression analysis of *mitf* gene relating to body color variation in *Amphilophus citrinellus*. *Prog. Fish. Sci.* **2021**, *42*, 107–118.
4. Ahi, E.P.; Lecaudey, L.A.; Ziegelbecker, A.; Steiner, O.; Glabonjat, R.; Goessler, W.; Hois, V.; Wagner, C.; Lass, A.; Sefc, K.M. Comparative transcriptomics reveals candidate carotenoid color genes in an East African cichlid fish. *BMC Genom.* **2020**, *21*, 54. [CrossRef]
5. Jiang, Y.; Song, H.; Liu, Y.; Wei, M.; Wang, X.; Hu, Y.; Luo, J. Cloning and expression analysis of the developing sequence and tissue expression of TYR gene in *Amphilophus citrinellus*. *J. Agric. Biotechnol.* **2016**, *24*, 697–707.
6. Andrade, P.; Pinho, C.; Pérez I De Lanuza, G.; Afonso, S.; Brejcha, J.; Rubin, C.; Wallerman, O.; Pereira, P.; Sabatino, S.J.; Bellati, A.; et al. Regulatory changes in pterin and carotenoid genes underlie balanced color polymorphisms in the wall lizard. *Proc. Natl. Acad. Sci. USA* **2019**, *116*, 5633–5642. [CrossRef] [PubMed]
7. Inaba, M.; Yamanaka, H.; Kondo, S. Pigment pattern formation by contact-dependent depolarization. *Science* **2012**, *335*, 677. [CrossRef] [PubMed]
8. Volkening, A.; Sandstede, B. Iridophores as a source of robustness in zebrafish stripes and variability in *Danio* patterns. *NAT Commun.* **2018**, *9*, 3231. [CrossRef]
9. Huang, Y.; Luo, Y.; Liu, J.; Gui, S.; Wang, M.; Liu, W.; Peng, L.; Xiao, Y. A light-colored region of caudal fin: A niche of melanocyte progenitors in crucian carp (*Cyprinus carpio* L.). *Cell Biol. Int.* **2017**, *41*, 42–50. [CrossRef]
10. Cal, L.; Suarez-Bregua, P.; Cerdá-Reverter, J.M.; Braasch, I.; Rotllant, J. Fish pigmentation and the melanocortin system. *Comp. Biochem. Physiol. Part A Mol. Integr. Physiol.* **2017**, *211*, 26–33. [CrossRef]
11. Kautt, A.F.; Kratochwil, C.F.; Nater, A.; Machado-Schiaffino, G.; Olave, M.; Henning, F.; Torres-Dowdall, J.; Härer, A.; Hulsey, C.D.; Franchini, P.; et al. Contrasting signatures of genomic divergence during sympatric speciation. *Nature* **2020**, *588*, 106–111. [CrossRef] [PubMed]
12. Huang, Z.; Ma, B.; Guo, X.; Wang, H.; Ma, A.; Sun, Z.; Wang, Q. Comparative transcriptome analysis of the molecular mechanism underlying the golden red colour in mutant Taiwanese loach. *Aquaculture* **2021**, *543*, 736979. [CrossRef]
13. Chen, Y.; Gong, Q.; Lai, J.; Song, M.; Liu, Y.; Wu, Y.; Ai, J.; Long, Z. Transcriptome analysis identifies candidate genes associated with skin color variation in *Triplophysa siluroides*. *Comp. Biochem. Physiol. Part D Genom. Proteom.* **2020**, *35*, 100682. [CrossRef] [PubMed]
14. Tang, S.; Janpoom, S.; Prasertlux, S.; Rongmung, P.; Ratdee, O.; Zhang, W.; Khamnamtong, B.; Klinbunga, S. Transcriptome comparison for identification of pigmentation-related genes in different color varieties of Siamese fighting fish *Betta splendens*. *Comp. Biochem. Physiol. Part. D Genom. Proteom.* **2022**, *43*, 101014. [CrossRef]
15. Rani, V.; Sengar, R.S. Biogenesis and mechanisms of microRNA-mediated gene regulation. *Biotechnol. Bioeng.* **2022**, *3*, 119. [CrossRef] [PubMed]
16. Gong, W.; Huang, Y.; Xie, J.; Wang, G.; Yu, D.; Sun, X. Genome-wide identification and characterization of conserved and novel microRNAs in grass carp (*Ctenopharyngodon idella*) by deep sequencing. *Comput. Biol. Chem.* **2017**, *68*, 92–100. [CrossRef] [PubMed]
17. He, L.; Zhang, A.; Chu, P.; Li, Y.; Huang, R.; Liao, L.; Zhu, Z.; Wang, Y. Deep Illumina sequencing reveals conserved and novel microRNAs in grass carp in response to grass carp reovirus infection. *BMC Genom.* **2017**, *18*, 195. [CrossRef] [PubMed]
18. Gong, W.; Huang, Y.; Xie, J.; Wang, G.; Yu, D.; Sun, X.; Zhang, K.; Li, Z.; Ermeng, Y.; Tian, J.; et al. Identification and expression analysis of miRNA in hybrid snakehead by deep sequencing approach and their targets prediction. *Genomics* **2019**, *111*, 1315–1324. [CrossRef] [PubMed]
19. Franchini, P.; Xiong, P.; Fruciano, C.; Meyer, A. The role of microRNAs in the repeated parallel diversification of lineages of Midas cichlid fish from Nicaragua. *Genome Biol. Evol.* **2016**, *8*, 1543–1555. [CrossRef]
20. Franchini, P.; Xiong, P.; Fruciano, C.; Schneider, R.F.; Woltering, J.M.; Hulsey, C.D.; Meyer, A. MicroRNA gene regulation in extremely young and parallel adaptive radiations of crater lake cichlid fish. *Mol. Biol. Evol.* **2019**, *36*, 2498–2511. [CrossRef]
21. Wu, S.; Huang, J.; Li, Y.; Zhao, L. Involvement of miR-495 in the skin pigmentation of rainbow trout (*Oncorhynchus mykiss*) through the regulation of *mc1r*. *Int. J. Biol. Macromol.* **2024**, *254*, 127638. [CrossRef] [PubMed]
22. Wang, L.; Song, F.; Yin, H.; Zhu, W.; Fu, J.; Dong, Z.; Xu, P. Comparative microRNAs expression profiles analysis during embryonic development of common carp, *Cyprinus carpio*. *Comp. Biochem. Physiol. Part D Genom. Proteom.* **2021**, *37*, 100754. [CrossRef] [PubMed]
23. Wang, P.; Zeng, D.; Xiong, G.; Zhou, X.; Wang, X. Integrated analysis of mRNA-seq and microRNA-seq depicts the potential roles of miRNA-mRNA networks in pigmentation of Chinese soft-shelled turtle (*Pelodiscus sinensis*). *Aquac. Rep.* **2021**, *20*, 100686. [CrossRef]
24. Guo, Y.; Wu, W.; Yang, X. Coordinated microRNA/mRNA Expression Profiles Reveal Unique Skin Color Regulatory Mechanisms in Chinese Giant Salamander (*Andrias davidianus*). *Animals* **2023**, *13*, 1181. [CrossRef] [PubMed]
25. Botchkareva, N.V. MicroRNA/mRNA regulatory networks in the control of skin development and regeneration. *Cell Cycle* **2012**, *11*, 468–474. [CrossRef]
26. Zhang, J.; Liu, Y.; Zhu, Z.; Yang, S.; Ji, K.; Hu, S.; Liu, X.; Yao, J.; Fan, R.; Dong, C. Role of microRNA508-3p in melanogenesis by targeting microphthalmia transcription factor in melanocytes of alpaca. *Animal* **2017**, *11*, 236–243. [CrossRef] [PubMed]
27. Dong, Z.; Luo, M.; Wang, L.; Yin, H.; Zhu, W.; Fu, J. MicroRNA-206 Regulation of Skin Pigmentation in Koi Carp (*Cyprinus carpio* L.). *Front. Genet.* **2020**, *11*, 47. [CrossRef]

28. Jiang, S.; Yu, X.; Dong, C. MiR-137 affects melanin synthesis in mouse melanocyte by repressing the expression of c-Kit and Tyrp2 in SCF/c-Kit signaling pathway. *J. Agric. Chem. Soc. Jpn.* **2016**, *80*, 2115–2121.
29. Zhu, Z.; Cai, Y.; Li, Y.; Li, H.; Zhang, L.; Xu, D.; Yu, X.; Li, P.; Lv, L. miR-148a-3p inhibits alpaca melanocyte pigmentation by targeting MITF. *Small Rumin. Res.* **2019**, *177*, 44–49. [CrossRef]
30. Wei, M.; Song, H.; Qi, B.; Liu, C.; Luo, J.; Hu, Y. Pigment cells development and body color variation of postembryonic development in *Amphilophus citrinellus* (Günther 1864). *J. Shanghai Ocean. Univ.* **2015**, *24*, 28–35.
31. Forsman, A.; Polic, D.; Sunde, J.; Betzholtz, P.E.; Franzén, M. Variable colour patterns indicate multidimensional, intraspecific trait variation and ecological generalization in moths. *Ecography* **2020**, *43*, 823–833. [CrossRef]
32. Henning, F.; Jones, J.C.; Franchini, P.; Meyer, A. Transcriptomics of morphological color change in polychromatic Midas cichlids. *BMC Genom.* **2013**, *14*, 171. [CrossRef] [PubMed]
33. Kratochwil, C.F.; Kautt, A.F.; Nater, A.; Harer, A.; Liang, Y.; Henning, F.; Meyer, A. An intronic transposon insertion associates with a trans-species color polymorphism in Midas cichlid fishes. *Nat. Commun.* **2022**, *13*, 296. [CrossRef] [PubMed]
34. Wu, G.; Mu, X.; Song, H.; Liu, Y.; Yang, Y.; Liu, C. Characterization and functional analysis of pax3 in body color transition of polychromatic Midas cichlids (*Amphilophus citrinellus*). *Comp. Biochem. Physiol. Part B: Biochem. Mol. Biol.* **2023**, *263*, 110779. [CrossRef] [PubMed]
35. Luo, J.; Liu, G.; Chen, Z.; Ren, Q.; Yin, H.; Luo, J.; Wang, H. Identification and characterization of microRNAs by deep-sequencing in *Hyalomma anatolicum anatolicum* (Acari: Ixodidae) ticks. *Gene* **2015**, *564*, 125–133. [CrossRef] [PubMed]
36. Betel, D.; Koppal, A.; Agius, P.; Sander, C.; Leslie, C. Comprehensive modeling of microRNA targets predicts functional non-conserved and non-canonical sites. *Genome Biol.* **2010**, *11*, R90. [CrossRef] [PubMed]
37. Agarwal, V.; Bell, G.W.; Nam, J.; Bartel, D.P. Predicting effective microRNA target sites in mammalian mRNAs. *eLife* **2015**, *4*, e05005. [CrossRef] [PubMed]
38. Mao, L.; Zhu, Y.; Yan, J.; Zhang, L.; Zhu, S.; An, L.; Meng, Q.; Zhang, Z.; Wang, X. Full-length transcriptome sequencing analysis reveals differential skin color regulation in snakeheads fish *Channa argus*. *Aquac. Fish.* **2023**. [CrossRef]
39. Tian, X.; Peng, N.; Ma, X.; Wu, L.; Shi, X.; Liu, H.; Song, H.; Wu, Q.; Meng, X.; Li, X. microRNA-430b targets scavenger receptor class B member 1 (*scarb1*) and inhibits coloration and carotenoid synthesis in koi carp (*Cyprinus carpio* L.). *Aquaculture* **2022**, *546*, 737334. [CrossRef]
40. Gong, W.; Huang, Y.; Xie, J.; Wang, G.; Yu, D.; Zhang, K.; Li, Z.; Yu, E.; Tian, J.; Zhu, Y. Identification and comparative analysis of the miRNA expression profiles from four tissues of *Micropterus salmoides* using deep sequencing. *Genomics* **2018**, *110*, 414–422. [CrossRef]
41. Zhang, R. miR-199a-5p Reduces Chondrocyte Hypertrophy and Attenuates Osteoarthritis Progression via the Indian Hedgehog Signal Pathway. *J. Clin. Med.* **2023**, *12*, 1313. [CrossRef] [PubMed]
42. Liang, L.; Zhang, Z.; Qin, X.; Gao, Y.; Zhao, P.; Liu, J.; Zeng, W. Gambogic Acid Inhibits Melanoma through Regulation of miR-199a-3p/ZEB1 Signalling. *Basic Clin. Pharmacol.* **2018**, *123*, 692–703. [CrossRef] [PubMed]
43. Stein, C.S.; McLendon, J.M.; Witmer, N.H.; Boudreau, R.L. Modulation of miR-181 influences dopaminergic neuronal degeneration in a mouse model of Parkinson's disease. *Mol. Ther. Nucleic Acids* **2022**, *28*, 1–15. [CrossRef] [PubMed]
44. Chen, B.; Deng, Y.N.; Wang, X.; Xia, Z.; He, Y.; Zhang, P.; Syed, S.E.; Li, Q.; Liang, S. miR-26a enhances colorectal cancer cell growth by targeting RREB1 deacetylation to activate AKT-mediated glycolysis. *Cancer Lett.* **2021**, *521*, 1–13. [CrossRef] [PubMed]
45. Yi, R.; O'Carroll, D.; Pasolli, H.A.; Zhang, Z.; Dietrich, F.S.; Tarakhovsky, A.; Fuchs, E. Morphogenesis in skin is governed by discrete sets of differentially expressed microRNAs. *Nat. Genet.* **2006**, *38*, 356–362. [CrossRef]
46. Zhu, Z.; He, J.; Jia, X.; Jiang, J.; Bai, R.; Yu, X.; Lv, L.; Fan, R.; He, X.; Geng, J.; et al. MicroRNA-25 functions in regulation of pigmentation by targeting the transcription factor MITF in alpaca (*Lama pacos*) skin melanocytes. *Domest. Anim. Endocrinol.* **2010**, *38*, 200–209. [CrossRef]
47. Zhang, H.; Fu, Y.; Su, Y.; Shi, Z.; Zhang, J. Identification and expression of HDAC4 targeted by miR-1 and miR-133a during early development in *Paralichthys olivaceus*. *Comp. Biochem. Physiol. Part. B Biochem. Mol. Biol.* **2015**, *179*, 1–8. [CrossRef] [PubMed]
48. Ceinos, R.M.; Guillot, R.; Kelsh, R.N.; Cerdá-Reverter, J.M.; Rotllant, J. Pigment patterns in adult fish result from superimposition of two largely independent pigmentation mechanisms. *Pigment Cell Melanoma Res.* **2015**, *28*, 196–209. [CrossRef]
49. Xu, F.; Zhang, H.; Su, Y.; Kong, J.; Yu, H.; Qian, B. Up-regulation of microRNA-183-3p is a potent prognostic marker for lung adenocarcinoma of female non-smokers. *Clin. Transl. Oncol.* **2014**, *16*, 980–985. [CrossRef]
50. Zhang, C.; Gu, H.; Liu, D.; Tong, F.; Wei, H.; Zhou, D.; Fang, J.; Dai, X.; Tian, H. The circ_FAM53B-miR-183-5p-CCDC6 axis modulates the malignant behaviors of papillary thyroid carcinoma cells. *Mol. Cell. Biochem.* **2022**, *477*, 2627–2641. [CrossRef]
51. Kaken, H.; Wang, S.; Zhao, W. P53 Regulates Osteogenic Differentiation Through miR-153-5p/miR-183-5p-X-Linked IAP (XIAP) Signal in Bone Marrow Mesenchymal Stem Cell (BMSC). *J. Biomater. Tissue Eng.* **2022**, *12*, 2427–2431. [CrossRef]
52. Jie, C.; Lin, G.; Jie, N.; Ping, H.; Kai, H.; Ying-Li, S. MiR-183 Regulates ITGB1P Expression and Promotes Invasion of Endometrial Stromal Cells. *Biomed. Res. Int.* **2015**, *2015*, 340218.
53. Zhu, Y.; Li, Q. Mitf involved in shell pigmentation by activating tyrosinase-mediated melanin synthesis in *Pacific oyster* (*Crassostrea gigas*). *Gene* **2024**, *897*, 148086. [CrossRef]
54. Yu, F.; Qu, B.; Lin, D.; Deng, Y.; Huang, R.; Zhong, Z. Pax3 Gene Regulated Melanin Synthesis by Tyrosinase Pathway in *Pteria penguin*. *Int. J. Mol. Sci.* **2018**, *19*, 3700. [CrossRef]

55. Kang, D.; Kim, H. Functional relation of agouti signaling proteins (ASIPs) to pigmentation and color change in the starry flounder, *Platichthys stellatus*. *Comp. Biochem. Physiol. Part A Mol. Integr. Physiol.* **2024**, *291*, 111524. [CrossRef] [PubMed]
56. Zhou, K.; Song, H.; Pan, X.; Liu, Y.; Jiang, Y.; Yang, Y.; Mou, X.; Liu, C.; Hu, Y.; Zhou, S. Expression analysis of *mc1r* gene relating to body color variation in *Amphilophus citrinellus*. *Chin. J. Zool.* **2019**, *54*, 45–56.
57. Motohiro, M.; Hiroyuki, T.; Akiko, S.; Tetsuaki, K.; Ikuko, W.; Hikaru, K.; Yusuke, N.; Kiyoshi, N.; Shin-Ichi, H.; Takashi, S. A gene regulatory network combining *Pax3/7*, *Sox10* and *Mitf* generates diverse pigment cell types in medaka and zebrafish. *Development* **2023**, *19*, dev202114.
58. Yin, H.; Luo, M.; Luo, W.; Wang, L.; Zhu, W.; Fu, J.; Dong, Z. miR-196a regulates the skin pigmentation of koi carp (*Cyprinus carpio* L.) by targeting transcription factormitfa. *Aquac. Res.* **2021**, *52*, 229–236. [CrossRef]
59. Yan, B.; Liu, B.; Zhu, C.D.; Li, K.L.; Yue, L.J.; Zhao, J.L.; Gong, X.L.; Wang, C.H. microRNA regulation of skin pigmentation in fish. *J. Cell Sci.* **2013**, *126*, 3401–3408.
60. Kanehisa, M.; Araki, M.; Goto, S.; Hattori, M.; Hirakawa, M.; Itoh, M.; Katayama, T.; Kawashima, S.; Okuda, S.; Tokimatsu, T.; et al. KEGG for linking genomes to life and the environment. *Nucleic Acids Res.* **2007**, *36*, D480–D484. [CrossRef]
61. Suzuki, N.; Mutai, H.; Miya, F.; Tsunoda, T.; Terashima, H.; Morimoto, N.; Matsunaga, T. A case report of reversible generalized seizures in a patient with Waardenburg syndrome associated with a novel nonsense mutation in the penultimate exon of SOX10. *BMC Pediatr.* **2018**, *18*, 171. [CrossRef] [PubMed]

fishes

Article

Does Size Matter? Small and Large Larvae of Pikeperch (*Sander lucioperca*) in a Comparative Gene Expression Analysis

Katrin Tönißen [1,*], George Philipp Franz [1], Alexander Rebl [2], Philipp Lutze [1,3] and Bianka Grunow [1,*]

[1] Fish Growth Physiology Workgroup, Research Institute for Farm Animal Biology (FBN), Wilhelm-Stahl-Allee 2, 18196 Dummerstorf, Germany; franz@fbn-dummerstorf.de (G.P.F.); philipp.lutze@med.uni-greifswald.de (P.L.)
[2] Fish Genetics Workgroup, Research Institute for Farm Animal Biology (FBN), Wilhelm-Stahl-Allee 2, 18196 Dummerstorf, Germany; rebl@fbn-dummerstorf.de
[3] Institute of Pathophysiology, University Medicine of Greifswald, Martin-Luther-Straße 6, 17489 Greifswald, Germany
* Correspondence: toenissen@fbn-dummerstorf.de (K.T.); grunow@fbn-dummerstorf.de (B.G.)

Abstract: Size differences are common in the aquaculture of fishes. In the larviculture of cannibalistic species such as pikeperch, they majorly influence mortality rates and consequently provoke losses in the aquaculture industry. With this study, we aim to reveal molecular differences between small and large pikeperch of the same age using a set of 20 genes associated with essential developmental processes. Hereby, we applied a general study design to early and late larval pikeperch before the onset of piscivory to explore the causes of growth differences in these developmental groups. The analysis of the expression levels showed developmental but not size-related differences in *PGC1A*, *TGFB1*, *MYOD1*, *MRF4*, and the collagens *COL1A1* and *COL1A2*. Furthermore, increased head lengths were found in larger late larvae compared to their smaller conspecifics. While no uniquely size-related expression differences were found, the expression patterns of *PGC1A* in combination with *TGFB1* as regulators of the citric acid cycle indicate a possible influence of mitochondrial energy metabolism. Furthermore, expression differences of *MYOD1* and *MRF4* point out possible temporal advantages of myogenetic processes in the larger late larval group and hypothesise growth advantages of the larger late larvae resulting from various influences, which provide a promising target for future research.

Keywords: fluidigm array; gene expression; ontogeny; larval fishes; Percidae

Key Contribution: This study demonstrates that developmental stage rather than size-dependent factors significantly influence growth in pikeperch larvae. This is supported by molecular differences in genes associated with key developmental processes and provides evidence for potential factors influencing larval growth.

1. Introduction

Fish larval development is a critical aspect of aquaculture, as it directly affects the success of commercial fish farming. Effective management during this stage is essential to ensure the survival and optimal growth of fish larvae, which will influence the profitability of aquaculture companies. Also in pikeperch aquaculture, the husbandry of early-stage larvae is a significant bottleneck in the rearing process [1,2]. One of the main challenges are size differences during the larval stages, which are connected to the cannibalism of larger individuals towards their smaller conspecifics when the animals start hunting independently [3–5]. This influences the overall mortality rate in pikeperch aquaculture and consequently affects its economic success. Previously, differences between cannibalistic and non-cannibalistic pikeperch were considered to result from developmental differences [6,7]. In pikeperch, the first cannibalism was found to occur among individuals attaining 15 mm total length under rearing conditions in aquaculture [8]. In a study by Colchen and colleagues [6], piscivorous pikeperch larvae exhibited a more advanced digestive system, characterised by increased digestive enzyme activity, along with larger heads and tails compared to their non-piscivorous counterparts.

Since size differences are the foremost influence on the occurrence of cannibalism, the understanding of growth phases and ontogenetic processes is crucial to aiding pikeperch larviculture. The growth of pikeperch larvae does not occur steadily during development. A previous study described different phases of growth occurring during the larval development of pikeperch reared in a semi-natural environment [9]. Similarly, a study with a focus on skeletal development demonstrated an intermediate reduction in size increase for individual pikeperch reared in environments with constant temperatures [10]. As a result of these studies, a distinct early larval growth phase with a low to no size increase can be differentiated from a later larval growth phase with increasing size variants [9,10]. Based on the occurrence of characteristic ontogenetic events, these groups can be attributed to the larval stages L1 to L2 (early larval group) as well as L3 to L6 (late larval group) described by Peňáz [11]. The early larvae are characterised by a straight notochord and mixed feeding, whereas in late larvae, the notochord is bent upwards and caudal fin skeletal elements are formed. The appropriate timing for the initial feeding, differentiated by the feed type and the feed quality, is a well-known factor that influences the growth of larval pikeperch [12,13]. However, it is not fully understood how these growth differences still occur in groups that are fed the same type of feed and ad libitum, as exemplarily carried out by Ott and colleagues [10]. This opens the question of whether there are also transcription-level differences that result in variations in the growth of same-age specimens, thus influencing the cannibalism rate of pikeperch in later stages within a batch.

With the present study, we aim to investigate possible differences in the gene expression patterns of same-age larval pikeperch of smaller and larger size. For this purpose, we comparatively investigated the expression patterns of genes relevant for developmental processes in larval pikeperch [14–16]. Hereby, we tried to exclude the influence of piscivory/cannibalism on the expression levels by sampling individuals before the onset of cannibalism based on the age and size of first-time cannibalistic individuals provided by the literature [3,6,8]. By combining size comparisons and gene expression analyses, our approach can contribute to elucidating potential relationships in the genetic regulation of growth differences in aquaculture pikeperch.

2. Materials and Methods

2.1. Sample Collection and Size Sorting

Pikeperch larvae were obtained from a fishery at Lake Hohen Sprenz in Mecklenburg-Vorpommern, Germany, that applies semi-controlled pikeperch reproduction (compare with [9,14]). From a local population of breeders of three females and five males, a mixed clutch nest from one fertilisation date was sampled. Eggs were kept in Zuger jars until the eye-point stage. Afterwards, they were transferred to net cages in flow-through channels. Following the hatch, lake zooplankton, collected with an additional light trap, was provided as a food source (*Daphnia* sp., *Bosmina* sp., copepod nauplii).

During the rearing process, lake water was used, having a variable temperature regime with a mean of 17.1 ± 3.9 °C (min 11.8 °C, max 23.8 °C). Data on water temperature and quality and air temperature were collected continuously (Maxim Integrated iButton MF1921G, Hanna Instruments hi9829). Besides the age in days post fertilisation (dpf), the age in degree-days (dd) was calculated based on the daily average water temperature. Due to the natural temperature variances and different sampling days, the calculation of dd is used to determine the developmental age of the poikilothermic organisms. In accordance with developmental [11] as well as described growth differences during larval development [9], the specimens were separated into an early and late larval group. In total, we collected six larval age stages (early larval: 16 dpf/227 dd, 18 dpf/256 dd, 21 dpf/306 dd; late larval: 31 dpf/512 dd, 35 dpf/600 dd, 38 dpf/666 dd). Representative light microscope images of the pikeperch larvae from the examined early and late larval groups, with their corresponding ages, are shown in Figure 1. All pikeperch individuals used in morphometric measurements and gene expression analysis were euthanised using MS222 (Serva) in a 0.25 g/L concentration. For each individual age stage, specimens were collected and stored in ethanol for measurements and photography, as well as in RNAlater for molecular analyses. All sampling procedures followed national and international animal welfare regulations (Directive 2010/63/EU and Act (§ 4(3) TierSchG)).

Figure 1. Exemplary light microscope images of pikeperch from the two distinguished larval developmental groups. (**a**) Larvae of the early larval group at the start of the exogenous feeding. (**b**) Late larval group specimen after the completed change to exogenous feeding but before the onset of cannibalism. Ages are given in days after fertilisation (dpf) and degree-days (dd). Scale: 1 mm.

For the morphometric measurements, the total length (TL) and head length (HL) were measured for 20 pikeperch of each age stage under a stereomicroscope (Leica SD9). Based on the individual TL in comparison to the overall sampling group mean TL and size range, the specimens were associated with the size class "small" or "large". For this, "large" specimens were collected from the largest third of the size range of each sampling group and "small" specimens from the smallest third. Per size class within a sampling group, two sample pools were created, each with $n = 15$. For gene expression analysis, the separated age stages and size classes of specimens were stored in RNAlater at −80 °C.

2.2. RNA Isolation

For RNA isolation from pikeperch larvae, pools of 15 animals from each group were used. After a mechanical disruption (Precellys Evolution, VWR), RNA was isolated in two steps, following Rebl et al. [17]. The first step consisted of a Trizol-chloroform precipitation, and in the second step, the precipitated total RNA was purified using the RNeasy Micro Kit (Qiagen, Hilden, Germany) and treated with a DNase (Qiagen, Hilden, Germany). The final RNA concentration was determined with the NanoDrop ND-1000 spectrophotometer (Peqlab, Erlangen, Germany) according to the manufacturer's instructions.

2.3. Design of Primer and Fluidigm Multiplex Real-Time PCR

A total of 20 genes of interest (GOI) and two reference genes (Ribosomal Protein L32 and Ribosomal Protein S5) were selected for analysis. The GOI and reference genes were established in previous studies on pikeperch and were selected to gain insight into the early ontogeny of pikeperch larvae, with a focus on general development and myogenesis [14,18].

The primers are based on pikeperch sequences (RefSeq NCBI: GCA_008315115) from the NCBI GenBank database. In the absence of sequence information, sequences of other perciform species, such as *Perca flavaescens* and *Epinephelus coioides*, were used. All primers were checked for homology via BLAST algorithms against the *Sander lucioperca* transcriptome. The polymerase chain reaction (PCR)-generated amplicons were evaluated by gel electrophoresis. Primer efficiency was also evaluated in quantitative PCR (qPCR) approaches at different temperatures in the LightCycler96 instrument (Roche) [14]. A full list of the primers for GOI and their properties is included in Supplementary Table S1. Expression analysis was performed on a 48.48 Gene Expression biochip (Standard BioTools) using the BioMark HD system (Standard BioTools) as described previously [17]. The data collection is based on two biological and two technical replicates per sample taken. Gene expressions within all experimental groups were output as normalised relative quantity (NRQ) using the delta-delta CT method [19].

2.4. Statistical Analyses

To check for the formation of significant size differences between the size classes per sampling, we compared their TL and HL using a Student's test. To allow for the comparison of the influence of the developmental stage or the size class, a two-tailed ANOVA was applied that included the two factors of size class (small and large) as well as the two larval age groups (early larval group and late larval group). To check test prerequisites, the normal distribution was determined by the Shapiro–Wilk test, and the homogeneity of variances was tested using the Levene test. All statistical analyses were done using R (version 4.2.3) with RStudio (version 2023.03.0) with a significance level of $p < 0.05$. Graph plotting was done using GraphPad Prism (version 10.0.2). In the graphs, gene expression data is presented as mean + SEM (standard error of the mean). To highlight the differences between the statistical models, asterisks were used for the Student's test and letters for the ANOVA and the Tukey test.

3. Results

3.1. Morphometry

Before the sorting procedures, all individuals in each early developmental group were similar in total lengths, i.e., 4.69 ± 0.17 mm at 227 dd, 4.20 ± 0.24 mm at 256 dd, and 4.94 ± 0.29 mm at 306 dd (linear regression, slope = 0.004). The late larval stages were characterised by an increase in total length and variance. This is reflected by the average total lengths of 4.98 ± 0.11 mm (512 dd), 7.30 ± 0.25 mm (600 dd), and 8.67 ± 0.29 mm (666 dd) (linear regression, slope = 0.024) (Figure 2).

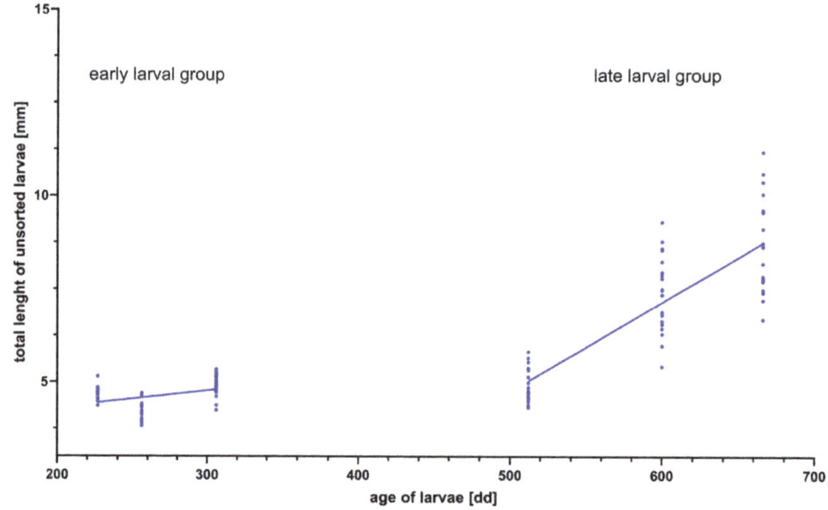

Figure 2. Total length (TL in mm) of pikeperch larval stages prior to the size separations (n = 20 per age stage) starting from 227 dd. Dots indicate the total length of single individuals. A linear regression line has been plotted within both developmental larval groups. The corresponding developmental groups for the expression analysis are added (early larval group: aged 227 to 306 dd, late larval group: aged 512 to 666 dd).

The size separation of the individuals into the groups of small and large specimens resulted in significantly different TL of the size groups within each age stage (Figure 3a,b). The characteristic for the late larval group was the significantly higher values in head length (HL) in the large size class compared to their small conspecifics ($p \leq 0.05$, Figure 3b).

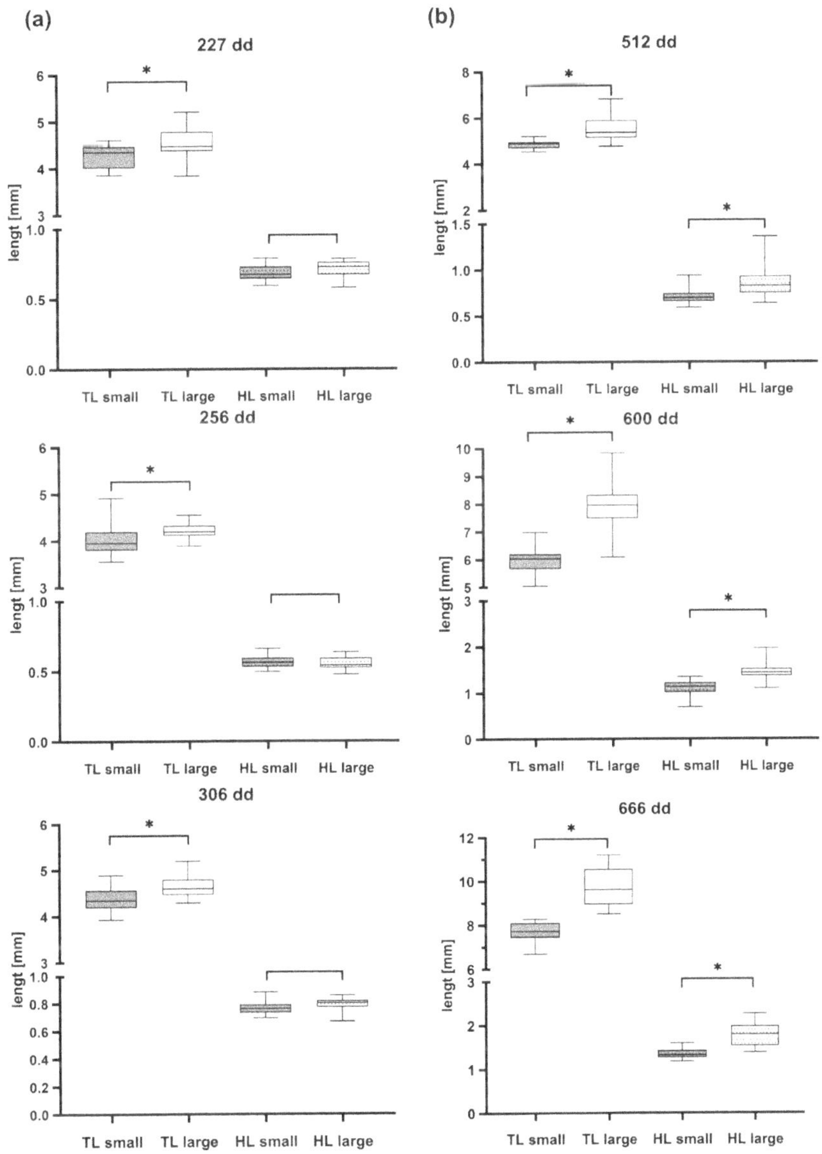

Figure 3. Total length (TL in mm) and head length (HL in mm) of the "small" (grey blots) and "large" (white blots) size class pikeperch larvae of the developmental groups. (**a**) Early larval group (blots without dots), age 227 to 306 dd. (**b**) Late larval group (blots with dots), age 512 to 666 dd. Abbreviations: * $p < 0.05$, using the Student's test.

3.2. Gene Expression Pattern

The variance analysis showed differences between the effects of the early and late larval developmental group but not between the small and large size classes (Table 1).

Overall, the factor developmental group showed significant influence on the expression of *MYOD1* (Myogenic Differentiation 1), *TGFB1* (Transforming growth factor beta1), *FGF4* (Fibroblast growth factor 4), *MRF4* (Myogenic regulatory factor 4), *PGC1A* (peroxisome proliferative activated receptor, gamma, coactivator 1 alpha) and the collagen-encoding genes *COL1A1* (Collagen type 1 alpha 1) and *COL1A2* (Collagen type 1 alpha 2). Furthermore, a combined effect of group and size class was determined for *MYOD1*. The size class alone, however, showed no influence on the expression of any of these genes.

Table 1. Resulting effect influences with their *p*-values of the calculated ANOVA. The effects of the factors developmental group (early larval, late larval), size class (small and large), and their effect interaction are given. *p*-values below 0.05 are highlighted in bold. *COL1A1*—Collagen type 1 alpha 1; *COL1A2*—Collagen type 1 alpha 2; *EN2*—Engrailed 2; *FGF4*—Fibroblast growth factor 2; *FGF6*—Fibroblast growth factor 6; *FGF8*—Fibroblast growth factor 8; *IGF1*—Insulin-like growth factor 1; *IGF2*—Insulin-like growth factor 2; *MEF2A*—Myocyte enhancer factor 2A; *MRF4*—Myogenic regulatory factor 4; *MSTN*—Myostatin; *MSX1*—Msh homeobox 1; *MYF5*—Myogenic factor 5; *MYH6*—Myosin heavy chain 6; *MYOD1*—Myogenic Differentiation 1; *MYOG*—Myogenin; *PAX3*—Paired box 3; *PAX7*—Paired box 7; *PGC1A*—Peroxisome proliferative activated receptor gamma coactivator 1 alpha; *TGFB1*—Transforming growth factor beta1.

Gene Symbol	Developmental Group *p* Value	Size Class *p* Value	Effect Interaction (Group × Size) *p* Value
COL1A1	**0.000**	0.196	0.100
COL1A2	**0.000**	0.065	0.130
EN2	0.195	0.445	0.423
FGF4	**0.032**	0.652	0.143
FGF6	0.274	0.691	0.594
FGF8	0.178	0.353	0.887
IGF1	0.064	0.201	0.237
IGF2	0.454	0.684	0.915
MEF2A	0.197	0.133	0.432
MRF4	**0.001**	0.161	0.353
MSTN	0.295	0.135	0.105
MSX1	0.544	0.254	0.837
MYF5	0.193	0.376	0.386
MYH6	0.939	0.799	0.362
MYOD1	**0.000**	0.977	**0.018**
MYOG	0.054	0.914	0.830
PAX3	0.813	0.901	0.838
PAX7	0.328	0.324	0.101
PGC1A	**0.005**	0.199	0.280
TGFB1	**0.004**	0.888	0.603

No significant differences in the expression levels between size classes within a developmental group were determined. However, in the comparison of the age/size combination groups, significant gene expression differences were found for six genes.

The expression pattern of genes associated with the activation of satellite cells and stem cells had no significantly different expression levels for three of the four genes analysed (Figure 4). *PAX3* (Paired box gene 3) was constantly expressed. A slightly elevated expression level of *PAX7* (Paired box gene 7) was found for the large early larval group. *MSX1* (Msh homeobox) had moderately higher levels of expression in the small larvae of the early and late developmental groups. In contrast, *PGC1A* had significantly higher expression levels in both early larval groups compared to the lowest expression in the large larval group.

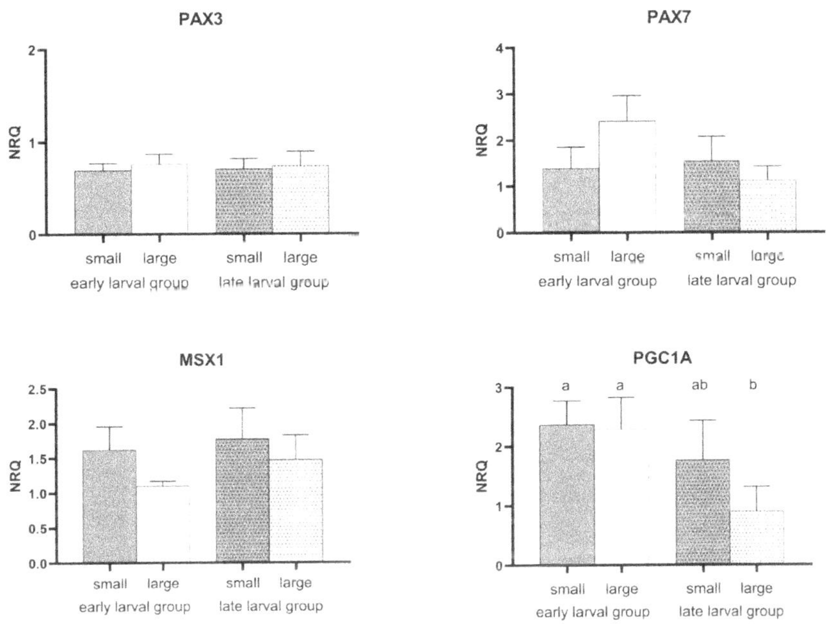

Figure 4. Expression patterns of selected genes during satellite cell and stem cell activation. Normalised relative quantity (NRQ) of pikeperch larvae of the early larval (blots without dots) and late larval groups (blots with dots) in the two size classes small (grey blots) and large (white blots) is illustrated by its mean and SEM. Small letters indicate significance $p \leq 0.05$ between age groups using ANOVA and Tukey's test. PAX3—Paired box 3; PAX7—Paired box 7; MSX1—Msh homeobox 1; PGC1A—Peroxisome proliferative activated receptor gamma coactivator 1 alpha.

The expression levels of the insulin-like growth factor-encoding genes *IGF1* and *IGF2*, as well as the fibroblast growth factor-encoding genes *FGF4*, *FGF6*, and *FGF8*, showed no significant differences across the samples (Figure 5). The expression of *TGFB1* (Transforming growth factor beta 1) was significantly increased between the early large larval and late large larval samples (Figure 5).

Two of the six genes coding for myogenic regulator factors and the muscle development regulators were significantly differentially expressed between the samples (Figure 6). The gene *MYOD1* had a significantly lower expression in the early larval group compared to the late larval group. Furthermore, in the early large larval stage, *MYOD1* is significantly less expressed than in the late large larval group. These differences in the expression pattern between small and large larvae could not be found for *MYF5* (Myogenic factor 5), which also initiates myoblast formation. However, the transcription factor *MRF4*, which is active later in myofibre formation, also showed a similar pattern to *MYOD1*, having significant lower expressions in both early larval sizes compared to the large late larval group. *MYOG* (Myogenin) had no significant expression-level differences. Nevertheless, a slight increase was observed between the small and large larvae within the developmental groups and between the early larval and late larval groups themselves. Of the genes that influence muscle development, *MEF2A* (Myocyte enhancer factor 2A) was uniformly expressed in all developmental groups and size classes. The muscle growth inhibitor-encoding gene *MSTN* (Myostatin) showed no significant differences in the examined developmental groups. However, there is a trend towards a higher *MSTN* expression in the large, late larvae.

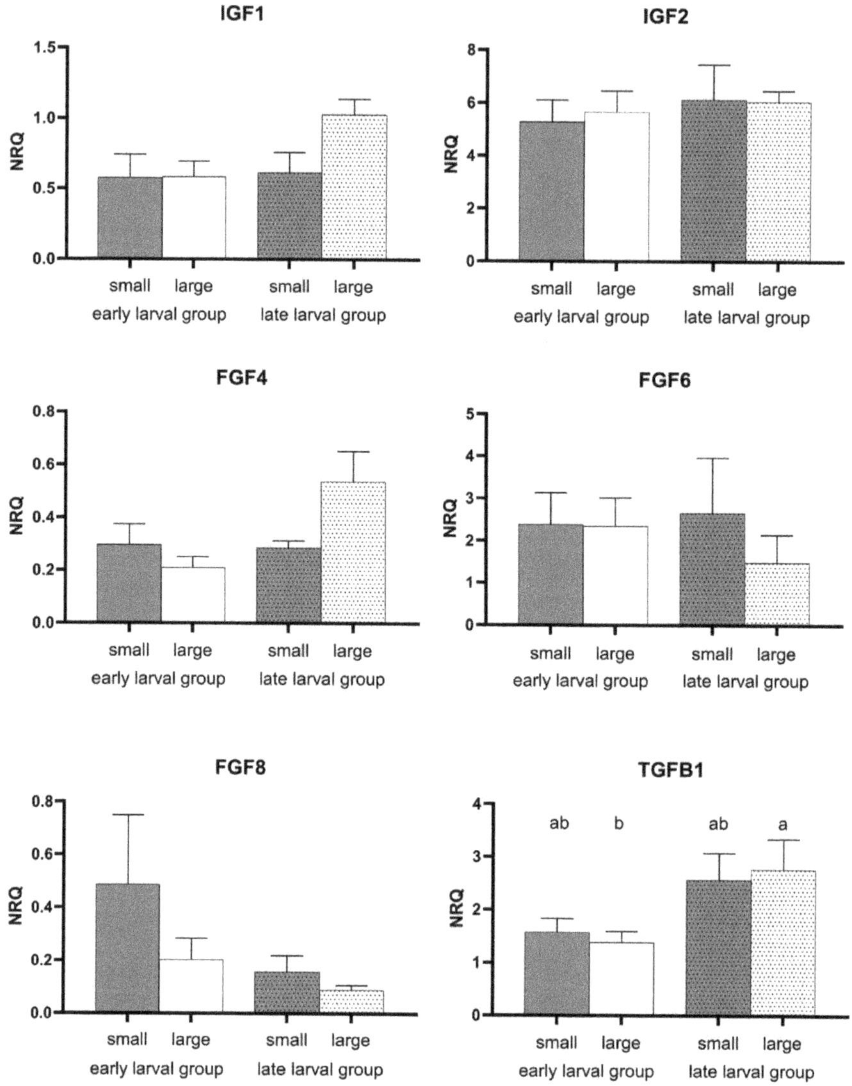

Figure 5. Expression patterns of the regulatory genes of growth and development. Normalised relative quantity (NRQ) of pikeperch larvae of the early larval (blots without dots) and late larval (blots with dots) groups in the two size classes small (grey blots) and large (white blots) is illustrated by its mean and SEM. Small letters indicate significance $p \leq 0.05$ between groups using ANOVA and Tukey's test. IGF1—Insulin-like growth factor 1; IGF2—Insulin-like growth factor 2; FGF4—Fibroblast growth factor 2; FGF6—Fibroblast growth factor 6; FGF8—Fibroblast growth factor 8; TGFB1—Transforming growth factor beta1.

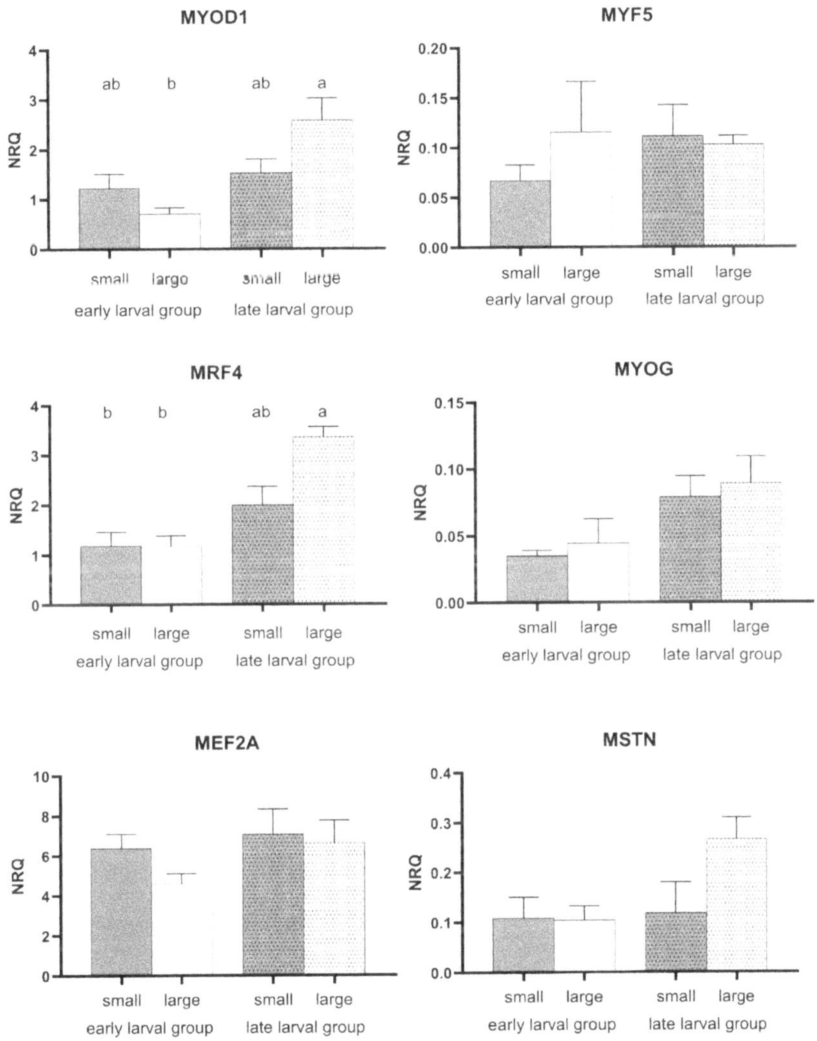

Figure 6. Expression patterns of genes in myogenic development. Normalised relative quantity (NRQ) of pikeperch larvae of the early larval (blots without dots) and late larval (blots with dots) groups in the two size classes small (grey blots) and large (white blots) is illustrated by its mean and SEM. Small letters indicate significance $p \leq 0.05$ between groups using ANOVA and Tukey's test. MYOD1—Myogenic Differentiation 1; MYF5—Myogenic factor 5; MRF4—Myogenic regulatory factor 4; MYOG—Myogenin; MEF2A—Myocyte enhancer factor 2A; MSTN—Myostatin.

The structural marker genes *COL1A1* and *COL1A2* were significantly differently expressed in the larval groups (Figure 7a). *COL1A1* was significantly more expressed in the large late larval samples compared to both early larval sizes. *COL1A2* was significantly more expressed in both late larval samples compared to the early larval sizes. The gene expression of *EN2* (Mandibular arch-muscle specific engrailed 2), responsible for the development of the jaw elements, had no significantly different expression between the samples (Figure 7b). This was also the case for the heart-specific transcription factor *MYH6* (Cardiac-specific myosin heavy chain 6) (Figure 7b).

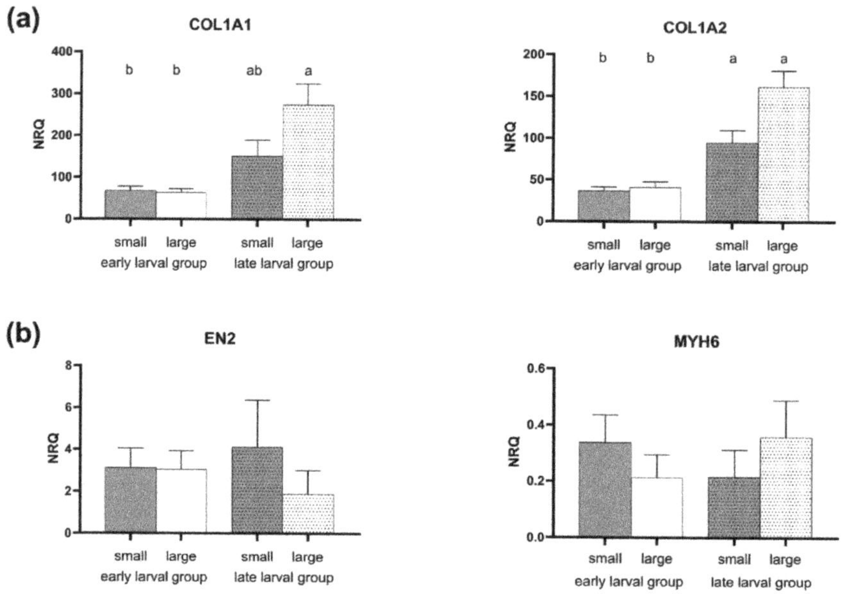

Figure 7. Expression patterns of genes for (**a**) structural markers and (**b**) regional markers. Normalised relative quantity (NRQ) of pikeperch larvae of the early larval (blots without dots) and late larval (blots with dots) groups in the two size classes small (grey blots) and large (white blots) is illustrated by its mean and SEM. Small letters indicate significance $p \leq 0.05$ between groups using ANOVA and Tukey's test. COL1A1—Collagen type 1 alpha 1; COL1A2—Collagen type 1 alpha 2; EN2—Engrailed 2; MYH6—Myosin heavy chain 6.

4. Discussion

4.1. Morphometric Distinctions of Size Classes and Growth

For the size comparisons, individuals were separated into either a small or a large size class based on the mean total length of the sampling groups. The pronounced distinction between small and large size classes during the ontogeny was already evident in the early larval group. Overall, general growth tendencies were positive in both developmental groups, with a higher regression slope in the late larval group. However, an intermediate decrease in the group mean total length was present in between the 227 dd and 256 dd stages, which we consider to be related to the sampling size for the number of measured individuals and the influence of temperature variation. It is worth noting that, in the subsequent late larval group, the increasing size differences were accompanied by significant size differences in head length. Given the interrelation between head development and the development of the jaw elements [20], we hypothesise that in pikeperch larvae, a large head is correlated with larger jaw bones, which could allow the larger larvae to take up a wider range of prey. This is supported by a strong correlation between fish length and mouth gape size in pikeperch [21] and further between head length and fish length [9]. Furthermore, previous studies have shown that pikeperch have size-dependent predator-prey relationships [22], with gape size being the limiting factor for prey ingestion [22,23]. Additionally, the size of the mouth has been shown to be associated with the occurrence of cannibalism, including the sister species *Sander vitreum* [24]. Consequently, the greater head length of the large pikeperch larvae would give these individuals a competitive advantage in the ingestion of prey. The sampling for this study, between mouth gape opening [25,26] and the start of piscivory [6,8], allowed to exclude potential influences of piscivory on the growth and gene expression. However, the differently sized natural zooplancton spectrum might have an effect on the growth and gene expression, especially for the later larval group with their differently sized head lengths.

Contradicting this, the expression of *EN2* is similar for all groups. *EN2* is expressed in the jaw muscles of vertebrates [27,28] and was previously found to be expressed more dominantly after the switch to external feeding [9]. However, since EN2 also influences midbrain development in fish [29,30] and generally determines muscle fibre type [31], the temporal and spatial expression changes of *EN2* cannot be differentiated by the grouped sample setup here. In the expression analysis of genes of interest, significant differences only occurred between the two developmental groups and not between the size classes within them. The expression patterns showed that initial transcription factors for general growth as well as muscle growth were active in the two developmental phases investigated.

4.2. Regulatory Genes Connected to Stem Cell Activation

Generally, considering genes related to the activation of satellite cells and stem cells, *PAX3* and *PAX7* are mainly expressed by cells in the dermomyotome [32]. In the dermomyotome of fishes, these regulatory genes are necessary for, but are also suppressed by, myogenesis [33,34]. Both transcription factors interact together with MSX1 in myogenic development [35], but are not affected by the developmental group or size class investigated in pikeperch larvae within this study. The complex interrelations of muscle development are reflected in the different expression patterns of the genes studied. The higher expression of *MRF4* and *MYOD1* in the older larval pikeperch stages, contrasted by the partly lower expression in the early larval group, indicates parallels to wave-like embryonic muscle development [36]. The expression of genes related to growth and muscle development exhibits a pattern that is dependent on development [14,37]. As already mentioned, in the present study, no significant difference was found in the expression of stem cell activation-related genes, with the exception of *PGC1A*. Here, significantly lower expressions in the large late larval stage compared to the early larval group were present. The transcription factor PGC1A is also a master regulator for mitochondrial function and organismal metabolism (reviewed in [38]). In interaction with other regulatory factors like TGFB1, an activation of PGC1A increases oxygen consumption in mammalian cells [39]. In the pikeperch larvae of the present study, no differences in gene expression were detected for general growth factors (see Figure 5). The transcript levels of the genes coding for the transcription factors PGC1A and TGFB1 were decreased as well as increased in specifically the large late larvae in comparison to the other larval groups. As shown by Nam and colleagues [40], the interaction of these two transcription factors can lead to regulation of the tricarboxylic acid (TCA) cycle of energy metabolism. Such findings suggest that the investigation into the factors contributing to the divergence in growth could be related to mitochondrial energy metabolism. However, an upregulation of genes related to energy metabolism pathways was previously found to occur in transcriptomic analyses of *Dicentrarchus labrax* larvae [41].

4.3. Genes of Growth and General Development

As shown by the stable gene expression of different growth factors (*IGF1, IGF2, FGF4, FGF6*, and *FGF8*), the genetic potential for growth of pikeperch larvae seems to be similar in the early and late larval groups, regardless of the size. The indications of a different utilisation in the energy metabolism (*PGC1A* and *TGFB1*) could lead to individual effects up to divergent growth. These individual strategies are not captured in our generalised experimental approach, which has been designed to examine rather general molecular mechanisms of size differences in pikeperch larvae. By using a mixed clutch approach (2–3 female spawners combined with five males), potential biases that may influence egg quality and subsequent development through both maternal and paternal effects were minimised [42,43].

4.4. Myogenic Development and Structural Marker Genes

Considering muscle development and based on the dynamic wave-like expression pattern in muscle development during pikeperch ontogeny [14,15], we detected significant differences in *MYOD1* and *MRF4* expression between the developmental groups. In contrast, the effects of other master regulators such as *MYF5* and *MYOG* and their interactions within the myogenesis cascade [44] could be overlaid by the selection of groups of pikeperch larvae that has been carried out in the present study. Nevertheless, the significant patterns of the expression levels might indicate possible differences in the developmental progress between the size classes of the same developmental group.

The generally higher expression of *MYOD1* and collagen genes (*COL1A1* and *COL1A2*) in the late larval developmental groups gives an indication of a temporal advance in ontogenesis and, furthermore, in the development of skin, bones, and scales through the formation of extracellular matrix proteins [45,46]. The development and interaction of muscles and bones together with other elements of the musculoskeletal system ensure the functionality of locomotion [47]. A resulting better locomotion could lead to a benefit for larvae with stronger and faster growth due to better hunting and escape abilities. This hypothesis is supported by the larger tail area of older pikeperch larvae in comparison to our studied age stages [3], as this area consists of the largest degree of locomotory muscle tissue.

5. Conclusions

The investigation of some of the genetic determinants contributing to growth differences in pikeperch rearing did not yield conclusive evidence. However, this in turn also means that the fundamental basis for our genes of interest is size-independent in the larval stages prior to the onset of piscivory. Following indications from the gene expression patterns, we hypothesise that influences on the individual level affect mitochondrial energy metabolism as well as myogenesis, resulting in size differences.

Our objective was to identify gene expression variations associated with development, growth, and muscle in early and late pikeperch larvae (227 dd to 666 dd), specifically comparing conspecifics of small and large size. The sampling strategy was designed to capture overall size effects across two development groups while mitigating the impact of later-onset cannibalism. Based on our data, we could not determine any size-related influence on gene expression patterns, either overall or within the developmental groups. Consequently, the capacity for growth appears to be largely independent of the expression of the analysed genes of interest. Nevertheless, variations in gene expression levels, particularly in the mitochondrial energy pathway (*PGC1A* and *TGFB1*) and myogenesis (*MYOD1* and *MRF4*), were observed, suggesting potential outcomes being influenced at the individual level.

Supplementary Materials: The following supporting information can be downloaded at: https://www.mdpi.com/article/10.3390/fishes9010033/s1, Table S1: Alphabetical primer list of genes of interest (GOI) for Fluidigm multiplex quantitative PCR.

Author Contributions: Conceptualisation, B.G.; methodology, K.T., G.P.F. and P.L.; validation, K.T., G.P.F. and A.R.; formal analysis, K.T.; investigation, K.T. and G.P.F.; resources, B.G.; data curation, K.T.; writing—original draft preparation, K.T.; writing—review and editing, K.T., G.P.F., A.R., P.L. and B.G.; visualisation, K.T.; supervision, B.G.; project administration, B.G.; funding acquisition, B.G. All authors have read and agreed to the published version of the manuscript.

Funding: This research was funded by the European Maritime and Fisheries Fund (EMFF, MV-II.1-LM-013). The publication of this article was funded by the Open Access Fund of the FBN.

Institutional Review Board Statement: This study does not contain any animal experiments. Fishes used in this study were euthanised according to the animal welfare law Directive 2010/63/EU and TierSchG § 4(3) before sampling.

Data Availability Statement: All relevant data are within the manuscript.

Acknowledgments: We would like to thank Julian Krinitskij for the technical support in performing the fluidigm multiplex real-time PCR.

Conflicts of Interest: The authors declare no conflicts of interest. The funders had no role in the design of the study; in the collection, analyses, or interpretation of data; in the writing of the manuscript; or in the decision to publish the results.

References

1. Policar, T.; Schaefer, F.J.; Panana, E.; Meyer, S.; Teerlinck, S.; Toner, D.; Zarski, D. Recent progress in European percid fish culture production technology-tackling bottlenecks. *Aquac. Int.* **2019**, *27*, 1151–1174.
2. Colchen, T.; Gisbert, E.; Krauss, D.; Ledoré, Y.; Pasquet, A.; Fontaine, P. Improving pikeperch larviculture by combining environmental, feeding and populational factors. *Aquac. Rep.* **2020**, *17*, 100337. [CrossRef]
3. Colchen, T.; Fontaine, P.; Ledoré, Y.; Teletchea, F.; Pasquet, A. Intra-cohort cannibalism in early life stages of pikeperch. *Aquac. Res.* **2019**, *50*, 915–924. [CrossRef]
4. Steenfeldt, S.; Lund, I.; Höglund, E. Is batch variability in hatching time related to size heterogeneity and cannibalism in pikeperch (*Sander lucioperca*)? *Aquac. Res.* **2011**, *42*, 727–732. [CrossRef]
5. Szczepkowski, M.; Zakęś, Z.; Szczepkowska, B.; Piotrowska, I. Effect of size sorting on the survival, growth and cannibalism in pikeperch (*Sander lucioperca* L.) larvae during intensive culture in RAS. *Czech. J. Anim. Sci.* **2011**, *56*, 483–489. [CrossRef]
6. Colchen, T.; Dias, A.; Gisbert, E.; Teletchea, F.; Fontaine, P.; Pasquet, A. The onset of piscivory in a freshwater fish species: Analysis of behavioural and physiological traits. *J. Fish Biol.* **2020**, *96*, 1463–1474. [CrossRef] [PubMed]
7. Pereira, L.S.; Agostinho, A.A.; Winemiller, K.O. Revisiting cannibalism in fishes. *Rev. Fish Biol. Fish.* **2017**, *27*, 499–513.
8. Szkudlarek, M.; Zakęś, Z. Effect of stocking density on survival and growth performance of pikeperch, *Sander lucioperca* (L.), larvae under controlled conditions. *Aquac. Int.* **2007**, *15*, 67–81.
9. Franz, G.P.; Lewerentz, L.; Grunow, B. Observations of growth changes during the embryonic-larval-transition of pikeperch (*Sander lucioperca*) under near-natural conditions. *J. Fish Biol.* **2021**, *99*, 425–436. [CrossRef]
10. Ott, A.; Löffler, J.; Ahnelt, H.; Keckeis, H. Early development of the postcranial skeleton of the pikeperch *Sander lucioperca* (Teleostei: Percidae) relating to developmental stages and growth. *J. Morphol.* **2012**, *273*, 894–908. [CrossRef]
11. Penaz, M. A general framework of fish ontogeny: A review of the ongoing debate. *Folia Zool.* **2001**, *50*, 241–256.
12. El Kertaoui, N.; Lund, I.; Assogba, H.; Dominguez, D.; Izquierdo, M.S.; Baekelandt, S.; Cornet, V.; Mandiki, S.N.M.; Montero, D.; Kestemont, P. Key nutritional factors and interactions during larval development of pikeperch (*Sander lucioperca*). *Sci. Rep.* **2019**, *9*, 7074. [CrossRef] [PubMed]
13. Imentai, A.; Gilannejad, N.; Martínez-Rodríguez, G.; López, F.J.M.; Martínez, F.P.; Pěnka, T.; Dzyuba, V.; Dadras, H.; Policar, T. Effects of First Feeding Regime on Gene Expression and Enzyme Activity in Pikeperch (*Sander lucioperca*) Larvae. *Front. Mar. Sci.* **2022**, *9*, 864536. [CrossRef]
14. Franz, G.P.; Tönißen, K.; Rebl, A.; Lutze, P.; Grunow, B. The expression of myogenic gene markers during the embryo-larval-transition in Pikeperch (*Sander lucioperca*). *Aquac. Res.* **2022**, *53*, 14. [CrossRef]
15. Lavajoo, F.; Falahatkar, B.; Perelló-Amorós, M.; Moshayedi, F.; Efatpanah, I.; Gutiérrez, J. The pattern of gene expression (IGF family, muscle growth regulatory factors and osteogenesis related genes) involved in growth of skeletal muscle in pikeperch (*Sander lucioperca*) during ontogenesis. *Preprint* **2023**. [CrossRef]
16. Schäfer, N.; Kaya, Y.; Rebl, H.; Stüeken, M.; Rebl, A.; Nguinkal, J.A.; Franz, G.P.; Brunner, R.M.; Goldammer, T.; Grunow, B.; et al. Insights into early ontogenesis: Characterization of stress and development key genes of pikeperch (*Sander lucioperca*) in vivo and in vitro. *Fish Physiol. Biochem.* **2021**, *47*, 515–532. [CrossRef]
17. Rebl, A.; Rebl, H.; Verleih, M.; Haupt, S.; Köbis, J.M.; Goldammer, T.; Seyfert, H.-M. At Least Two Genes Encode Many Variants of Irak3 in Rainbow Trout, but Neither the Full-Length Factor Nor Its Variants Interfere Directly With the TLR-Mediated Stimulation of Inflammation. *Front. Immunol.* **2019**, *10*, 2246. [CrossRef]
18. Swirplies, F.; Wuertz, S.; Baßmann, B.; Orban, A.; Schäfer, N.; Brunner, R.M.; Hadlich, F.; Goldammer, T.; Rebl, A. Identification of molecular stress indicators in pikeperch Sander lucioperca correlating with rising water temperatures. *Aquaculture* **2019**, *501*, 260–271. [CrossRef]
19. Hellemans, J.; Mortier, G.; De Paepe, A.; Speleman, F.; Vandesompele, J. qBase relative quantification framework and software for management and automated analysis of real-time quantitative PCR data. *Genome Biol.* **2007**, *8*, R19. [CrossRef]
20. Löffler, J.; Ott, A.; Ahnelt, H.; Keckeis, H. Early development of the skull of *Sander lucioperca* (L.) (Teleostei: Percidae) relating to growth and mortality. *J. Fish Biol.* **2008**, *72*, 233–258. [CrossRef]
21. Mehner, T.; Hülsmann, S.; Worischka, S.; Plewa, M.; Benndorf, J. Is the midsummer decline of Daphnia really induced by age-0 fish predation? Comparison of fish consumption and M Daphnia mortality and life history parameters in a biomanipulated reservoir. *J. Plankton Res.* **1998**, *20*, 1797–1811. [CrossRef]
22. Dörner, H.; Hülsmann, S.; Hölker, F.; Skov, C.; Wagner, A. Size-dependent predator–prey relationships between pikeperch and their prey fish. *Ecol. Freshw. Fish* **2007**, *16*, 307–314. [CrossRef]
23. Vehanen, T.; Hyvärinen, P.; Huusko, A. Food consumption and prey orientation of piscivorous brown trout (*Salmo trutta*) and pikeperch (*Stizostedion lucioperca*) in a large regulated lake. *J. Appl. Ichthyol.* **1998**, *14*, 15–22. [CrossRef]

24. Baras, E.; Jobling, M. Dynamics of intracohort cannibalism in cultured fish. *Aquac. Res.* **2002**, *33*, 461–479. [CrossRef]
25. Xu, Z.; Li, C.; Ling, Q.; Gaughan, S.; Wang, G.; Han, X. Early development and the point of no return in pikeperch (*Sander lucioperca* L.) larvae. *Chin. J. Oceanol. Limnol.* **2017**, *35*, 1493–1500. [CrossRef]
26. Güralp, H.; Pocherniaieva, K.; Blecha, M.; Policar, T.; Psenicka, M.; Saito, T. Development, and effect of water temperature on development rate, of pikeperch *Sander lucioperca* embryos. *Theriogenology* **2017**, *104*, 94–104. [CrossRef] [PubMed]
27. Hatta, K.; Schilling, T.F.; BreMiller, R.A.; Kimmel, C.B. Specification of Jaw Muscle Identity in Zebrafish: Correlation with engrailed-Homeoprotein Expression. *Science* **1990**, *250*, 802–805. [CrossRef]
28. Knight, R.D.; Mebus, K.; Roehl, H.H. Mandibular arch muscle identity is regulated by a conserved molecular process during vertebrate development. *J. Exp. Zool. Part B Mol. Dev. Evol.* **2008**, *310B*, 355–369. [CrossRef]
29. Vecino, E.; Ekström, P. Expression of the homeobox engrailed gene during the embryonic development of the nervous system of the trout (*Salmo fario* L.). *Neurosci. Lett.* **1991**, *129*, 311–314. [CrossRef]
30. Scholpp, S.; Brand, M. Morpholino-induced knockdown of zebrafish engrailed genes eng2 and eng3 reveals redundant and unique functions in midbrain--hindbrain boundary development. *Genesis* **2001**, *30*, 129–133. [CrossRef]
31. Degenhardt, K.; Sassoon, D.A. A role for Engrailed-2 in determination of skeletal muscle physiologic properties. *Dev. Biol.* **2001**, *231*, 175–189. [CrossRef] [PubMed]
32. Relaix, F.; Rocancourt, D.; Mansouri, A.; Buckingham, M. A Pax3/Pax7-dependent population of skeletal muscle progenitor cells. *Nature* **2005**, *435*, 948–953. [CrossRef] [PubMed]
33. Hammond, C.L.; Hinits, Y.; Osborn, D.P.S.; Minchin, J.E.N.; Tettamanti, G.; Hughes, S.M. Signals and myogenic regulatory factors restrict pax3 and pax7 expression to dermomyotome-like tissue in zebrafish. *Dev. Biol.* **2007**, *302*, 504–521. [CrossRef] [PubMed]
34. Akolkar, D.B.; Asaduzzaman, M.; Kinoshita, S.; Asakawa, S.; Watabe, S. Characterization of Pax3 and Pax7 genes and their expression patterns during different development and growth stages of Japanese pufferfish Takifugu rubripes. *Gene* **2016**, *575*, 21–28. [CrossRef] [PubMed]
35. Bendall, A.J.; Ding, J.; Hu, G.; Shen, M.M.; Abate-Shen, C. Msx1 antagonizes the myogenic activity of Pax3 in migrating limb muscle precursors. *Development* **1999**, *126*, 4965–4976. [CrossRef]
36. Rossi, G.; Messina, G. Comparative myogenesis in teleosts and mammals. *Cell Mol. Life Sci.* **2014**, *71*, 3081–3099. [CrossRef] [PubMed]
37. Franz, A.C.; Faass, O.; Kollner, B.; Shved, N.; Link, K.; Casanova, A.; Wenger, M.; D'Cotta, H.; Baroiller, J.F.; Ullrich, O.; et al. Endocrine and Local IGF-I in the Bony Fish Immune System. *Biology* **2016**, *5*, 9. [CrossRef]
38. Puigserver, P. Tissue-specific regulation of metabolic pathways through the transcriptional coactivator PGC1-alpha. *Int. J. Obes.* **2005**, *29* (Suppl. S1), S5–S9. [CrossRef]
39. Yu, E.; Foote, K.; Bennett, M. Mitochondrial function in thoracic aortic aneurysms. *Cardiovasc. Res.* **2018**, *114*, 1696–1698. [CrossRef]
40. Nam, H.; Kundu, A.; Karki, S.; Brinkley, G.J.; Chandrashekar, D.S.; Kirkman, R.L.; Liu, J.; Liberti, M.V.; Locasale, J.W.; Mitchell, T.; et al. The TGF-β/HDAC7 axis suppresses TCA cycle metabolism in renal cancer. *JCI Insight* **2021**, *6*, e148438. [CrossRef]
41. Darias, M.J.; Zambonino-Infante, J.L.; Hugot, K.; Cahu, C.L.; Mazurais, D. Gene Expression Patterns During the Larval Development of European Sea Bass (Dicentrarchus Labrax) by Microarray Analysis. *Mar. Biotechnol.* **2008**, *10*, 416–428. [CrossRef] [PubMed]
42. Bobe, J.; Labbé, C. Egg and sperm quality in fish. *Gen. Comp. Endocrinol.* **2010**, *165*, 535–548. [CrossRef] [PubMed]
43. Sullivan, C.V.; Chapman, R.W.; Reading, B.J.; Anderson, P.E. Transcriptomics of mRNA and egg quality in farmed fish: Some recent developments and future directions. *Gen. Comp. Endocrinol.* **2015**, *221*, 23–30. [CrossRef] [PubMed]
44. Schnapp, E.; Pistocchi, A.S.; Karampetsou, E.; Foglia, E.; Lamia, C.L.; Cotelli, F.; Cossu, G. Induced early expression of mrf4 but not myog rescues myogenesis in the myod/myf5 double-morphant zebrafish embryo. *J. Cell Sci.* **2009**, *122*, 481–488. [CrossRef]
45. Gelse, K.; Pöschl, E.; Aigner, T. Collagens—Structure, function, and biosynthesis. *Adv. Drug Deliv. Rev.* **2003**, *55*, 1531–1546. [CrossRef]
46. Rescan, P.Y. Development of myofibres and associated connective tissues in fish axial muscle: Recent insights and future perspectives. *Differentiation* **2019**, *106*, 35–41. [CrossRef]
47. Sefton, E.M.; Kardon, G. Chapter Five—Connecting muscle development, birth defects, and evolution: An essential role for muscle connective tissue. In *Current Topics in Developmental Biology*; Wellik, D.M., Ed.; Academic Press: Cambridge, MA, USA, 2019; Volume 132, pp. 137–176.

fishes

MDPI

Article

Molecular Cloning and Characterization of Scavenger Receptor Class B Type 1 in Grass Carp (*Ctenopharyngodon idellus*) and Its Expression Profile following Grass Carp Reovirus Challenge

Yang Zhang [1,2,†], Jiayuan Shi [1,3,†], Yuntao Lu [1,2], Qing Luo [1], Pengfei Chu [3], Rong Huang [4], Kunci Chen [1], Jian Zhao [1], Yaping Wang [1,*] and Mi Ou [1,2,3,*]

1 Key Laboratory of Tropical and Subtropical Fishery Resources Application and Cultivation, Ministry of Agriculture and Rural Affairs, Pearl River Fisheries Research Institute, Chinese Academy of Fishery Sciences, Guangzhou 510380, China; dy211004@163.com (Y.Z.); 19825305074@163.com (J.S.); 17633537502@163.com (Y.L.); luoqing@prfri.ac.cn (Q.L.); chenkunci@prfri.ac.cn (K.C.); zhaojian@prfri.ac.cn (J.Z.)

2 School of Life and Health Sciences, Hunan University of Science and Technology, Xiangtan 411201, China

3 College of Animal Science and Technology, Yangzhou University, Yangzhou 225000, China; chupf@yzu.edu.cn

4 State Key Laboratory of Freshwater Ecology and Biotechnology, Institute of Hydrobiology, Chinese Academy of Sciences, Wuhan 430072, China; huangrong@ihb.ac.cn

* Correspondence: wangyp@ihb.ac.cn (Y.W.); om1990@prfri.ac.cn (M.O.); Tel.: +86-020-81174522 (M.O.)

† These authors contributed equally to this work.

Abstract: As a member of the pattern recognition receptor (PRR) class, scavenger receptor class B type 1 (SRB1) plays a key role in innate immunity. Grass carp (*Ctenopharyngodon idellus*) ranks among the most extensively cultivated freshwater aquaculture species in China. However, little is known about the function of SRB1 in *C. idellus*. In this research study, a *SRB1* gene was identified in *C. idellus*, named *CiSRB1*. The full-length cDNA of *CiSRB1* is 2486 bp long, with an open reading frame (ORF) of 2486 bp encoding a 497 amino acid (aa) protein containing a conserved CD36 domain. The identified genomic DNA length of *CiSRB1* is 20,042 bp, including 12 exons and 11 introns. The predictive analysis of protein interactions revealed that *Ci*SRB1 could interact with the outer capsid proteins of typical GCRV strains. The tissue distribution of *CiSRB1* exhibited age-dependent characteristics. *CiSRB1* displayed the highest expression in the intestines and moderate levels in muscle, spleen, liver, and brain of one-year-old grass carp while maintaining relatively low levels in three-year-old grass carp. Following grass carp reovirus (GCRV) infection, notable upregulation of *CiSRB1* transcripts was observed in major immune tissues (gills, intestines, spleen, and liver). Furthermore, significant differences were found between one-year-old and three-year-old grass carp, with lower *CiSRB1* expression levels being detected in the older group. Additionally, a distinct response to GCRV infection was observed in one-year-old and three-year-old grass carp. It was found that one-year-old individuals had a mortality rate of up to 84% 6 days post-infection (dpi), whereas all three-year-old counterparts survived after GCRV infection. The analysis of GCRV copy numbers across tissues revealed substantially higher levels in one-year-old grass carp compared with their older counterparts, confirming the existence of age-dependent susceptibility to GCRV infection in grass carp. Combined with these results, it was speculated that the decline in cell-surface *CiSRB1* expression with age may impede reovirus binding to host cells, potentially explaining why older grass carp demonstrated enhanced resistance to GCRV infection. This observation accentuates the importance of *CiSRB1* in the context of GCRV infection and provides insights into age-dependent susceptibility to reovirus.

Keywords: grass carp; GCRV; SRB1; gene expression; age-dependent susceptibility

Key Contribution: The *CiSRB1* gene, identified in economically important grass carp (*Ctenopharyngodon idellus*), exhibited age-dependent expression patterns. In one-year-old grass carp, *CiSRB1* was highly expressed in the intestines and moderately so in muscle, spleen, liver, and brain, while levels were relatively low in three-year-old carp. Upon GCRV infection, there was a significant increase in *CiSRB1* transcripts in immune tissues, with lower levels observed being in three-year-old fish compared with one-year-olds. Notably, higher mortality and GCRV copy numbers were seen in one-year-old fish, indicating age-related susceptibility to GCRV. The predictive analysis of protein interactions revealed that *Ci*SRB1 could interact with the outer capsid proteins of typical GCRV strains. This suggests that age may impact the cell-surface expression of CiSRB1 post-GCRV infection, potentially affecting reovirus binding to target cells. These findings shed light on age-related restrictions in reovirus infection in grass carp.

1. Introduction

Grass carp (*Ctenopharyngodon idellus*) is one of the most extensively cultivated freshwater aquaculture species in China, significantly contributing to the agricultural economy [1]. However, the production of *C. idellus* has experienced a decline due to various infectious diseases, leading to substantial financial losses annually. One of the most severe and prevalent diseases is grass carp hemorrhage disease, caused by grass carp reovirus (GCRV) [2]. GCRV, a double-stranded RNA virus, typically induces intestinal and muscle bleeding, often resulting in death upon infection. It is estimated that GCRV accounts for an economic loss of approximately 20% of the total grass carp production [3]. Unfortunately, there are currently no optimal methods for preventing GCRV infection or curing hemorrhagic disease in *C. idellus* [4]. Hence, the identification of factors in *C. idellus* linked to host immune defense is crucial to developing antiviral drugs and improving fish breeding programs.

In teleosts, the innate immune system heavily relies on pattern recognition receptors (PRRs) to identify conserved domains [5]. Among these PRRs, scavenger receptors (SRs) play a pivotal role in the immune system, encompassing functions such as phagocytosis, antigen presentation, clearance of apoptotic cells, and activation of various signaling pathways [6]. The expansive family of scavenger receptors (SRs) is categorized into ten classes (classes A-J) based on structural characteristics and functional domains, and the majority of SRs are present on macrophages to mediate lipid transport and immune defense [7].

Within the SRBs, there are three members: cluster differentiation-36 (CD36), scavenger receptor class B type 1 (SRB1), and lysosomal integral membrane protein type 2 (LIMP2) [6]. SRB1, a high-density lipoprotein (HDL) receptor crucial for HDL metabolism [8], has been associated with the entry of multiple viruses into host cells, including hepatitis C virus (HCV) [9–11], dengue virus (DV) [12], malaria parasite [13], SARS-CoV-2 [14], and GCRV [15,16]. Extensive research on SRB1 has been conducted across different species, such as humans (*Homo sapiens*) [17], mice (*Mus musculus*) [18], orange-spotted grouper (*Epinephelus coioides*) [19], rare minnow (*Gobiocypris rarus*) [16], and turbot (*Scophthalmus maximus*) [20]. Transcriptome analysis of GCRV-infected *C. idellus* kidney (CIK) cells reveals the upregulation of SRB1 between 8 and 24 h post-infection (hpi), indicating its potential role in GCRV cell entry [15]. Despite these insights, the specific action of *SRB1* during GCRV infection in *C. idellus* remains unknown. This study aims to clone the *SRB1* gene and explore its functions during grass carp hemorrhage, shedding light on the role of *SRB1* in virus entry into host cells and contributing to a more comprehensive understanding of fish *SRB1* function.

2. Materials and Methods

2.1. Experimental Fish and Sampling

Healthy one-year-old (n = 200; weight, 80.3 ± 10.1 g; length, 18.1 ± 3.5 cm) and three-year-old (n = 150; weight, 1500.8 ± 500.4 g; length, 70.7 ± 10.3 cm) grass carp were adapted to 4 × 4 × 2 m^3 indoor concrete ponds with dechlorinated and aerated water at 28.0 ± 1.0 °C at the Fangcun Experimental Station, Pearl River Fisheries Research Institute (Guangzhou, China). These fish were fed commercial feed from Guangdong Bairong Aquatic Varieties Group Co., Ltd. (FoShan, China), twice daily. After a one-week acclimatization period with no observed abnormalities, the grass carp were prepared for further experimentation.

For tissue distribution experiments, nine tissues (gills, head kidney, heart, intestines, brain middle kidney, liver, muscle, and spleen) were collected from healthy one-year-old and three-year-old grass carp ($n = 3$ per age group). All tissues were promptly homogenized in TRIzol reagent (Invitrogen, Waltham, MA, USA) and stored at −80 °C until RNA extraction. Further, tail tissues were obtained and fixed in 95% ethanol.

2.2. Gene Cloning and Sequence Analysis of SRB1 in C. idellus

Total RNA was extracted from healthy samples of one-year-old fish by using TRIzol reagent (Invitrogen, Carlsbad, CA, USA) according to the manufacturer's instruction. The ReverTra Ace kit (Toyobo, Osaka, Japan) was used to synthesize the first-strand cDNA by using an oligo dT primer (Takara, Kyoto, Japan). Fragments of *SRB1* in *C. idellus*, designated as *CiSRB1*, were obtained by blasting the sequence of *SRB1* in *G. rarus* (*GrSRB1*, GenBank accession No. MK436208) against the draft genome of grass carp (SRA accession No. PRJEB5920) [21]. Partial cDNA fragments of the *CiSRB1* gene were amplified by specific primers (Table 1). Subsequently, specific and adaptor primers (Table 1) were designed to obtain the 5′ and 3′ untranslated regions (UTRs) by using the SMART^{TM} RACE cDNA Amplification Kit (Clontech, Mountain View, CA, USA). The full-length cDNA sequence of *CiSRB1* was amplified by using primers CiSRB1-F2 and CiSRB1-R2 (Table 1) targeting the 5′ and 3′ UTRs, respectively. Then, a Gel Extraction Kit (Omega Bio-Tek, Norcross, GA, USA) was used to purify the PCR products, which were then ligated into pMD19-T vectors (Takara, Japan) and transformed into competent *Escherichia coli* DH5α cells (TransGen, Beijing, China). Positive colonies of the target fragment were sequenced by a commercial company (Tianyi huiyuan, Guangzhou, China).

Table 1. Primers for full-length cDNA cloning and qPCR.

Primer Name	Sequence (5′→3′)	Application
CiSRB1-F1	GGTTTGGCAGTTTTGTTCGG	Partial sequence obtaining
CiSRB1-R1	GAATGGTTGCGAGTCCGAGA	
CiSRB1-5′R-out	GTTCGTTCTTCGGGTTTATCTCTACATTC	5′-Race PCR amplification
CiSRB1-5′R-in	CAAACACCACGACAGTCCCGAAC	
CiSRB1-3′F-out	ACAGAAACAGGAAAGATAACAGAGGTG	3′-Race PCR amplification
CiSRB1-3′F-in	ATCTTCATCGGTCTCGGACTCGCA	
CiSRB1-F2	GTGTGAAGTAAGGATGGCGGT	ORF qualifying
CiSRB1-R2	GTGGATATTGGCTCTAGCTCGT	
CiSRB1-qF	GGGAGATGAATCCGATGTGGTC	qPCR amplification
CiSRB1-qR	GACCTTCAACGAGGGACCTTTC	
Ciβ-actin-qF	GGATGATGAAATTGCCGCACTGG	qPCR amplification
Ciβ-actin-qR	ACCGACCATGACGCCCTGATGT	
HZ08F	AGCGCAGCAGGCAATTACTATCT	qPCR amplification
HZ08R	ATCTGCTGGTAATGCGGAACG	

The sequences of *CiSRB1* were analyzed by using Sequence Manipulation Suite (STS) (http://www.bio-soft.net/sms/, accessed on 15 March 2024), and Simple Modular Architecture Research Tool (SMART) (http://smart.embl-heidelberg.de/, accessed on 15 March 2024) was utilized to predict the protein domains. The NetNglyc 1.0 server (http://www.cbs.dtu.dk/services/NetNGlyc/, accessed on 16 March 2024) predicted potential N-glycosylation sites. SRB1 sequences from other species were acquired from BLASTP (https://blast.ncbi.nlm.nih.gov/Blastp.cgi, accessed on 16 March 2024), and multiple-sequence alignments were conducted by using ClustalW 2.1 (http://www.ebi.ac.uk/tools/clustalw2.1, accessed on 16 March 2024). The phylogenetic tree based on the amino acid sequences was constructed by utilizing Mega 7.0 (http://www.megasoftware.net/index.html, accessed on 16 March 2024) with the neighbor-joining (NJ) algorithm and a bootstrapping procedure with a minimum of 1000 bootstraps.

The genomic DNA sequences of *CiSRB1* were predicted from the grass carp draft genome based on cDNA sequences, and specific primers (Table 2) were designed accordingly. Genomic DNA was extracted from the tail tissue of one-year-old fish by using the Universal Genomic DNA Kit (CWBio, Beijing, China). PCR amplification was performed by using the extracted genomic DNA as a template, and the resulting PCR products were sequenced to obtain the genomic DNA sequences of *CiSRB1*. Exons and introns were determined by comparing the genomic DNA sequences with the cDNA sequences.

Table 2. Primers for the amplification of genomic DNA sequences.

Primer Name	Sequence (5′→3′)	Length (bp)
CiSRB1-D-F1	GTGTGAAGTAAGGATGGCGGT	1310 bp
CiSRB1-D-R1	CTTTTAGGATTTCGTCAGGATTG	
CiSRB1-D-F2	CACTATGTGGAAGGACATCCC	1040 bp
CiSRB1-D-R2	GTTTCGCTTCAGAAGACCTCTAT	
CiSRB1-D-F3	GAGAGTCAAAAAATACATATACAGGC	1223 bp
CiSRB1-D-R3	ATACAGCACTGTTACGCTTTGG	
CiSRB1-D-F4	TCAAAAACGGTCCACGAGC	1424 bp
CiSRB1-D-R4	CCCTCATGTTTGCATAAGTCTAGAT	
CiSRB1-D-F5	CTGTTCCAAACGTAGCTGCCT	1432 bp
CiSRB1-D-R5	TTCATTTGGGCTCAATCCGT	
CiSRB1-D-F6	CCATCCACACTTTTTTGCGG	1384 bp
CiSRB1-D-R6	TTGATCGAATAGGACCAGAGGG	
CiSRB1-D-F7	ATGACATTTCTGCCCCACTATGG	1321 bp
CiSRB1-D-R7	AAACACCGAAACTAACACGACCC	
CiSRB1-D-F8	CTGTCCCTTTTATAGCATTGG	1187 bp
CiSRB1-D-R8	TAAACATTTGAAATATATCAGTCTG	
CiSRB1-D-F9	GGTGCTGGTCATATAATTAGAAT	1216 bp
CiSRB1-D-R9	GACGACCATTGACACCTT	
CiSRB1-D-F10	TCCTTCCCTTCGCCTCTC	1311 bp
CiSRB1-D-R10	TATTCCTGTGATGCAAAGCTG	
CiSRB1-D-F11	GAGCAGCAAATCAGCATATTAG	1253 bp
CiSRB1-D-R11	CACCACCTCTGTTATCTTTCCT	
CiSRB1-D-F12	CTGTACAATCTCTCATGGACTCAT	1252 bp
CiSRB1-D-R12	CAAACGTTACTGCGGCTC	
CiSRB1-D-F13	TAGGATTACTCTTGTTGCCTAGC	1297 bp
CiSRB1-D-R13	TGAGGTATAATATCACGGCTCC	
CiSRB1-D-F14	CCGTTCTCAACACGTTCCG	1229 bp
CiSRB1-D-R14	ACATTAGAAGGAACATTGCCCAC	
CiSRB1-D-F15	ACTTTATCTCCTTCACTGATCTGT	1355 bp
CiSRB1-D-R15	AGTGTTTGGCATAGTAATGGAT	
CiSRB1-D-F16	ACGAAGGTCTTTCAGGTTTGGA	1538 bp
CiSRB1-D-R16	AGGAAGAAACATCCCCACTCAA	

AlphaFold Server (https://alphafoldserver.com, accessed on 30 June 2024) [22] was used to predict the potential interaction between *Ci*SRB1 and GCRV particles, which play key roles in viral attachment and infection [23,24]. The outer capsid proteins of several typical GCRV strains were analyzed, including the VP5 protein from the GCRV-873 strain (GenBank accession No. AAG17823.1), the VP7 protein from the GCRV-873 strain (GenBank accession No. AAM92742.1), the VP5 protein from the HZ08 strain (GenBank accession No. ADJ75337.1, designated as VP5H in this paper), and the VP7 protein from the GCRV-GD108 strain (GenBank accession No. ADT79738.1, designated as VP7G in this paper). UCSF ChimeraX (http://www.cgl.ucsf.edu/chimera/, accessed on 15 March 2024) [25] was then utilized to visualize the results predicted by AlphaFold Server.

2.3. GCRV Challenge and Sampling

The acquisition of the GCRV and GCRV challenge was based on [26] with modifications. In brief, two hundred fish were divided into four groups (Groups I–IV), each consisting of 50 fish. Group I, containing one-year-old fish, and Group II, containing three-year-old fish, were intraperitoneally (i.p.) injected with 20 μL/g body weight of 0.7% NaCl as the control treatments. Samples of gills, spleen, intestines, and liver were collected from three individuals in each group, respectively. Meanwhile, Group III, containing one-year-old fish, and Group IV, containing three-year-old fish, were i.p. injected with 20 μL/g body weight of GCRV (HZ08 strain) at a titer of 2.97×10^3 copy/μL. These injected fish were maintained under the same conditions as mentioned above. Afterwards, three individuals from each challenge group were euthanized, and samples of gills, spleen, intestines, and liver were collected 6, 12, 24, 48, 72, 96, 120, and 144 h post-injection (hpi). All samples were immediately homogenized in TRIzol reagent and stored at −80 °C until RNA extraction for further analysis. The total mortality of these four groups was assessed by recording the number of dead fish daily. The experiment ended when no deaths were recorded for two consecutive weeks, and the total mortality was calculated at that point.

2.4. Quantification of Gene Expression

Total RNAs were extracted from various tissues of healthy one-year-old and three-year-old grass carp, including gills, head kidney, heart, intestines, brain middle kidney, liver, muscle, and spleen. First-strand cDNA synthesis was performed by using ReverTra Ace® qPCR RT Master Mix with gDNA Remover (Toyobo, Japan). Quantitative real-time PCR (qPCR) was conducted by using SYBR® Green Realtime PCR Master Mix (Toyobo, Japan) on the StepOnePlus™ Real-Time PCR System (ABI, Los Angeles, CA, USA), with three replicates performed per sample. The *β-actin* gene in *C. idellus* (GenBank accession No. M25013.1) served as an internal control for cDNA normalization. Gene expression levels were calculated by using the $2^{-\Delta\Delta Ct}$ method [16]. The expression level in the gills from healthy one-year-old grass carp was set as the baseline (1.00) for tissue distribution, and the relative gene expression was calculated as the ratio of gene expression in each tissue relative to that in the gills from one-year-old grass carp.

To determine the effects of viral infection on *CiSRB1* transcripts, three individuals were sampled from each group. Four representative immune tissues, gills, spleen, intestines, and liver were collected 6, 12, 24, 48, 72, 96, 120, and 144 hpi. The mRNA expression levels of *CiSRB1* at 0 h in healthy one-year-old fish were set to 1.00, and *β-actin* was used as an internal control to normalize the relative *CiSRB1* transcripts in different tissues in response to GCRV infection. To determine dynamic changes in GCRV levels in the infected fish, the relative GCRV copy numbers were examined by specific primers (Table 1) for the M6 segment of the GCRV-HZ08 strain [27]. For convenience, the relative copy number of GCRV 1 dpi in the gills of one-year-old infected fish was used for normalization, and GCRV relative copy numbers of one-year-old and three-year-old infected fish in gills, spleen, intestines, and liver were calculated 1, 2, 3, 4, 5, and 6 dpi.

2.5. Statistical Analysis

The experimental data were expressed as the means ± standard deviation (SD) of three replicates. Significant differences were subjected to one-way ANOVA followed by Dunnett's multiple comparisons post-test by using SPSS Statistics 22.0, with $p < 0.05$ being considered statistically significant.

3. Results

3.1. Molecular Features of CiSRB1

The full-length cDNA sequence of *CiSRB1* (GenBank accession No. MT643909) was obtained via RT-PCR and RACE, measuring 2486 bp in length. It comprised a 1494 bp ORF encoding 497 amino acids (aa), a 310 bp 5′-UTR and a 682 bp 3′-UTR with three RNA instability motifs (ATTTA), five polyadenylation signal sequences (ATTAA), and a poly(A) tail (Figure 1). Structure analysis revealed that the *Ci*SRB1 protein contained two primary putative transmembrane domains of 23 aa each (residues 7~29 and 442~464) and six putative N-glycosylation sites (residues 101, 107, 211, 309, 329, and 382). Additionally, a CD36 domain spanning aa 14 to 462 (E-value = 9.5×10^{-155}) was discovered.

```
   1 ctctggctcataggtggcgctttgagcaccgtaaacatggtaatcacacagcaagagcgctataaaacctgcaggattagagagccggac
  91 tttaacatcactcacagctgtaatgaacatagccataaagtcagacgagcattaataactcaatgaccagagttttatggctttatctcg
 181 tataagtgatagtttgttcagctgttgaggactcacgcgtgctttgagttctgtgcgctttctagaagaattaactttattgttctcaaa
   1                                           M  A  V  S  K  S  T  L  A  I  V  F  L  V  L  G  G
 271 cctttgcattaatttgtatcgcgaactgtgtgaagtaaggATGGCGGTGTCTAAATCTACATTAGCGATCGTTTTCTTAGTTCTGGGAGG
  18 L  A  V  L  F  G  T  V  V  V  F  V  G  P  I  I  I  D  D  Q  I  V  K  N  V  E  I  N  P  K
 361 TTTGGCAGTTTTGTTCGGGACTGTCGTGGTGTTTGTTGGACCTATTATAATAGACGATCAAATAGTAAAGAATGTAGAGATAAACCCGAA
  48 N  E  L  S  Y  T  M  W  K  D  I  P  V  P  F  F  M  S  V  Y  F  F  H  I  V  N  P  D  E  I
 451 GAACGAACTCTCCTACACTATGTGGAAGGACATCCCGGTTCCCTTTTTTATGTCTGTATATTTCTTTCATATTGTCAATCCTGACGAAAT
  78 L  K  G  E  K  P  M  V  I  Q  R  G  P  Y  V  Y  R  E  N  R  W  K  D  N  I  T  F  H  D  N
 541 CCTAAAAGGAGAAAAGCCCATGGTGATACAGAGGGGGCCATATGTGTACCGTGAAAACCGCTGGAAGGACAACATCACATTCCATGACAA
 108 N  T  V  S  Y  K  E  F  R  Q  Y  F  F  E  E  S  M  S  V  G  D  E  S  D  V  V  T  I  P  N
 631 CAACACAGTTTCGTATAAGGAATTTCGGCAGTATTTCTTTGAGGAGAGTATGTCTGTGGGAGATGAATCCGATGTGGTCACCATCCCTAA
 138 M  L  V  L  G  A  S  V  M  M  E  N  M  P  F  P  I  R  V  L  L  S  A  T  F  K  T  F  N  E
 721 CATGCTAGTGCTGGGCGCATCAGTAATGATGGAGAATATGCCGTTTCCTATACGCGTTTTGCTCAGCGCCACGTTCAAGACCTTCAACGA
 168 G  P  F  L  T  K  P  V  G  E  L  M  W  G  Y  D  S  K  L  V  D  F  L  N  K  Y  L  P  G  M
 811 GGGACCTTTCTTGACAAAACCAGTAGGAGAACTCATGTGGGGCTACGACAGCAAGTTGGTGGACTTCCTGAACAAATATCTCCCTGGCAT
 198 L  P  S  S  G  K  F  G  L  F  A  E  F  N  N  S  N  T  G  Q  F  T  V  F  T  G  Q  D  D  I
 901 GCTTCCATCCAGCGGCAAGTTTGGCCTATTTGCTGAGTTTAACAACTCAAACACTGGGCAGTTCACCGTCTTCACTGGCCAAGATGACAT
 228 R  K  V  H  K  V  D  S  W  N  G  L  K  S  V  D  Y  W  R  S  D  Q  C  N  M  I  N  G  T  A
 991 CCGAAAAGTTCATAAGGTGGACTCTTGGAATGGCCTAAAAAGTGTGGATTACTGGAGGTCTGACCAGTGTAACATGATCAATGGTACAGC
 258 G  Q  M  W  P  P  F  M  T  T  E  S  T  L  P  F  Y  S  P  D  A  C  R  S  M  E  L  V  Y  Q
1081 GGGTCAAATGTGGCCTCCGTTCATGACCACAGAGTCGACGCTGCCCTTCTACAGCCCTGATGCGTGCAGGTCCATGGAGCTAGTGTACCA
 288 R  P  G  V  S  Q  G  I  P  V  F  R  F  V  A  P  K  T  L  F  A  N  G  T  D  Y  P  P  N  E
1171 AAGGCCAGGAGTGTCTCAGGGGATTCCAGTTTTCCGCTTTGTGGCCCCAAAGACTCTTTTTGCCAACGGTACAGATTATCCTCCCAATGA
 318 G  F  C  P  C  R  Q  S  G  L  L  N  V  S  T  C  R  H  N  S  P  V  F  I  S  H  P  H  F  F
1261 GGGCTTCTGTCCCTGTCGGCAGTCCGGCCTTCTCAACGTCAGCACCTGCAGACACAATTCCCCTGTGTTCATCTCCCATCCACACTTTTT
 348 A  A  D  P  V  L  L  D  T  V  N  G  L  S  P  N  E  D  E  H  G  L  F  I  D  I  H  P  E  T
1351 TGCGGCTGATCCCGTCCTTTTGGACACTGTTAACGGATTGAGCCCAAATGAAGATGAACATGGACTTTTTATCGACATCCACCCGGAGAC
 378 G  V  P  M  N  V  S  I  R  L  Q  L  N  L  L  M  K  R  V  S  G  I  T  E  T  G  K  I  T  E
1441 TGGAGTGCCGATGAACGTTTCCATACGGCTGCAGCTCAATCTGCTCATGAAGAGAGTTTCAGGCATCACAGAAACAGGAAAGATAACAGA
 408 V  V  M  P  M  I  W  F  E  E  S  G  Y  I  D  G  P  V  L  N  T  F  R  T  N  L  V  V  L  P
1531 GGTGGTGATGCCCATGATCTGGTTTGAAGAGAGTGGCTACATTGACGGTCCCGTTCTCAACACGTTCCGCACTAATCTGGTGGTGCTGCC
 438 M  V  M  E  Y  M  Q  Y  I  F  I  G  L  G  L  A  T  I  L  G  A  V  I  L  Y  L  S  D  K  V
1621 CATGGTCATGGAGTACATGCAGTACATCTTCATCGGTCTCGGACTCGCAACCATTCTGGGAGCCGTGATATTATACCTCAGCGACAAGGT
 468 K  S  K  K  C  G  Q  P  C  T  D  V  D  P  S  S  S  A  S  E  K  T  P  L  L  Q  A  S  T  S
1711 AAAAAGTAAGAAGTGTGGCCAGCCCTGCACAGATGTGGATCCATCCAGCTCCGCCAGTGAAAAGACCCCATTACTACAGGCCTCAACGAG
 498 *
1801 CTAGagccaatatccacactgcagcttactttcattagaaactcacaatcccacaatccactggccagcaaataaaagttcttttattaa
1891 gcgttgtggttgaatggactggatcaagaacagaccatcagcaacaggattatgaacactttgagtggggatgtttcttcctgtggttt
1981 ttttgcactgcagttgggtttaaagtgctctgtggggttaatactacattttctcgttctttttatttgtatgaaaaagtgagcagcgtc
2071 tgcaatgtggtacaggtagatatttttttgacccatttaaaattaaataattgataaaatagtagacaatgaattttcaaatggtgtgtt
2161 cactaatggtcaaaacactccaaaccttacttcctcagagggaaaaataaagagatttaacctttatcctgtaaataaggaactgttatg
2251 aacactttgtaacaatcacgtgctgactgaaatatcctccgaaagcagaaaaaatgctgagccaacacatttctcagtatgctgaataca
2341 cttgttgcttgtattctgtccgtaaaggcacttgtatttaatgaatgtaaaaaattctttaatatgtcattgtttgcattctcttgatgt
2431 tcagtataaaataaaatgggattatttcaaaaaaaaaaaaaaaaaaaaaaaaaaa
```

Figure 1. Nucleotide and putative amino acid sequences of *Ci*SRB1. Nucleotide (lower row) and putative amino acid (upper row) sequence numbers are shown on the left. The start codon (ATG) and stop codon (TAG) are highlighted in bold red. The mRNA instability motif (ATTTA) is double-underscored, and the poly-adenylation signal sequence (AATAA) is shown with a wavy line. Transmembrane regions are marked with yellow backgrounds. The CD36 domain is underscored. Asn-Xaa-Ser/Thr sequons are highlighted in green background, and Asn residues predicted to be N-linked glycosylation sites are highlighted in red. * indicates that amino acids stop being produced.

The full-length genomic DNA sequence of *CiSRB1* was 20,042 bp in length, with a schematic diagram depicted in Figure 2. *CiSRB1* contained 12 exons and 11 introns, adhering to the consensus GT/AG rule. Comparison with previous research indicated conserved lengths of most exons (1~7) between *CiSRB1* (Figure 2a) and *GrSRB1* (Figure 2b), with distinctions starting from exon 8. Furthermore, significant differences were observed in the intron lengths, with *CiSRB1* spanning 20,042 bp and *GrSRB1* only 10,792 bp.

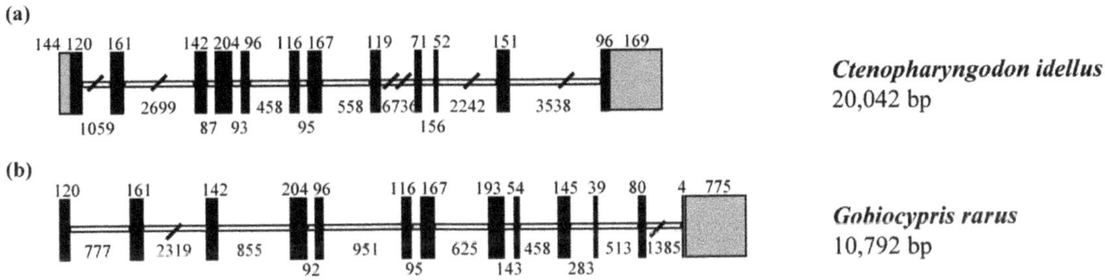

Figure 2. Genomic structure of *SRB1* genes. (**a**) *CiSRB1* and (**b**) *GrSRB1*. Exons are represented by dark bars and introns by white boxes, while 5′-UTR and 3′-UTR are indicated by dark bars with a diagonal line. The number of nucleotides in each exon or intron is shown above or below the corresponding element.

3.2. Homology Analysis of CiSRB1

The deduced amino acid sequence of *Ci*SRB1 showed 86.56%, 80.32%, 80.15%, 72.19%, 54.06%, 54.04%, 52.28%, 50.39%, and 50.00% identity with the SRB1 homologs from *G. rarus*, *D. rerio*, *Cyprinus carpio*, *Oreochromis niloticus*, *Gallus gallus*, *H. sapiens*, *Xenopus laevis*, *Bos Taurus*, and *M. musculus*, respectively (Figure 3).

The phylogenetic tree indicated that homologous SRB1 proteins from various vertebrate species clustered into four groups: mammals, birds, amphibians, and fish. All fish SRB1 proteins clustered together, with SRB1 from *C. idellus* branching most closely to SRB1 from *G. rarus* (Figure 4), reflecting a genetic relationship consistent with the evolution of species.

3.3. Predictive Analysis of Protein Interactions between CiSRB1 and GCRV Particles

In AlphaFold3, the modified local distance difference test (pLDDT) is a per-atom confidence estimate on a scale of 0–100, and the predicted aligned error (PAE) indicates model confidence in the relative orientations of the protein parts. Higher pLDDT or PAE values indicate greater confidence in the prediction [22]. High pLDDT and PAE values were observed between *Ci*SRB1 and VP5, VP7, VP5H, and VP7G (Figure S1), confirming the reliability of our predictions. These results, predicted by AlphaFold3, were then visualized by using ChimeraX. In ChimeraX, three methods are used to analyze the interactions between protein structures: "Clashes" (unfavorable interactions due to close proximity), "Contacts" (various polar and nonpolar interactions), and "H-Bonds" (hydrogen bond analysis) [25]. The interaction between the *Ci*SRB1 protein and the outer capsid protein of GCRV depends on amino acid residues at their interaction interface. There were 3 hydrogen bonds and 63 contacts between *Ci*SRB1 and VP5, 8 hydrogen bonds and 108 contacts between *Ci*SRB1 and VP7, 3 hydrogen bonds and 66 contacts between *Ci*SRB1 and VP5H, and 11 hydrogen bonds and 687 contacts between *Ci*SRB1 and VP7G (Figure 5).

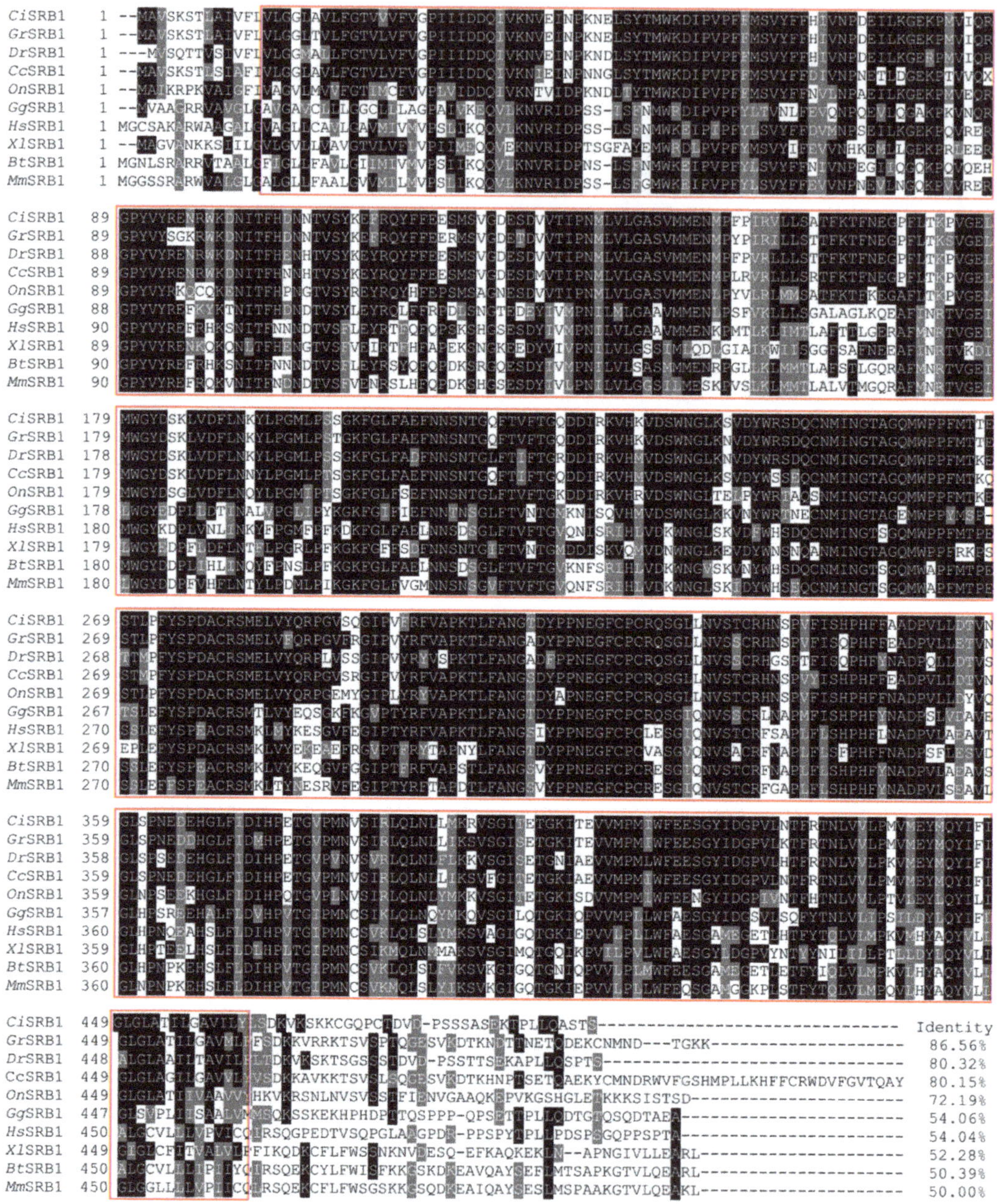

Figure 3. Multiple-sequence alignments of SRB1 proteins from different species. The dark shade indicates 100% identity, and the gray shade indicates 50% identity. The CD36 domain is highlighted in a red box. *Ci*SRB1 denotes SRB1 in *C. idellus*, *Gr*SRB1 denotes SRB1 in *G. rarus*, *Dr*SRB1 denotes SRB1 in *D. rerio*, *Cc*SRB1 denotes SRB1 in *C. carpio*, *On*SRB1 denotes SRB1 in *O. niloticus*, *Gg*SRB1 denotes SRB1 in *G. gallus*, *Xl*SRB1denotes SRB1 in *X. laevis*, *Hm*SRB1 denotes SRB1 in *H. sapiens*, *Bt*SRB1 denotes SRB1 in *B. taurus*, and *Mm*SRB1 denotes SRB1 in *M. musculus*.

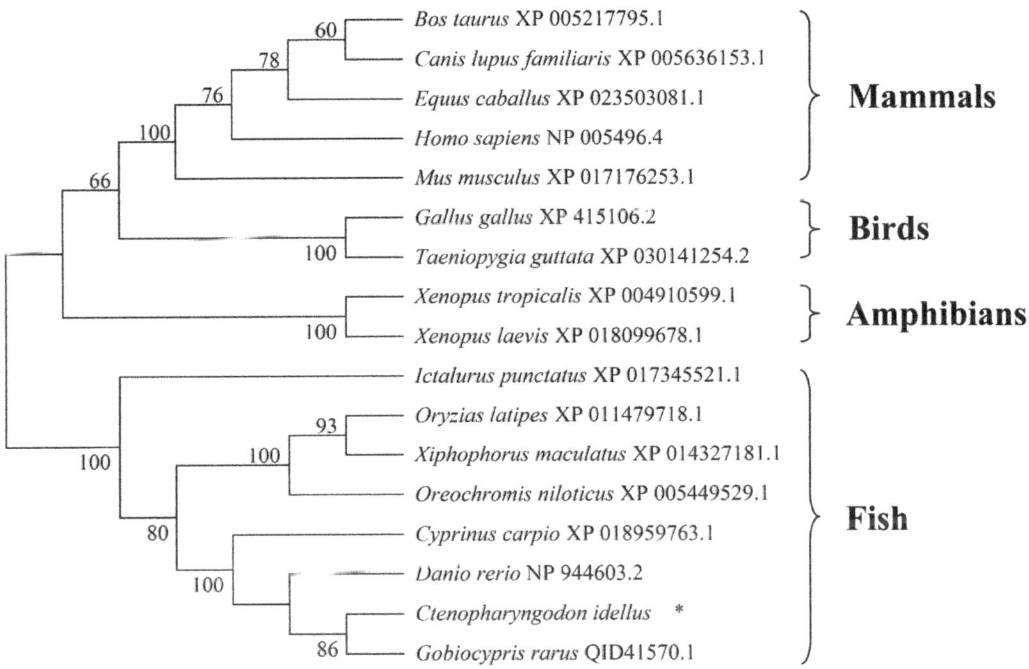

Figure 4. Neighbor-joining phylogenetic tree analysis of SRB1 proteins in vertebrates. The tree was constructed based on amino acid sequences of grass carp SRB1 and 16 orthologs by using MEGA 7.0 software. The GenBank accession numbers of the SRB1 proteins are given after the species names in the tree. * represents SRB1 in *C. idellus*.

3.4. Tissue Distribution of CiSRB1 in Healthy Grass Carp

In healthy one-year-old grass carp, *CiSRB1* expression was detected in all tested tissues, with the highest expression being in the intestines (16.48 $\pm$ 1.33 fold change), followed by muscle (6.06 $\pm$ 0.15 fold change), spleen (5.98 $\pm$ 0.06 fold change), liver (5.14 $\pm$ 0.32 fold change), and brain (4.68 $\pm$ 0.36 fold change). Lower expression levels were observed in the middle kidney (3.10 $\pm$ 0.11 fold change), heart (2.75 $\pm$ 0.04 fold change), and head kidney (2.03 $\pm$ 0.11 fold change), with the lowest being in the gills (1.00 $\pm$ 0.06 fold change). Conversely, in healthy three-year-old grass carp, *CiSRB1* exhibited high expression levels in the intestines (2.72 $\pm$ 0.10 fold change) and gills (2.20 $\pm$ 0.04 fold change), intermediate levels in the liver (1.89 $\pm$ 0.05 fold change), and low levels in the other tissues, with the lowest expression being observed in the spleen (0.17 $\pm$ 0.00 fold change). Overall, *CiSRB1* transcripts in the tested tissues of three-year-old individuals were significantly lower compared with one-year-old individuals, except for gills ($p < 0.05$) (Figure 6).

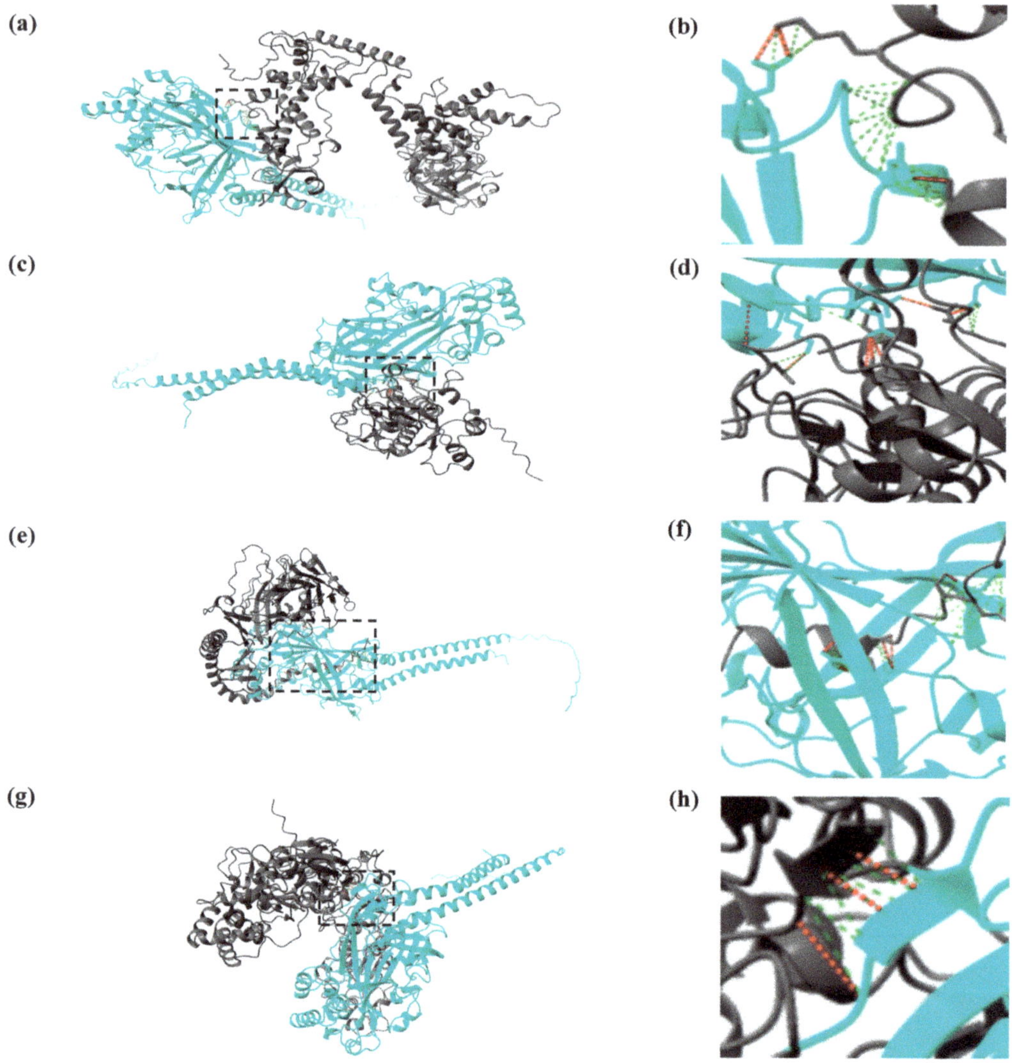

Figure 5. Visualized results of protein interactions between *Ci*SRB1 and GCRV particles by UCSF ChimeraX. The interactions are depicted as follows: (**a**) *Ci*SRB1 and VP5, (**c**) *Ci*SRB1 and VP7, (**e**) *Ci*SRB1 and VP7G, and (**g**) *Ci*SRB1 and VP5H; panels (**b**), (**d**), (**f**), and (**h**) are magnifications of panels (**a**), (**c**), (**e**), and (**g**), respectively. In the visualizations, the blue protein chain represents the *Ci*SRB1 protein, and the gray protein chain represents the outer capsid protein from different typical GCRV strains. The red dashed lines between the two protein chains represent hydrogen bonds, and the green dashed lines indicate various polar and nonpolar interactions between the molecules.

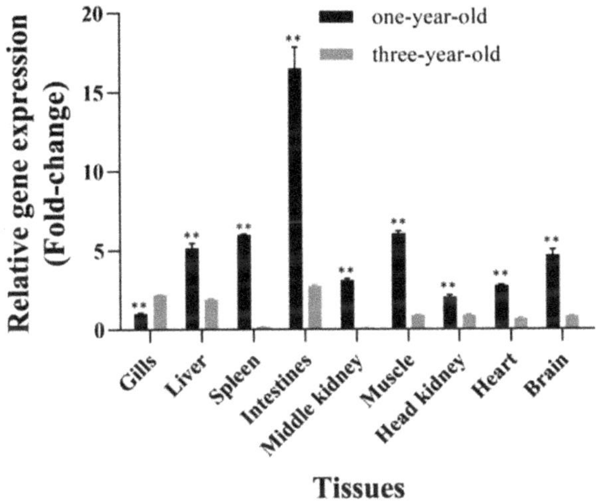

Figure 6. Tissue distribution of *CiSRB1* in healthy one-year-old and three-year-old grass carp. Data are expressed as the ratios of *CiSRB1* mRNA expression in each tissue relative to its expression in the gills from one-year-old fish. Significant differences in *CiSRB1* expression between samples from one-year-old and three-year-old fish were analyzed by one-way ANOVA followed by Dunnett's multiple comparisons post-test. ** indicates $p < 0.01$.

3.5. Mortality of Grass Carp after GCRV Infection

One-year-old and three-year-old grass carp were injected with either 0.7% NaCl or GCRV virus, and the mortality of both control (Groups I and II) and experimental (Groups III and IV) groups was evaluated after infection. No clinical disease symptoms were detected in the control fish (Group I: one-year-old fish; Group II: three-year-old fish). However, among the one-year-old grass carp injected with GCRV (Group III), mortality was observed from 6 dpi and continued up to 13 dpi (Figure 7). The total mortality of one-year-old grass carp infected with GCRV was 84% (42/50), with all deceased fish exhibiting typical symptoms of muscular or intestinal hemorrhage. In contrast, throughout the entire infection period, three-year-old grass carp infected with GCRV (Group IV) remained asymptomatic, with no mortality recorded.

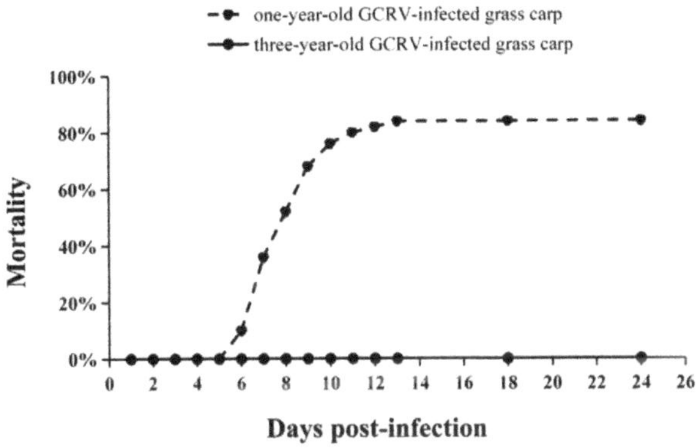

Figure 7. Mortality rates of one-year-old and three-year-old grass carp after GCRV infection. The number of deceased grass carp in each group was monitored daily until no deaths were recorded for two consecutive weeks.

3.6. Expression Profiles of *CiSRB1* following GCRV Infection

To assess the potential involvement of the *CiSRB1* gene in GCRV cell entry across different age groups, the mRNA levels of *CiSRB1* in immune-related tissues, including gills, spleen, intestines and liver, were determined.

In the gills of one-year-old grass carp, robust upregulation of *CiSRB1* expression was observed upon GCRV stimulation (Figure 8a). *CiSRB1* mRNA levels peaked 12 hpi (3.48 ± 0.15 fold change) and then quickly dropped back to the original levels by 48 hpi, with slight fluctuations thereafter. In three-year-old individuals, *CiSRB1* exhibited a different expression pattern, with transcript levels steadily declining from the peak 0 hpi (2.85 ± 0.08 fold change) to the lowest point 120 hpi (0.72 ± 0.05 fold change).

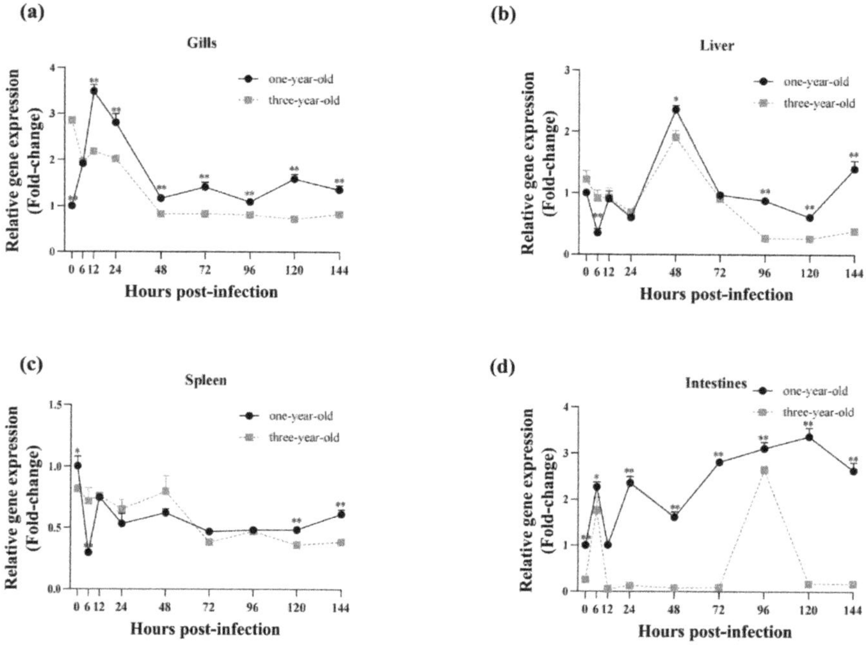

Figure 8. Temporal expression analysis of *CiSRB1* in immune tissues from one-year-old and three-year-old carp after GCRV infection. The mRNA levels of *CiSRB1* in the gills (**a**), liver (**b**), spleen (**c**), and intestines (**d**) 0~144 hpi were subjected to qPCR analysis. Expression levels of *CiSRB1* in the healthy samples from one-year-old fish (0 hpi) were set to 1.00, and *β-actin* served as an internal control to normalize the relative expression levels of the target genes. The results are based on three independent experiments and are expressed as means ± SD. Significant differences in *CiSRB1* expression between the samples from the one-year-old and three-year-old fish were analyzed by one-way ANOVA followed by Dunnett's multiple comparisons post-test. ** indicates $p < 0.01$, and * indicates $p < 0.05$.

In the liver of one-year-old grass carp, *CiSRB1* expression reached its lowest level 6 hpi (0.35 ± 0.07 fold change), followed by a gradual upregulation to peak levels (2.36 ± 0.07 fold change) 48 hpi, before declining again from 72 to 120 hpi (0.61 ± 0.07 fold change). A similar trend was observed in three-year-old grass carp, with *CiSRB1* transcripts gradually decreasing to a nadir 24 hpi (0.69 ± 0.06 fold change) (Figure 8b), followed by an increase to peak levels (1.90 folds ± 0.11 fold change). This was succeeded by a significant decline to the lowest point (0.26 ± 0.03 fold change) 120 hpi, followed by a gradual increase 144 hpi.

In the spleen of one-year-old grass carp, *CiSRB1* mRNA expression exhibited rapid downregulation, reaching its lowest level 6 hpi (0.30 ± 0.01 fold change), followed by slight variations and fluctuating patterns from 12 to 144 hpi. This fluctuating expression pattern was also observed in the spleen of three-year-old individuals infected with GCRV (Figure 8c).

In the intestines of one-year-old grass carp, *CiSRB1* mRNA expression levels demonstrated unstable changes following GCRV challenge. After GCRV exposure, *CiSRB1* expression gradually increased 6 hpi (2.36 ± 0.14 fold change), then slowly declined to baseline levels 12 hpi. Subsequently, *CiSRB1* expression was upregulated again, reaching its peak levels (3.37 ± 0.19 fold change) 120 hpi. In three-year-old grass carp stimulated with GCRV, a 1.76 to 2.65 fold-change increase in *CiSRB1* expression was observed 6 hpi and 96 hpi in the intestines, with lower transcription levels of *CiSRB1* being observed at other tested timepoints (Figure 8d).

3.7. Relative Copy Numbers of GCRV RNA in Infected Grass Carp

The relative GCRV copy numbers in four tissues (gills, liver, spleen, and intestines) were examined by qPCR by using specific primers for the M6 segment of the GCRV-HZ08 strain to determine the GCRV infection status of GCRV-infected grass carp. Minor differences in the GCRV copy numbers of gills, liver, and spleen were observed between one-year-old and three-year-old grass carp 1 dpi, indicating that the GCRV dose used for injection was appropriate (Figure 9). In one-year-old infected fish, the GCRV relative copy numbers in gills (Figure 9a), spleen (Figure 9c), and intestines (Figure 8d) showed an increasing trend and peaked 6 dpi during the tested period. However, the GCRV relative copy numbers in the liver followed a parabolic trend, reaching a maximum 3 dpi and then decreasing to the original level 6 dpi (Figure 9b). In three-year-old carp, the GCRV levels remained quite low in the gills throughout the infection (Figure 9a), and a similar parabolic curve was observed for the liver (Figure 9b). The GCRV copy numbers in the spleen increased after infection in the three-year-old fish, although they were significantly lower than those in the one-year-old fish. Similarly, the GCRV RNA levels gradually increased to their peak in the intestines 5 dpi before declining 6 dpi (Figure 9d). In general, the relative copy numbers of GCRV in the tissues of the three-year-old grass carp were significantly lower than those in the one-year-old grass carp after GCRV infection.

(a)

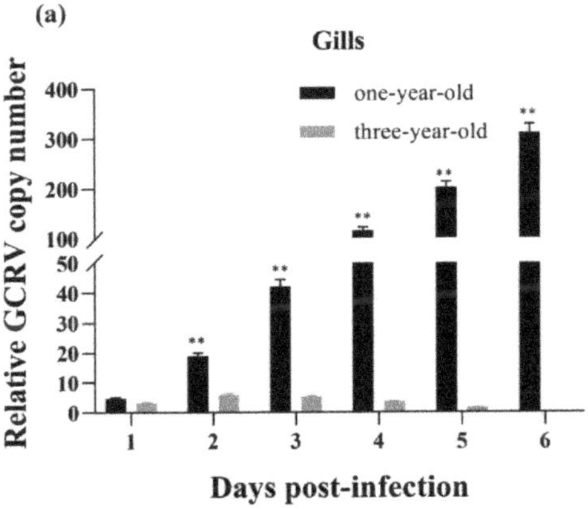

(b)

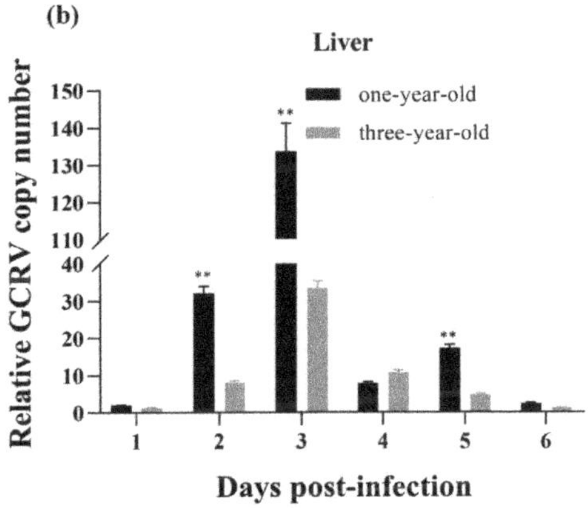

Figure 9. *Cont.*

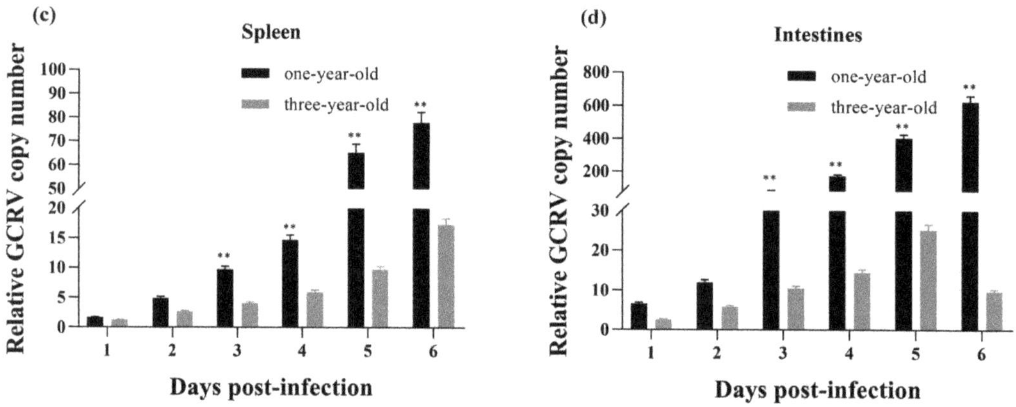

Figure 9. Relative GCRV copy numbers in infected one-year-old and three-year-old grass carp. The relative GCRV copy numbers were examined by using specific primers targeting the M6 segment of the GCRV-HZ08 strain. (**a**) The relative GCRV copy number in gills; (**b**) The relative GCRV copy number in liver; (**c**) The relative GCRV copy number in spleen; (**d**) The relative GCRV copy number in intestines. The copy number of GCRV 1 dpi in the gills of one-year-old infected fish was used for normalization. The data (expressed as means ± SD) were analyzed by one-way ANOVA followed by Dunnett's multiple comparisons post-test. ** indicates $p < 0.01$.

4. Discussion

SRs are pivotal for immune defense, functioning as PRRs. Among them, SRB1, renowned as an HDL receptor, performs vital functions in viral entry into host cells [9–12,14–16], immune regulation [7], phagocytosis of apoptotic cells [28], and more. In this study, the *SRB1* gene was discerned from *C. idellus*, designated as *CiSRB1*. The *Ci*SRB1 protein contained a CD36 domain, two transmembrane regions, and two cytoplasmic tails (Figure 1), resembling observations in other species [17]. Alignment and structural analysis of diverse SRB1 orthologs from different species unveiled that fish SRB1 proteins share notable similarities of at least 72%, clustering together in a distinct clade within the phylogenetic tree.

The genomic architecture of the *SRB1* gene varies across species. In *G. rarus* [16], *C. carpio* (Gene ID: 109090404), and *S. maximus* [20], the *SRB1* gene comprises 13 exons and 12 introns. However, *CiSRB1*, as discovered in this study, consists of 12 exons and 11 introns, akin to the genomic structure observed in *H. sapiens* (Gene ID: 949) and *Bos taurus* (Gene ID: 282346). Comparing the sequences of *GrSRB1* and *CiSRB1*, the cDNA sequences demonstrated substantial similarity, sharing up to 88.76%. Notably, differences in the size of the *CiSRB1* genomic sequence compared with that of *GrSRB1* primarily stemmed from variations in intron length. Research suggests that introns are involved in regulating gene expression, with DNA methylation within introns playing an important role in this regulatory process [29]. The role of introns in gene expression regulation, particularly in the context of GCRV mortality, warrants further investigation.

CiSRB1 exhibited a widespread distribution in various tissues with varying expression levels in healthy grass carp, predominantly expressed in the intestines, akin to reports in *S. maximus* [20], *E. coioides* [19], and Atlantic salmon (*Salmo salar*) [30]. However, a novel paralog of *SCARB1* (*SCARB1-2*) in *S. salar* was equally expressed in muscle, liver, and midgut [31]. Furthermore, it was reported that *SRB1* exhibited the highest expression in the liver and steroidogenic tissues in mammals or nonmammals, such as *M. musculus* [32], *D. rerio* [33], goldfish (*Carassius auratus*) [34], and *G. rarus* [16]. This suggests that the tissue distribution of *SRB1* varies across different species. It is possible that *SRB1* emerged relatively early in vertebrate evolution, possessing several metabolic functions [35].

Lower levels of *CiSRB1* transcript were detected in most tested tissues of three-year-old grass carp compared with one-year-old individuals, except for the gills. These findings paralleled the expression pattern observed for *JAM-A* in grass carp, a potential receptor for GCRV infection, where *JAM-A* exhibited higher expression levels in one-year-old grass carp compared with three-year-old individuals [26]. Furthermore, it was found that one-year-old grass carp had a mortality rate of up to 84% 6 dpi, whereas all three-year-old grass carp survived after GCRV infection (Figure 6), consistently with previous research [26,27,35]. Similar age-dependent patterns of susceptibility to viral infection have been documented in various species, such as spring viremia of carp virus (SVCV) challenge in North American fish species [36] and European common carp (*Cyprinus carpio carpio*) [37], as well as nervous necrosis virus (NNV) challenge in barramundi (*Lates calcarifer*) [38] and gilthead seabream (*Sparus aurata*) [39]. It is possible that the immunity of young individuals, including the expression of antiviral, inflammatory, and cell-mediated cytotoxicity genes, renders them more susceptible to viral infection [35,39].

SRB1 serves as an essential factor in the entry of multiple viruses. It can interact directly with HCV particles or indirectly via ApoB or HDL to promote HCV entry into cells [9–11]. The interaction of ApoA-I and SRB1 facilitates the attachment and entry of DV into cells [12]. SR-B1 mediates the interaction of cholesterol with the spike of SARS-CoV-2, thereby facilitating the entry of SARS-CoV-2 into host cells [14]. SRB1 also plays a crucial role in the innate immunity of teleost fish. In large yellow croakers (*Larimichthys crocea*), notable upregulation of *SRB1* was observed after *Pseudomonas plecoglossicida* infection [40]. Similarly, *SRB1* is implicated in the immune response of Japanese flounder (*Paralichthys olivaceus*) against *Vibrio anguillarum* infection [41]. In *S. salar*, two distinct paralogues (*SRB1-a* and *SRB1-b*) show similar responses to viral mimic (dsRNA: pIC) [42]. This suggests that *SRB1* influences both bacterial and viral infection. To explore the role of *CiSRB1* during GCRV infection in grass carp, challenges were conducted across different age groups to evaluate responses. Notable upregulation of *CiSRB1* transcripts was observed in major immune tissues (gills, liver, spleen, and intestines) following GCRV challenge. This result corroborates our previous transcriptome analysis, which demonstrates that the expression of phagosome pathway-related gene *SRB1* in CIK cells is upregulated from 8 to 24 h after GCRV infection [15]. Similar findings have been confirmed for *CiSRB2a* and *CiSRB2b*, the alternative splicing products of *SRB1*, whose expression levels are elevated in the gills and spleen following GCRV infection [43]. This suggests that SRB1 plays a crucial role in the immune regulation of *C. idellus* in response to GCRV challenge. In *G. rarus*, a model fish used to analyze the mechanism of GCRV infection, *GrSRB1* expression is upregulated in the gills, spleen, liver, and intestines after GCRV infection. Furthermore, co-immunoprecipitation assays reveal that *Gr*SRB1 can interact with GCRV particles, such as VP5 and VP7 proteins from the GCRV-873 strain and the VP7 protein from the GCRV-GD108 strain [16]. The protein sequence similarity between *Gr*SRB1 and *Ci*SRB1 was approximately 90%, indicating that both likely play a similar role in GCRV infection. Additionally, protein interactions between *Ci*SRB1 and GCRV particles were predicted by AlphaFold3 and visualized by using ChimeraX. These predictions indicated that *Ci*SRB1 interacts with the outer capsid proteins of typical GCRV strains (VP5 and VP7 proteins from the GCRV-873 strain, VP5 protein from the HZ08 strain, and VP7 protein from the GCRV-GD108 strain) through hydrogen bonding as well as various polar and nonpolar interactions. These findings preliminarily infer that SRB1 may interact with the outer capsid proteins of GCRV and then facilitates the entry of GCRV into host cells via endocytosis, similarly to other reported viral attachment protein [44–46]. These research findings provide support for our subsequent studies on the specific mechanisms of *Ci*SRB1 in GCRV infection. Further research is needed to determine whether *Ci*SRB1 can be directly or indirectly associated with GCRV particles to facilitate entry into host cells, such as knockdown or overexpression of *CiSRB1* in CIK cells.

Significant differences were observed between one-year-old and three-year-old grass carp after GCRV challenge, with lower *CiSRB1* expression levels being detected in the older group. Similarly, the relative copy numbers of GCRV across tissues revealed substantially higher levels in one-year-old grass carp compared with their older counterparts, consistently with prior findings [26,35]. This indicates that *CiSRB1* is positively correlated with the degree of GCRV infection in one-year-old grass carp. The overexpression analysis of *CiSRB2a* and *CiSRB2b* in CIK cells also supports this, showing increased GCRV copy numbers following overexpression of *CiSR-B2a* and *CiSR-B2b* [43]. As a potential receptor for GCRV infection, higher levels of endogenous SRB1 may facilitate increased interactions with the viral outer capsid proteins, thereby promoting the rapid entry of GCRV into host cells via endocytosis, resulting in a greater number of GCRV copies within the host cells. Our previous study finds that the accumulation of GCRV copy numbers up to a certain threshold may lead to hemorrhagic disease, leading to organ malfunction and death [27,35]. In this study, the GCRV copy numbers peaked 6 dpi, coinciding with observed mortality from 6 dpi in one-year-old grass carp, indicating that the viral load had reached a fatal threshold. Conversely, the GCRV copy numbers in three-year-old grass carp were far lower those in the one-year-old individuals. Moreover, multi-omics sequencing indicates that three-year-old grass carp recognize the virus immediately, rapidly activate the immune response, and enhance host translation machinery to defend against viruses [35]. This coincides with our finding that GCRV copy numbers decreased from 2 dpi in the gills, 3 dpi in the liver, and 5 dpi in the intestines in the three-year-old individuals. Overall, although SRB1 is also expressed in three-year-old grass carp, allowing GCRV to enter, their immune system rapidly recognizes the virus and activates the immune response against it. Consequently, the viral load is not enough to cause hemorrhagic disease, preventing organ malfunction and death, allowing three-year-old grass carp to survive after GCRV infection. This observation accentuates the importance of *CiSRB1* in the context of GCRV infection and provides insights into age-dependent susceptibility to reovirus.

5. Conclusions

In this study, *CiSRB1* was identified, and its gene structure and expression levels were thoroughly analyzed. It was unveiled that *Ci*SRB1 closely resembles SRB1 in other vertebrates, exhibiting the conserved CD36 domain. The predictive analysis of protein interactions revealed that *Ci*SRB1 interact with the outer capsid proteins of typical GCRV strains through hydrogen bonding as well as various polar and nonpolar interactions. Tissue distributions exhibited age-dependent characteristics, with *CiSRB1* being highly expressed in the intestines and moderately in the muscle, spleen, liver, and brain of one-year-old grass carp, while maintaining relatively low levels in three-year-old grass carp. Notable upregulation of *CiSRB1* transcripts was observed in major immune tissues (gills, spleen, liver, and intestines) following GCRV challenge. Furthermore, significant differences were found between one-year-old and three-year-old grass carp, with lower *CiSRB1* expression levels being detected in the older group. Additionally, higher mortality and more GCRV copy numbers were observed in one-year-old grass carp compared with their three-year-old counterparts, confirming the existence of age-related restriction to GCRV in grass carp. These results suggest that the cell-surface expression of *CiSRB1* may decrease with age following GCRV infection, potentially acting as the reovirus receptor and reducing reovirus binding to target cells. This sets the stage for further investigations into the mechanisms underlying viral infection in fish.

Supplementary Materials: The following supporting information can be downloaded at: https://www.mdpi.com/article/10.3390/fishes9070276/s1, Figure S1: The predicted results of protein interactions between *Ci*SRB1 and GCRV particles by AlphaFold3. The interactions are depicted as follows: (**a**) *Ci*SRB1 and VP5, (**b**) *Ci*SRB1 and VP7, (**c**) *Ci*SRB1 and VP5H, and (**d**) *Ci*SRB1 and VP7G. Predictions colored according to pLDDT (orange: pLDDT $\leq$ 50; yellow: 50 < pLDDT $\leq$ 70; light blue: 70 < pLDDT $\leq$ 90; and dark blue: 90 $\leq$ pLDDT < 100) and PAE matrix of same prediction (darker color means more confident).

Author Contributions: Conceptualization, Y.Z. and J.S.; Methodology, Y.L. and Q.L.; Investigation, Y.Z., P.C. and R.H.; Data Curation, M.O.; Writing—Original Draft Preparation, K.C., J.Z. and J.S.; Writing—Review and Editing, Y.Z., Y.W. and Y.L.; Visualization, Q.L., P.C. and R.H.; Supervision,

K.C., J.Z., Y.W. and M.O.; Project Administration, M.O.; Funding Acquisition, J.Z. All authors have read and agreed to the published version of the manuscript.

Funding: This work was supported by the Central Public-interest Scientific Institution Basal Research Fund, CAFS (2023TD37), China-ASEAN Maritime Cooperation Fund (CAMC-2018F), and National Freshwater Genetic Resource Center (FGRC18537).

Institutional Review Board Statement: All fish experiments in the present study were approved by the Pearl River Fisheries Research Institute and the Chinese Academy of Fishery Sciences under contract LAEC-PRFRI-2020-01-01, and the experimental process complied with the protocols of international guidelines for the ethical use of animals in research.

Informed Consent Statement: Not applicable.

Data Availability Statement: Data are contained within the article and Supplementary Materials.

Conflicts of Interest: The authors declare no conflicts of interest.

References

1. Yang, L.; Wang, C.; Huang, Y.; Xu, B.; Liu, Y.; Yu, J.; Xiong, L.; Xiao, T.; Liu, Q. Identification of the *C1qDC* gene family in grass carp (*Ctenopharyngodon idellus*) and the response of *C1qA*, *C1qB*, and *C1qC* to GCRV infection in vivo and in vitro. *Fish Shellfish Immunol.* **2024**, *148*, 109477. [CrossRef] [PubMed]
2. Yang, L.; Su, J. Type II grass carp reovirus infects leukocytes but not erythrocytes and thrombocytes in grass carp (*Ctenopharyngodon idella*). *Viruses* **2021**, *13*, 870. [CrossRef] [PubMed]
3. Lu, Y.; Zhao, W.; Ji, N.; Xu, D.; Li, Y.; Xiao, T.; Wang, J.; Zou, J. Analysis of tissue tropism of GCRV-II infection in grass carp using a VP35 monoclonal antibody. *Dev. Comp. Immunol.* **2024**, *157*, 105189. [CrossRef] [PubMed]
4. Xiao, F.; Liao, L.; Xu, Q.; He, Z.; Xiao, T.; Wang, J.; Huang, J.; Yu, Y.; Wu, B.; Yan, Q. Host–Microbiota Interactions and Responses to Grass Carp Reovirus Infection in *Ctenopharyngodon idellus*. *Environ. Microbiol.* **2021**, *23*, 431–447. [CrossRef] [PubMed]
5. Ouyang, G.; Yuan, L.; Xia, X.-Q.; Zhang, W.; Shi, M. Transcriptomes of zebrafish in early stages of multiple viral invasions reveal the role of sterols in innate immune switch-on. *Int. J. Mol. Sci.* **2023**, *24*, 4427. [CrossRef]
6. Ravi, S.; Martin, L.C.; Krishnan, M.; Kumaresan, M.; Manikandan, B.; Ramar, M. Interactions between macrophage membrane and lipid mediators during cardiovascular diseases with the implications of scavenger receptors. *Chem. Phys. Lipids* **2024**, *258*, 105362. [CrossRef]
7. Vogel, A.; Brunner, J.S.; Hajto, A.; Sharif, O.; Schabbauer, G. Lipid scavenging macrophages and inflammation. *Biochim. Biophys. Acta BBA Mol. Cell Biol. Lipids* **2022**, *1867*, 159066. [CrossRef]
8. Huby, T.; Le Goff, W. Macrophage SR-B1 in atherosclerotic cardiovascular disease. *Curr. Opin. Lipidol.* **2022**, *33*, 167–174. [CrossRef]
9. Barth, H.; Schnober, E.K.; Neumann-Haefelin, C.; Thumann, C.; Zeisel, M.B.; Diepolder, H.M.; Hu, Z.; Liang, T.J.; Blum, H.E.; Thimme, R.; et al. Scavenger receptor class B is required for hepatitis C virus uptake and cross-presentation by human dendritic cells. *J. Virol.* **2008**, *82*, 3466–3479. [CrossRef]
10. Colpitts, C.C.; Baumert, T.F. SCARB1 variants and HCV infection: Host susceptibility is lost in translation. *J. Hepatol.* **2017**, *67*, 211–213. [CrossRef]
11. Jennelle, L.T.; Magoro, T.; Angelucci, A.R.; Dandekar, A.; Hahn, Y.S. Hepatitis C virus alters macrophage cholesterol metabolism through interaction with scavenger receptors. *Viral Immunol.* **2022**, *35*, 223–235. [CrossRef]
12. Li, Y.; Kakinami, C.; Li, Q.; Yang, B.; Li, H. Human apolipoprotein A-I is associated with dengue virus and enhances virus infection through SR-BI. *PLoS ONE* **2013**, *8*, e70390. [CrossRef]
13. Langlois, A.-C.; Marinach, C.; Manzoni, G.; Silvie, O. *Plasmodium* sporozoites can invade hepatocytic cells independently of the ephrin receptor A2. *PLoS ONE* **2018**, *13*, e0200032. [CrossRef]
14. Tang, Y.; Hu, L.; Liu, Y.; Zhou, B.; Qin, X.; Ye, J.; Shen, M.; Wu, Z.; Zhang, P. Possible mechanisms of cholesterol elevation aggravating COVID-19. *Int. J. Med. Sci.* **2021**, *18*, 3533–3543. [CrossRef]
15. Chen, G.; He, L.; Luo, L.; Huang, R.; Liao, L.; Li, Y.; Zhu, Z.; Wang, Y. Transcriptomics sequencing provides insights into understanding the mechanism of grass carp reovirus infection. *Int. J. Mol. Sci.* **2018**, *19*, 488. [CrossRef] [PubMed]
16. Ou, M.; Huang, R.; Luo, Q.; Xiong, L.; Chen, K.; Wang, Y. Characterisation of scavenger receptor class B type 1 in rare minnow (*Gobiocypris rarus*). *Fish Shellfish Immunol.* **2019**, *89*, 614–622. [CrossRef]
17. Shen, W.-J.; Asthana, S.; Kraemer, F.B.; Azhar, S. Thematic review series: Lipid transfer proteins scavenger receptor B type 1: Expression, molecular regulation, and cholesterol transport function. *J. Lipid Res.* **2018**, *59*, 1114–1131. [CrossRef]
18. Cai, L.; Ji, A.; de Beer, F.C.; Tannock, L.R.; van der Westhuyzen, D.R. SR-BI protects against endotoxemia in mice through its roles in glucocorticoid production and hepatic clearance. *J. Clin. Investig.* **2008**, *118*, 364–375. [CrossRef]
19. Han, H.; Wang, L.; Xu, S.; Wang, S.; Yang, M.; Qin, Q.; Wei, S. Identification and characterization of scavenger receptor class B type 1 in orange-spotted grouper, *Epinephelus coioides*. *Aquaculture* **2022**, *546*, 737366. [CrossRef]

20. Li, C.; Ge, X.; Su, B.; Fu, Q.; Wang, B.; Liu, X.; Ren, Y.; Song, L.; Yang, N. Characterization of class B scavenger receptor type 1 (*SRB1*) in turbot (*Scophthalmus maximus* L.). *Fish Shellfish Immunol.* **2020**, *100*, 358–367. [CrossRef]
21. Wang, Y.; Lu, Y.; Zhang, Y.; Ning, Z.; Li, Y.; Zhao, Q.; Lu, H.; Huang, R.; Xia, X.; Feng, Q.; et al. The draft genome of the grass carp (*Ctenopharyngodon idellus*) provides insights into its evolution and vegetarian adaptation. *Nat. Genet.* **2015**, *47*, 625–631. [CrossRef] [PubMed]
22. Abramson, J.; Adler, J.; Dunger, J.; Evans, R.; Green, T.; Pritzel, A.; Ronneberger, O.; Willmore, L.; Ballard, A.J.; Bambrick, J.; et al. Accurate structure prediction of biomolecular interactions with AlphaFold 3. *Nature* **2024**, *630*, 493–500. [CrossRef]
23. Cheng, L.; Fang, Q.; Shah, S.; Atanasov, I.C.; Zhou, Z.H. Subnanometer-resolution structures of the grass carp reovirus core and virion. *J. Mol. Biol.* **2008**, *382*, 213–222. [CrossRef] [PubMed]
24. Zhang, F.; Guo, H.; Zhang, J.; Yan, L.; Chen, Q.; Yan, S.; Fang, Q. VP5 autocleavage is required for efficient infection by in vitro-recoated aquareovirus particles. *J. Gen. Virol.* **2015**, *96*, 1795–1800. [CrossRef]
25. Meng, E.C.; Goddard, T.D.; Pettersen, E.F.; Couch, G.S.; Pearson, Z.J.; Morris, J.H.; Ferrin, T.E. UCSF ChimeraX: Tools for structure building and analysis. *Protein Sci.* **2023**, *32*, e4792. [CrossRef]
26. Du, F.; Su, J.; Huang, R.; Liao, L.; Zhu, Z.; Wang, Y. Cloning and preliminary functional studies of the *JAM-A* gene in grass carp (*Ctenopharyngodon idellus*). *Fish Shellfish Immunol.* **2013**, *34*, 1476–1484. [CrossRef] [PubMed]
27. Chen, G.; Xiong, L.; Wang, Y.; He, L.; Huang, R.; Liao, L.; Zhu, Z.; Wang, Y. Different responses in one-year-old and three-year-old grass carp reveal the mechanism of age restriction of GCRV infection. *Fish Shellfish Immunol.* **2019**, *86*, 702–712. [CrossRef]
28. Osada, Y.; Sunatani, T.; Kim, I.-S.; Nakanishi, Y.; Shiratsuchi, A. Signalling pathway involving GULP, MAPK and Rac1 for SR-BI-induced phagocytosis of apoptotic cells. *J. Biochem.* **2009**, *145*, 387–394. [CrossRef]
29. Walker, C.G.; Littlejohn, M.D.; Meier, S.; Roche, J.R.; Mitchell, M.D. DNA methylation is correlated with gene expression during early pregnancy in *Bos taurus*. *Physiol. Genom.* **2013**, *45*, 276–286. [CrossRef]
30. Kleveland, E.J.; Syvertsen, B.L.; Ruyter, B.; Vegusdal, A.; Jørgensen, S.M.; Gjøen, T. Characterization of scavenger receptor class B, type I in Atlantic salmon (*Salmo salar* L.). *Lipids* **2006**, *41*, 1017–1027. [CrossRef]
31. Sundvold, H.; Helgeland, H.; Baranski, M.; Omholt, S.W.; Våge, D. Characterisation of a novel paralog of scavenger receptor class B member I (SCARB1) in Atlantic salmon (*Salmo salar*). *BMC Genet.* **2011**, *12*, 52. [CrossRef] [PubMed]
32. Brundert, M.; Ewert, A.; Heeren, J.; Behrendt, B.; Ramakrishnan, R.; Greten, H.; Merkel, M.; Rinninger, F. Scavenger receptor class B type I mediates the selective uptake of high-density lipoprotein–associated cholesteryl ester by the liver in mice. *Arterioscler. Thromb. Vasc. Biol.* **2005**, *25*, 143–148. [CrossRef] [PubMed]
33. Verwilligen, R.A.F.; Mulder, L.; Araújo, P.M.; Carneiro, M.; Bussmann, J.; Hoekstra, M.; Van Eck, M. Zebrafish as outgroup model to study evolution of scavenger receptor class B type I functions. *Biochim. Biophys. Acta BBA Mol. Cell Biol. Lipids* **2023**, *1868*, 159308. [CrossRef]
34. Duggan, A.E.; Marie, R.S.; Callard, I.P. Expression of SR-BI (scavenger receptor class B type I) in turtle (*Chrysemys picta*) tissues and other nonmammalian vertebrates. *J. Exp. Zool.* **2002**, *292*, 430–434. [CrossRef]
35. He, L.; Zhu, D.; Liang, X.; Li, Y.; Liao, L.; Yang, C.; Huang, R.; Zhu, Z.; Wang, Y. Multi-Omics sequencing provides insights into age-dependent susceptibility of grass carp (*Ctenopharyngodon idellus*) to reovirus. *Front. Immunol.* **2021**, *12*, 694965. [CrossRef] [PubMed]
36. Emmenegger, E.J.; Sanders, G.E.; Conway, C.M.; Binkowski, F.P.; Winton, J.R.; Kurath, G. Experimental infection of six North American fish species with the North Carolina strain of spring viremia of carp virus. *Aquaculture* **2016**, *450*, 273–282. [CrossRef]
37. Embregts, C.W.E.; Rigaudeau, D.; Veselý, T.; Pokorová, D.; Lorenzen, N.; Petit, J.; Houel, A.; Dauber, M.; Schütze, H.; Boudinot, P.; et al. Intramuscular DNA vaccination of juvenile carp against spring viremia of carp virus induces full protection and establishes a virus-specific B and T cell response. *Front. Immunol.* **2017**, *8*, 1340. [CrossRef]
38. Jaramillo, D.; Hick, P.; Whittington, R.J. Age dependency of nervous necrosis virus infection in barramundi *Lates Calcarifer* (Bloch). *J. Fish Dis.* **2017**, *40*, 1089–1101. [CrossRef] [PubMed]
39. García-Álvarez, M.Á.; Arizcun, M.; Chaves-Pozo, E.; Cuesta, A. Profile of innate immunity in gilthead seabream larvae reflects mortality upon betanodavirus reassortant infection and replication. *Int. J. Mol. Sci.* **2022**, *23*, 5092. [CrossRef]
40. Sun, X.; Bao, N.; Rui, C.; Xue, Y.; Fang, Q.; Zheng, T.; Lin, Z.; Liu, X.; Wang, X. Identification of large yellow croakers (*Larimichthys crocea*) scavenger receptor genes: Involvement in immune response to *Pseudomonas plecoglossicida* infection and hypoxia-exposure experiments. *Fish Shellfish Immunol.* **2024**, *144*, 109307. [CrossRef]
41. Ning, X.; Sun, L. Gene network analysis reveals a core set of genes involved in the immune response of Japanese flounder (*Paralichthys olivaceus*) against vibrio anguillarum infection. *Fish Shellfish Immunol.* **2020**, *98*, 800–809. [CrossRef] [PubMed]
42. Eslamloo, K.; Xue, X.; Hall, J.R.; Smith, N.C.; Caballero-Solares, A.; Parrish, C.C.; Taylor, R.G.; Rise, M.L. Transcriptome profiling of antiviral immune and dietary fatty acid dependent responses of Atlantic salmon macrophage-like cells. *BMC Genom.* **2017**, *18*, 706. [CrossRef] [PubMed]
43. Li, Y.; Huang, R.; Chen, L.; Li, Y.; Li, Y.; Liao, L.; He, L.; Zhu, Z.; Wang, Y. Characterization of *SR-B2a* and *SR-B2b* genes and their ability to promote GCRV infection in grass carp (*Ctenopharyngodon idellus*). *Dev. Comp. Immunol.* **2021**, *124*, 104202. [CrossRef]
44. Wang, H.; Liu, W.; Yu, F.; Lu, L. Disruption of clathrin-dependent trafficking results in the failure of grass carp reovirus cellular entry. *Virol. J.* **2016**, *13*, 25. [CrossRef] [PubMed]

45. Zhang, F.; Guo, H.; Zhang, J.; Chen, Q.; Fang, Q. Identification of the caveolae/raft-mediated endocytosis as the primary entry pathway for aquareovirus. *Virology* **2018**, *513*, 195–207. [CrossRef]
46. Sutherland, D.M.; Strebl, M.; Koehler, M.; Welsh, O.L.; Yu, X.; Hu, L.; Dos Santos Natividade, R.; Knowlton, J.J.; Taylor, G.M.; Moreno, R.A.; et al. NgR1 binding to reovirus reveals an unusual bivalent interaction and a new viral attachment protein. *Proc. Natl. Acad. Sci. USA* **2023**, *120*, e2219404120. [CrossRef]

fishes

Article

Mechanistic Insights into Nonylphenol Stress on *BMP2* and *BMP4* Gene Expression in Red Crucian Carp (*Carassius auratus* Red var.)

Die Li [1,†], Xiaojuan Cui [1,†], Shuailin Chen [1], Jia Xu [1], Yujing Li [1], Qiongyu Zhang [2] and Yuandong Sun [1,*]

[1] School of Life and Health Sciences, Hunan University of Science and Technology, Xiangtan 411201, China
[2] School of Fundamental Sciences, YongZhou Vocational Technical College, Yongzhou 425100, China
* Correspondence: syd@hnust.edu.cn; Tel.: +86-13786280789
† These authors contributed equally to the work.

Abstract: Nonylphenol (NP) is a known endocrine-disrupting chemical (EDC) that has been shown to affect bone development in mammals. However, the detrimental impacts of NP on the skeletal growth and development of aquatic species, especially bony fish, remain poorly understood. Bone morphogenic proteins (BMPs), essential for bone formation and osteoblast differentiation, act through the BMP-Smad signaling pathway. In this study, two *BMP* genes, *BMP2* and *BMP4*, were cloned and characterized in the red crucian carp (*Carassius auratus* red var.). The full-length cDNAs of *BMP2* and *BMP4* were 2029 bp and 2095 bp, respectively, encoding polypeptides of 411 and 433 amino acids, and share a typical TGF-β domain with other BMPs. The tissue expression patterns of both genes were identified, showing ubiquitous expression across all studied tissues. Additionally, the exposure of embryos or adult fish to NP stress resulted in a downregulation of *BMP2*, *BMP4*, and other genes associated with the BMP-Smad signaling pathway. Moreover, the combined treatment of adult fish with NP and the specific BMP receptor inhibitor significantly reduced these genes' expression. These findings elucidate the mechanism of NP stress on *BMP2* and *BMP4*, suggesting a role for the BMP-Smad signaling pathway in the response to endocrine-disrupting chemicals in fish.

Keywords: Nonylphenol; endocrine-disrupting chemicals; BMPs; toxic effects; *Carassius auratus* red var.

Key Contribution: This study suggests the crucial role of *BMP2* and *BMP4* in the adaptive responses of teleost fish to NP and highlights the significance of safeguarding aquatic ecosystems against EDC pollution.

1. Introduction

Nonylphenol (NP), a prominent endocrine-disrupting chemical (EDC) and the primary degradation product of alkylphenol ethoxylates, exhibits estrogenic activity in a variety of wildlife species [1]. Widespread in industrial applications and consumer products, human and animal exposure to NP occurs through multiple routes, including latex coatings, adhesives, paper products, detergents, and cosmetics [2]. Research on mammals has revealed that perinatal exposure to NP can affect brain function [3], cardiac function [4], and bone development [5] in the offspring of exposed individuals. In particular, the impact of NP on bone development has been well established. For instance, exposure to NP significantly compromises bone integrity by affecting osteoblasts and osteoclasts, which are vital for the formation and homeostasis of bone tissue [6,7]. Additionally, NP may contribute to abnormal skeletal development by interfering with critical signaling pathways, such as Wingless/Int-1 (Wnt) and β-catenin, either directly or indirectly, thereby disrupting the normal processes of skeletal growth and maturation [8]. NP also exerts toxic effects on aquatic organisms, resulting in multi-organ damage [9–11]. It has been shown that exposure to NP is correlated with the development of skeletal abnormalities in fish embryos.

Abnormalities were observed in developing Asian stinging catfish (*Heteropneustes fossilis*) larvae exposed to NP 5 h after hatching in [12]. In zebrafish (*Danio rerio*), 24 h of embryonic exposure to NP resulted in the observation of tail skeletal malformations in [13]. Similarly, in goldfish (*Carassius auratus*), early blastula-stage embryos were exposed to NP and exhibited skeletal malformations between 24 and 72 h after fertilization following exposure to varying concentrations of NP in [14].

During skeletal development, the majority of the bones in the body are established by the endochondral bone formation process [15]. Chondrocyte maturation and the endochondral bone development process are tightly regulated by a series of growth factors and transcription factors, including bone morphogenetic proteins (BMPs), which play a crucial role [16]. BMPs activate the BMP-Smad signaling pathway, which is essential for osteogenesis, skeletal development, and bone formation [17]. Specifically, BMP2 and BMP4 act as secreted ligands that engage serine–threonine kinase-type II receptors, leading to the activation of type I receptors and the subsequent phosphorylation of Smad proteins. This cascade regulates the expression of key bone formation markers such as runt-related transcription factor 2 (Runx2) and Osterix [18]. Multiple endocrine-disrupting chemicals (EDCs), such as bisphenol A (BPA) [19] and polychlorinated biphenyls (PCBs) [20], have been shown to interfere with bone development by inhibiting genes associated with the BMP-Smad pathway. Studies of various mammalian species, including mice [21], pigs [22], and goats [23], have highlighted the significant role of BMP2 and BMP4 in bone development. Additionally, extensive research has been conducted on BMP2 and BMP4 in various fish species. The cDNA sequences of *BMP2* and *BMP4* have been identified in early investigations focusing on zebrafish (*Danio rerio*) [24] and Japanese flounder (*Paralichthys olivaceus*) [25]. Subsequently, studies delving into the expression of *BMP2* were carried out on Jian carp (*Cyprinus carpio* var. Jian) [26] and barbel steed (*Hemibarbus labeo*) [27]. Similarly, *BMP4* expression studies were conducted on mandarin fish (*Siniperca chuatsi*) [28], providing further insights into the significance of *BMP2* and *BMP4* in fish skeletal development.

Carassius auratus red var. accounts for an important proportion of freshwater aquaculture production worldwide, but it is susceptible to various factors during production [29]. A previous study by our research group showed that NP affects *C. auratus* red var. and leads to the development of abnormal skeletons [30]; however, the molecular mechanism of NP's effect on the skeletal development of *C. auratus* red var. remains unclear. BMPs act as the key genes of the BMP-Smad pathway, which are known to be involved in the regulation of skeletal development. In the present study, two key members of the BMP family (*BMP2* and *BMP4*) were successfully cloned and characterized from *C. auratus* red var. The expression levels of the *BMP2* and *BMP4* transcripts in different tissues were analyzed by real-time fluorescence quantitative PCR (qRT-PCR). Furthermore, the temporal patterns of *BMP2* and *BMP4* in response to NP exposure were investigated in embryos and adult fish. The results of this study can help broaden the understanding of the roles of the BMP-Smad pathway in response to environmental endocrine disruptors.

2. Materials and Methods

2.1. Fish and Sampling

Two-year-old healthy *C. auratus* red var. were obtained from the Engineering Research Center of Polyploid Fish Breeding and Reproduction of the State Education Ministry at Hunan Normal University. The fish were acclimatized in an indoor freshwater tank at 25 ± 1 °C and fed a commercial diet (crude protein, 32.2%; crude lipid, 6.5%; ash, 10.4%; and gross energy, 18.5 MJ/kg) twice daily at 9:00 and 16:00 for one week. After no abnormal symptoms were observed, the *C. auratus* red var. fish were subjected to further study.

Three healthy fish were sacrificed as one group, and samples from the gill (G), caudal fin (C), heart (H), intestine (I), kidney (K), liver (L), muscle (M), brain (B), and spleen (S) were collected. All samples were immediately homogenized in TRIzol reagent (Invitrogen, Carlsbad, CA, USA) and stored at −80 °C until needed for RNA extraction for cloning and the detection of tissue differential expression in *BMP2* and *BMP4* genes.

2.2. NP Treatments of Embryos

A previous study by our research group found that the embryos of *C. auratus* red var. treated with 3 μmol/L of NP exhibited relatively obvious skeletal malformations, and the mortality rate was low [30]. Therefore, 3 μmol/L NP was selected as the experimental concentration. The eggs and sperm from three sexually mature male and female *C. auratus* red var. were artificially fertilized using the semi-dry method. The fertilized eggs were then placed in a 25 cm sterile, dry Petri dish containing aerated water for incubation. After fertilization for 2 min, all the embryos were exposed to NP with concentrations of 0 μmol/L (blank control, 0.01% ethanol) or 3 μmol/L. Each treatment group was employed for 5 parallel replicates (5 Petri dishes), with approximately 300 embryos in each Petri dish. Embryo incubation and NP exposure were carried out in the plates at 25 ± 1 °C. The water was changed approximately every 4 h. Intact embryos were collected at six stages: 6 hpf (hours post-fertilization), 12 hpf, 24 hpf, 48 hpf, 72 hpf, and 96 hpf. From each group, we collected 3 tubes of 30 embryos per tube from 5 Petri dishes, for a total of 6 tubes at each time point (Table S1). We used liquid nitrogen to stop embryo development and then stored the samples at −80 °C until RNA extraction.

2.3. NP Exposure and Inhibitor Intraperitoneal Injection of Adult Fish

Before the initiation of the study, all fish were acclimatized to laboratory conditions for a week. For NP exposure, *C. auratus* red var. fish (approximately 50 g in weight) were divided into three groups randomly. Each group had three replicates with 15 fish. Three identical round fiberglass tanks (200 L) were used to rear the fish (n = 15 per tank), with continuous aeration. The rearing conditions were as follows: water temperature was 25 ± 1 °C, dissolved oxygen was higher than 5 mg/L, the concentration of ammonia nitrogen was <0.5 mg/kg, and pH was maintained at 6.5–7.5. The experimental water was tap water after aeration for 3 days. Adult fish in the experimental groups were exposed to NP at concentrations of 251.3 μg/L or 753.9 μg/L dissolved in ethanol. The control group was treated with an equal concentration of ethanol in the tanks (0.01% ethanol, *v*/*v*). The exposure was conducted using a semi-static water system, where half of the water was replaced daily, and then, the reagent was added to maintain the original concentration. The selected NP concentrations were based on the 96 h LC50 value for *C. auratus* red var., which is 251.3 μg/L [31]. Three fish from each group were sampled randomly at 12, 24, 48, 72, and 96 h post-exposure to NP. The caudal fin of *C. auratus* red var. at each time point was collected, frozen immediately in liquid nitrogen, and stored at −80 °C until RNA isolation.

For the combination treatment with the BMPRI-specific inhibitor LDN193189 and NP, *C. auratus* red var. fish were selected (n = 12) and randomly divided into four groups. Experiments were carried out in a 50 L round fiberglass tank. The four groups were divided into control, LDN193189, NP, and NP + LDN193189 treatments. The experimental conditions and the experimental water were the same as those used in the previously described NP exposure experiment. Each fish received an intraperitoneal injection of 200 μL of the respective treatment, with three replicates per group. Both the control and NP groups were injected with 0.01% dimethyl sulfoxide (DMSO), while the LDN193189 and NP + LDN193189 groups received LDN193189 at a dose of 0.625 mg/kg body weight, freshly prepared in dimethyl sulfoxide. This dose was based on the effective dose used in a previous study with sailfin molly (*Poecilia latipinna*) [32]. At 12 h after injection, the control and LDN193189 groups were exposed to 0.01% ethanol, whereas the NP and NP + LDN193189 groups were exposed to a concentration of 251.3 μg/L NP. Samples of the caudal fin were collected from 12 live fish at 48 h post-exposure. Fish were not fed throughout the entire exposure process.

2.4. RNA Extraction

Total RNA from the collected samples was extracted individually with Trizol reagent under RNase-free conditions. RNA purity was estimated using a nucleic acid protein analyzer (BioPhotometer Eppendorf, Hamburg, Germany) based on the A260/A280 ratio, and the quality was assessed by electrophoresis on a 1.0% agarose gel.

2.5. Cloning the cDNA Sequence of BMP2 and BMP4

All primers (Table 1) were designed using the Primer Premier 5 software (https://www.premierbiosoft.com/primerdesign/index.html (accessed on 5 July 2022)) and synthesized by the Sangon Biotech Co., Ltd. (Shanghai, China). To amplify partial cDNA fragments of *BMP2* and *BMP4* in *C. auratus* red var., degenerate primers of *C. auratus* red var. *BMP2* were designed from homologous regions of the *BMP2* sequences of other teleosts (GenBank accession no. AB265811.1), and degenerate primers of *BMP4* were designed from homologous regions of the *BMP4* sequences of other teleosts (GenBank accession no. AB874478.1). The PCR template for the *BMP2*/*BMP4* genes was, respectively, synthesized from 1 μg of brain and intestine RNA of *C. auratus* red var. with a PrimeScript™ 1st Strand cDNA Synthesis Kit (Takara, Kusatsu, Japan) according to the manufacturer's instructions. PCR was performed in a total reaction volume of 20 μL containing 2 μL of 10× LA Taq Buffer II (Mg^{2+} plus), 1 μL of cDNA template, 2 μL of dNTP mixture (2.5 mM each), 0.5 μL of each primer (10 μM), 13.8 μL of ddH_2O, and 0.2 μL of LA-Taq DNA Polymerase (Takara, Kusatsu, Japan). The PCR conditions were 94 °C for 1 min; 35 cycles of 30 s at 98 °C; and 1 min 10 s at 61 °C, followed by a final extension for 10 min at 72 °C. The PCR products were purified using the Agarose Gel DNA Purification Kit (Accurate Biotech Co., Ltd., Changsha, China) and were inserted into the pMD18-T vector using a TA cloning kit (Accurate Biotech Co., Ltd., Changsha, China). The recombinant plasmid vector was transformed into *Escherichia coli DH5α* competent cells (Accurate Biotech Co., Ltd., Changsha, China). The positive clones were sequenced by Beijing Tsingke Biotechnology Co., Ltd. (Beijing, China), and the results were verified by the BLAST program (https://blast.ncbi.nlm.nih.gov/Blast.cgi (accessed on 12 September 2022)). All the PCRs in this study were repeated, and the results were consistent.

Table 1. The sequences of primers used in this study.

Primer Names	Sequence (5′ to 3′)	Application
BMP2-F	GCTGTTGCTCGGTCAGGTGT	Partial sequence obtaining
BMP2-R	CAGCCCTCCACCACCATGT	Partial sequence obtaining
BMP4-F	GAAGGGAAGAAGAAAGCGTCG	Partial sequence obtaining
BMP4-R	GACCTCTTTGCTTCGGGCTG	Partial sequence obtaining
BMP2-5′GSP	GTGGAGCACCTCAACCAGAAGCCCG	5′-RACE PCR
BMP2-3′GSP	AGCATGGCCCCTTCCAAAGAGCCTC	3′-RACE PCR
BMP4-5′GSP	CTCCCGTGGGTTGGGGATCTGAGAC	5′-RACE PCR
BMP4-3′GSP	CGAGTCGAGCGAACACCGTGAGAGG	3′-RACE PCR
BMP2-5′NGSP	AGCCCGTGGTTGTGCTGGGATTCGC	Nested 5′-RACE PCR
BMP2-3′NGSP	TGTTCAGGCCAGCATGGCCCCTTCC	Nested 3′-RACE PCR
BMP4-5′NGSP	TGGGGATCTGAGACTGCATCATCTATCT	Nested 5′-RACE PCR
BMP4-3′NGSP	CACCGTGAGAGGATTCCATCATGAAGAG	Nested 3′-RACE PCR
BMP2-QF	GCGATCCGATATTAACTTCCTG	qRT-PCR
BMP2-QR	GCTTTCCCATAGTGCTCCTTG	qRT-PCR
BMP4-QF	TGAACTGCTGCGGGACTTTG	qRT-PCR
BMP4-QR	GACTCGTGGACCTCTCGGGAT	qRT-PCR
Runx2-QF	CACAGAGCCATAAAGGTCACGG	qRT-PCR
Runx2-QR	GGAGTTGGGGTTGCTAAGCG	qRT-PCR
Osterix-QF	CAAACCCGTCCCATCTTCTG	qRT-PCR
Osterix-QR	GCACCAAGCCTCTCCAACTC	qRT-PCR
BMPRI-QF	TGGCGTACTCTGCAGCCTGT	qRT-PCR
BMPRI-QR	TGGGATGTCCACTTCATTTGTG	qRT-PCR
β-actin-QF	TCCCTTGCTCCTTCCACCA	qRT-PCR
β-actin-QR	GGAAGGGCCAGACTCATCGTA	qRT-PCR

To obtain the 5′ and 3′ ends of each cDNA, a number of gene-specific primers (Table 1) were designed for the target genes to replicate sense or antisense regions of their amplified partial fragments, sequenced above. 5′-RACE and 3′-RACE were performed with the RACE cDNA Amplification Kit (Vazyme Biotech Co., Ltd., Nanjing, China) using brain or intestine RNA following the manufacturer's instructions. Nested 5′- and 3′-RACE PCR products of the expected size were handled and sequenced as described above. The full-length cDNAs of the target genes were assembled by aligning the partial cDNA fragments and the 5′- and 3′-RACE fragments with the aid of the ContigExpress program in the Vector NTI Advance 11 sequence analysis software packages (Invitrogen).

2.6. Sequence Identification of BMP2 and BMP4

The ORF finder program (https://www.ncbi.nlm.nih.gov/orffinder/ (accessed on 5 April 2023)) and ExPASy (http://web.expasy.org/translate/ (accessed on 5 April 2023)) were used to deduce the amino acid sequences of BMP2/BMP4. The BLASTX and BLASTP programs (http://blast.ncbi.nlm.nih.gov/Blast.cgi (accessed on 11 April 2023)) were used to analyze the nucleotide and deduced protein sequences, respectively. The isoelectric point (pI) and molecular weight (Mw) were calculated by the pI/Mw tool (http://www.expasy.ch/tools/pi_tool.html (accessed on 11 April 2023)). The protein motif features were predicted by the Simple Modular Architecture Research Tool (http://smart.cmbl-heidelberg.de/ (accessed on 11 April 2023)). The BMP2 and BMP4 amino acid sequences of other vertebrates were searched in the NCBI protein database, and the GenBank accession numbers are listed in Table 2. Multiple sequence alignment of BMP2 and BMP4 was performed with the MegAlign (https://www.dnastar.com/software/lasergene/megalign-pro/ (accessed on 3 June 2023)) and GeneDoc (https://nrbsc.org/gfx/genedoc/ (accessed on 3 June 2023)) programs, and the identity was analyzed by the NCBI BLASTP program. Phylogenetic analysis was constructed using the neighbor-joining (NJ) method implemented in the MEGA 5.0 package (http://www.megasoftware.net/(accessed on 3 June 2023)) based on sequence alignment using the Clustal W method with 1000 bootstrap replicates.

Table 2. Amino acids used to construct the phylogenetic tree of the BMP2/BMP4 genes.

Protein Name	GenBank Access. No.
Carassius auratus red variety BMP2	PP411940
Carassius auratus BMP2	BAN17326.1
Danio rerio BMP2	AAI14257.1
Cyprinus carpio BMP2	XP_042602743.1
Megalops cyprinoides BMP2	XP_036398004.1
Anguilla Anguilla BMP2	XP_035277091.1
Astyanax mexicanus BMP2	KAG9268738.1
Puntigrus tetrazona BMP2	XP_043074619.1
Sinocyclocheilus anshuiensis BMP2	XP_016337425.1
Anabarilius grahami BMP2	ROL52452.1
Salmo trutta BMP2	XP_029587557.1
Homo sapiens BMP2	ACV32596.1
Mus musculus BMP2	AAI00345.1
Gorilla gorilla gorilla BMP2	XP_004061840.1
Bos taurus BMP2	AAI42130.1
Gallus gallus BMP2	NP_001385099.1
Columba livia BMP2	XP_021150021.1
Carassius auratus red variety BMP4	PP411941
Carassius auratus BMP4	XP_026085110.1
Cyprinus carpio BMP4	XP_042630442.1
Sinocyclocheilu rhinocerous BMP4	XP_016424235.1
Puntigrus tetrazona BMP4	XP_043118272.1
Danio rerio BMP4	NP_571417.1
Salmo salar BMP4	XP_014066471.1
Sparus aurata BMP4	XP_030298905.1
Homo sapiens BMP4	NP_001334843.1
Mus musculus BMP4	AAH13459.1
Gorilla gorilla gorilla BMP4	XP_030857865.1
Gallus gallus BMP4	NP_990568.4
Bos taurus BMP4	NP_001039342.1
Columba livia BMP4	XP_005510342.1
Xenopus laevis BMP4	NP_001081501.1

2.7. Gene Expression Analysis by RT-qPCR

Total RNA was reverse-transcribed into cDNA using the HiScript II Q RT SuperMix for qPCR (+gDNA wiper) kit (Vazyme Biotech Co., Ltd., Nanjing, China), and the cDNA was diluted five times for a quantitative template. Gene-specific primers for quantitative real-time PCR (qRT-PCR) were designed to determine the mRNA expression levels of *BMP2*, *BMP4*, *Runx2*, *Osterix*, and *BMPRI* in various samples (Table 2). *C. auratus* red var. β-actin was used as an internal control to verify successful transcription and to calibrate the cDNA template for corresponding samples. qRT-PCR was carried out on the Quant-Studio™ 3 Real-Time PCR System (Thermo Fisher, Waltham, MA, USA) in a total volume of 20 μL, containing 1 μL of cDNA template, 0.3 μL of each primer (10 μM), 10 μL of 2× ChamQ Universal SYBR qPCR Master Mix (Vazyme Biotech Co., Ltd., Nanjing, China), and 8.4 μL of RNase-Free ddH_2O. The PCR cycling procedure was as follows: initial denaturation at 95 °C for 30 s, followed by 40 cycles of amplification at 95 °C for 10 s and 60 °C for 30 s; a melting curve was obtained at the end of the reaction, and a single peak was observed, demonstrating the absence of dimeric primers or the generation of non-specific amplification. Relative expression levels were calculated by the $2^{-\Delta\Delta Ct}$ method.

2.8. Statistical Analysis

Data obtained from qRT-PCR were expressed as mean ± SEM (n = 3) and were analyzed with the SPSS 26.0 software (Chicago, IL, USA). Differences between means were assessed by one-way analysis of variance (ANOVA) followed by Duncan's multiple comparison tests. Differences were considered significant when $p < 0.05$. Relative expressions of the *BMP2* and *BMP4* genes were plotted using the GraphPad Prism 8.0 software (San Diego, CA, USA).

3. Results

3.1. Cloning and Sequence Characteristics of BMP2 and BMP4 cDNA

The full length of *BMP2* cDNA that we isolated from *C. auratus* red var. was 2029 bp, which contained a 5′-untranslated region (UTR) of 375 bp, a 1236-bp open reading frame (ORF), and a 418 bp 3′-UTR with a 31 bp polyA tail. The ORF encoded a 411-amino-acid peptide with a predicted molecular weight of 46.97 kDa and a theoretical isoelectric point of 7.72 (Figure 1A). The complete cDNA sequence of *BMP4* was 2095 bp, and the predicted ORF was 1302 bp, encoding a protein with 433 amino acids. It included a 359 bp 5′-UTR and a 434 bp 3′-UTR with a 26bp polyA tail (Figure 1B). The calculated molecular mass and theoretical isoelectric point of BMP4 were 49.56 kDa and 8.56, respectively. The sequences of *BMP2* and *BMP4* were submitted to NCBI GenBank, and the accession numbers are listed in Table 2. Domain architecture analysis of BMP2/BMP4 revealed three conserved and key structural features, including a signal peptide, a TGF-β propeptide, and a mature TGF-β domain (Figure 2).

3.2. Homology and Phylogenetic Analysis

Multiple sequence alignments of BMP2 and BMP4 are presented in Figures 3 and 4, respectively, showing their homology at the protein level. BMP2 exhibited the highest homology with BMP2 from other cyprinids including goldfish (*Carassius auratus*) (BAN17326.1), sharing 98.30% sequence identity. It also showed significant homology with blind cavefish (*Sinocyclocheilus anshuiensis*) (XP_016337425.1), common carp (*Cyprinus carpio*) (XP_042602743.1), and tiger barb (*Puntigrus tetrazona*) (XP_043074619.1), and the sequence identity was about 93%. Similarly, *C. auratus* red var. BMP4 shared the highest homology with *C. auratus* (XP_026085110.1), and sequence identity was 98.21%. It has higher homology with *C. carpio* (XP_042630442.1), *S. anshuiensis* (XP_016424235.1), and *P. tetrazona* (XP_043118272.1), and the sequence identities were 96.43%, 94.90%, and 93.89%, respectively.

The phylogenetic analysis in Figure 5A revealed that the BMP2 sequences formed two distinct branches, one representing bony fish, while the other was the branch of birds and mammals, indicating a closer relationship between *C. auratus* red var. BMP2 and *C. auratus* BMP2. The phylogenetic tree of BMP4 encompassed sequences from various species, including fish, amphibians, birds, and mammals, while *C. auratus* red var. BMP4 had the highest homology with *C. auratus* BMP4 (Figure 5B). These relationships depicted in the phylogenetic tree align well with traditional taxonomic affinities.

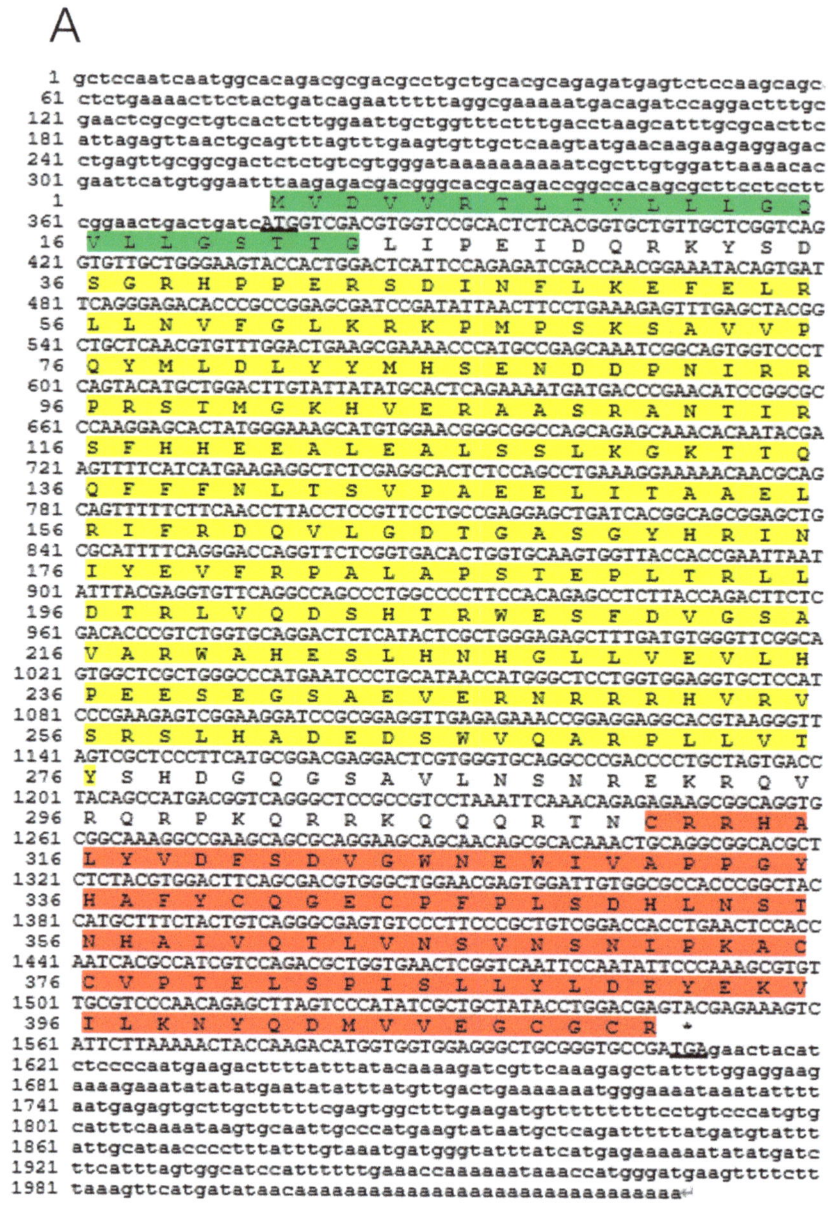

Figure 1. *Cont.*

B

```
   1 ggagaactggatgcgggttcagacggtgatgtcctccgtgctcgactcttcctggagata
  61 cacagatccgcgttcgcatgggactcttttggcttcttctacgtgatgcggaactaatga
 121 tctaaacattcgccttagggttggacttttgctgtctgaggtgaactcctttaaaaactc
 181 ctttttaccgtcttctcatccaggaataccgcgcgcatctctaccaatggaaaggaatct
 241 ccccgctgttgttaaacagccttccaggagcacgtaacattcactcgaggaactcgtcct
   1                                                            M
 301 ggatcagctctcgacaaacatggactgttttcccatgctttatttctgtcgagacatcA
   2    I  P  G  N  R  M  P  M  V  L  L  L  C  Q  V  L  L  G  F  S
 361 TGATTCCTGGTAATCGAATGCCGATGGTCCTTTTATTATGCCAAGTCCTACTGGGAGAGA
  22    S  Y  A  S  L  I  P  E  E  G  K  K  K  A  S  A  Q  H  L  G
 421 GCAGCTATGCTAGTCTAATACCTGAGGAAGGGAAGAAGAAAGCGTCGGCTCAGCACCTGG
  42    Q  S  H  E  L  L  R  D  F  E  A  T  L  L  H  M  F  G  L  Q
 481 GTCAGAGTCATGAACTGCTGCGGGACTTTGAAGCCACGCTGCTGCACATGTTTGGGCTGC
  62    R  R  P  R  P  S  Y  S  A  V  V  P  Q  Y  L  L  D  L  Y  R
 541 AGAGGCGACCGCGACCCAGCTACTCGGCTGTGGTGCCCCAGTATCTGCTGGACCTGTATC
  82    L  Q  S  G  E  A  E  E  A  G  A  H  D  V  S  F  H  Y  P  E
 601 GCCTGCAGTCAGGGGAGGCCGAAGAGGCCGGGGCCCACGACGTGAGCTTCCATTATCCCG
 102    R  S  T  S  R  A  N  T  V  R  G  F  H  H  E  E  H  M  E  E
 661 AGAGGTCCACGAGTCGAGCGAACACCGTGAGAGGATTCCATCATGAAGAGCACATGGAGG
 122    L  Q  S  D  G  S  E  T  P  L  R  F  V  F  N  L  S  S  I  P
 721 AGCTGCAGTCAGACGGCTCTGAGACTCCCCTACGCTTCGTCTTCAATCTCAGCAGCATCC
 142    E  E  E  L  I  S  T  A  E  L  R  L  Y  R  K  Q  I  D  D  A
 781 CAGAGGAGGAACTCATTTCTACCGCAGAGCTTCGTCTCTACAGGAAACAGATAGATGATG
 162    I  S  D  P  Q  P  T  G  D  E  G  L  H  R  I  N  I  Y  E  V
 841 CAATCTCAGATCCCCAACCCACGGGAGACGAAGGTCTACACAGGATAAACATATACGAGG
 182    L  K  P  P  R  A  G  Q  L  I  T  Q  L  L  D  T  R  L  V  R
 901 TGCTGAAGCCCCCGCGGGCCGGACAGCTCATCACACAGCTGCTGGACACTCGACTGGTGC
 202    H  N  T  T  R  W  E  S  F  D  V  S  P  A  V  L  R  W  T  R
 961 GCCACAACACGACTCGATGGGAGAGCTTCGACGTGAGTCCGGCCGTGCTGCGCTGGACCC
 222    D  K  S  S  N  H  G  L  A  V  E  V  V  P  L  N  R  T  S  R
1021 GTGACAAGAGCTCCAATCACGGCCTGGCTGTGGAGGTGGTGCCCCTGAACCGAACCTCGC
 242    H  Q  G  R  H  V  R  V  S  R  S  L  H  P  L  P  D  A  D  W
1081 GTCACCAGGGACGCCACGTACGCGTCAGTCGCTCCTTGCATCCGCTTCCCGACGCAGACT
 262    S  Q  L  R  P  L  L  V  T  F  G  H  D  G  K  S  H  P  L  T
1141 GGAGCCAGCTTCGCCCCCTGCTGGTCACGTTTGGACACGACGGAAAAAGTCACCCGCTGA
 282    R  R  A  K  R  S  P  K  Q  R  G  R  K  R  N  R  N  C  R  R
1201 CGCGGCGAGCGAAGCGCAGCCCGAAGCAAAGAGGTCGCAAGCGCAACCGAAACTGTCGGC
 302    H  A  L  Y  V  D  F  S  D  V  G  W  N  D  W  I  V  A  P  P
1261 GACACGCGCTTTATGTGGACTTCAGTGATGTGGGCTGGAACGACTGGATCGTGGCGCCGC
 322    G  Y  Q  A  Y  Y  C  H  G  E  C  P  F  P  L  A  D  H  L  N
1321 CCGGATATCAGGCGTACTACTGTCACGGGGAGTGTCCGTTCCCTTTGGCCGATCACCTGA
 342    S  T  N  H  A  I  V  Q  T  L  V  N  S  V  N  T  N  I  P  K
1381 ACTCCACCAATCACGCTATTGTGCAGACGCTGGTGAACTCGGTGAACACCAATATTCCCA
 362    A  C  C  V  P  T  E  L  S  A  I  S  M  L  Y  L  D  E  T  G
1441 AGGCCTGCTGCGTTCCCACAGAGCTCAGTGCCATCTCCATGCTCTACCTGGACGAGACGG
 382    T  G  W  V  L  K  N  Y  Q  E  R  W  W  W  G  K  R  M  C  A
1501 GAACAGGGTGGGTGCTGAAGAACTATCAGGAGAGGTGGTGGTGGGGAAAGAGGATGTGTG
 402    A  A  D  R  D  T  A  T  P  W  T  L  I  Y  N  I  W  V  L  G
1561 CTGCCGCTGACCGGGACACAGCTACACCCTGGACACTTATTTATAACATCTGGGTCTTGG
 422    R  C  V  F  L  V  S  S  I  H  I  F  *
1621 GAAGATGTGTGTTTCTTGTCTCCTCGATACATATATTTTGAcaaaaaaagtatatataaa
1681 tatatatttatatcaacataattaaacaaataaaaccgtccttattttaattatatgtca
1741 gctgtacagcttttcgtaatgtatatcagactgtataagggtttttctatgagctagaag
1801 aattagccactaaaaagtttatttactggtagtatttatttctttaatacaatatccatt
1861 caccaaatgtgtagccgagcgacccccctcataaagcatcgtttgtgtacttaaaatagag
1921 cactgggtgaaatgctaggaaacctgattgtcttgctatatcagacggagagatcagagg
1981 cgagttcaatgttctgcatcagcatgtgatttacatggtcccaaaatttttgttttgttg
2041 ttaaataaattttgaaacattccaacaccaaaaaaaaaaaaaaaaaaaaaaaaaa
```

Figure 1. Complete coding sequences and deduced protein sequences of BMP2 (**A**) and BMP4 (**B**). The conserved TGF-β propeptide and mature TGF-β domain are indicated by yellow and red shading, respectively. The signal peptide is marked with a green background, and the initiation codons (ATG) and termination codons (TGA) are underlined.

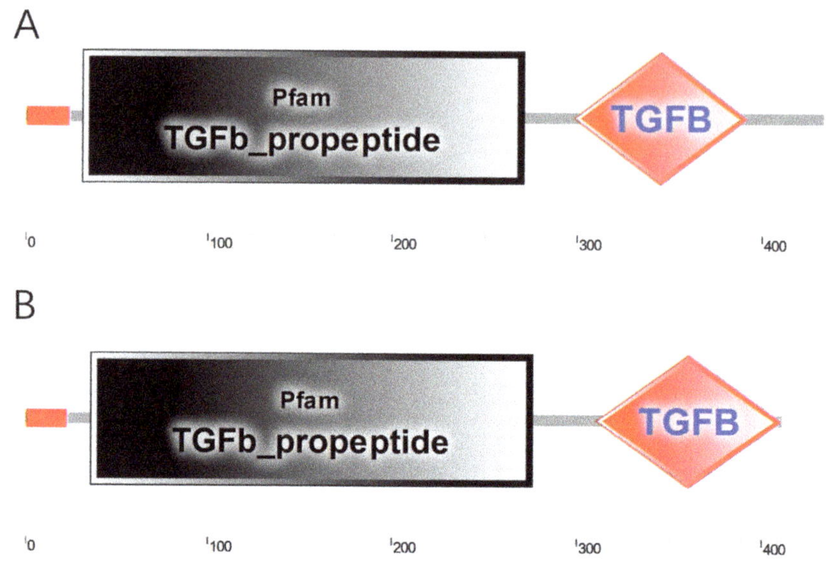

Figure 2. BMP2 (**A**) and BMP4 (**B**) domain structures in *C. auratus* red var. Three key structural features are shown, including a signal peptide (the first region marked in red), a TGF-β propeptide, and a mature TGF-β domain.

3.3. Expression Profiles of BMP2 and BMP4 in the Adult Fish

The expression patterns of *BMP2* and *BMP4* genes were detected in different tissues of adult *C. auratus* red var. fish by qRT-PCR. The findings revealed the widespread expression of *BMP2* and *BMP4* in all nine tissues examined: gill (G), caudal fin (C), heart (H), intestine (I), kidney (K), liver (L), muscle (M), brain (B), and spleen (S). Specifically, *BMP2* exhibited high expression levels in the muscle, gill, liver, and caudal fin, with lower expression in the heart. Moderate expression was observed in the brain, intestine, kidney, and spleen (Figure 6A). On the other hand, *BMP4* showed relatively high expression in muscle and the liver, followed by the caudal fin and brain, but lower expression in the gill, intestine, spleen, heart, and kidney (Figure 6B).

3.4. BMP2 and BMP4 Expression in Different Developmental Stages after NP Treatment

Under the NP treatment, abnormal tailbone formation was observed in embryos (Figure S1), which is consistent with the results of our group's previous research [30]. To investigate the effect of NP exposure on the expression of *BMP2* and *BMP4*, the levels of *BMP2* and *BMP4* mRNA were examined at six developmental stages of *C. auratus* red var. embryos, including 6 hpf, 12 hpf, 24 hpf, 48 hpf, 72 hpf, and 96 hpf (Figure 7). qRT-PCR analysis showed that the expressions of *BMP2* and *BMP4* in the 3μmol/L NP treatment group were significantly downregulated at 6 hpf, 12 hpf, and 96 hpf compared with the control group, and *BMP2* expression was also significantly decreased at 72 hpf ($p < 0.05$). There were non-significant changes in the expression levels of *BMP2* at 24 hpf and 48 hpf and *BMP4* at 24 hpf, 48 hpf, and 72 hpf after the NP treatment ($p > 0.05$).

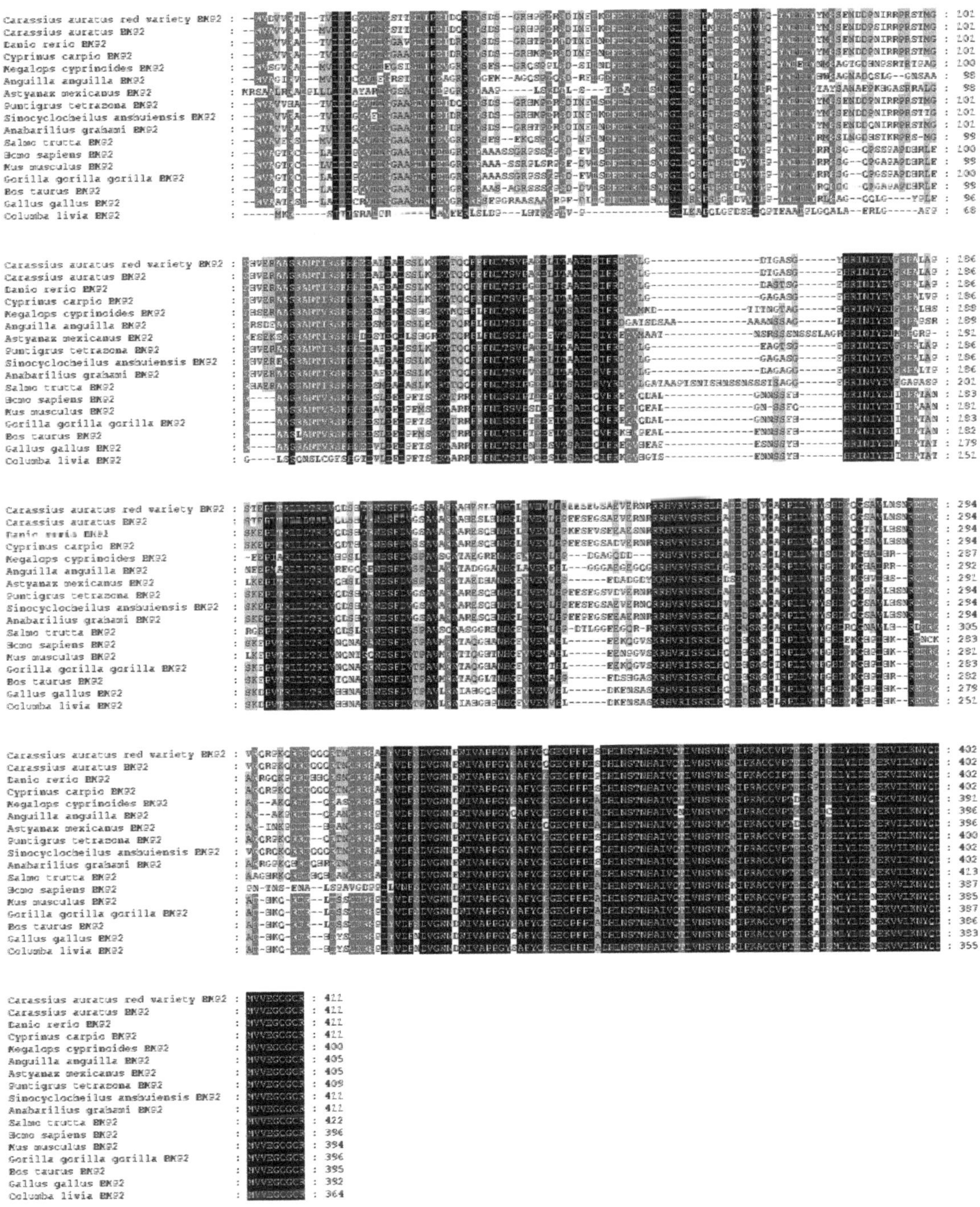

Figure 3. Multiple amino acid sequence alignment of BMP2 from *C. auratus* red var. and other species. The same amino acids are marked with black, and amino acids with more than 75% and 50% similarity are, respectively, shown in dark gray and light gray.

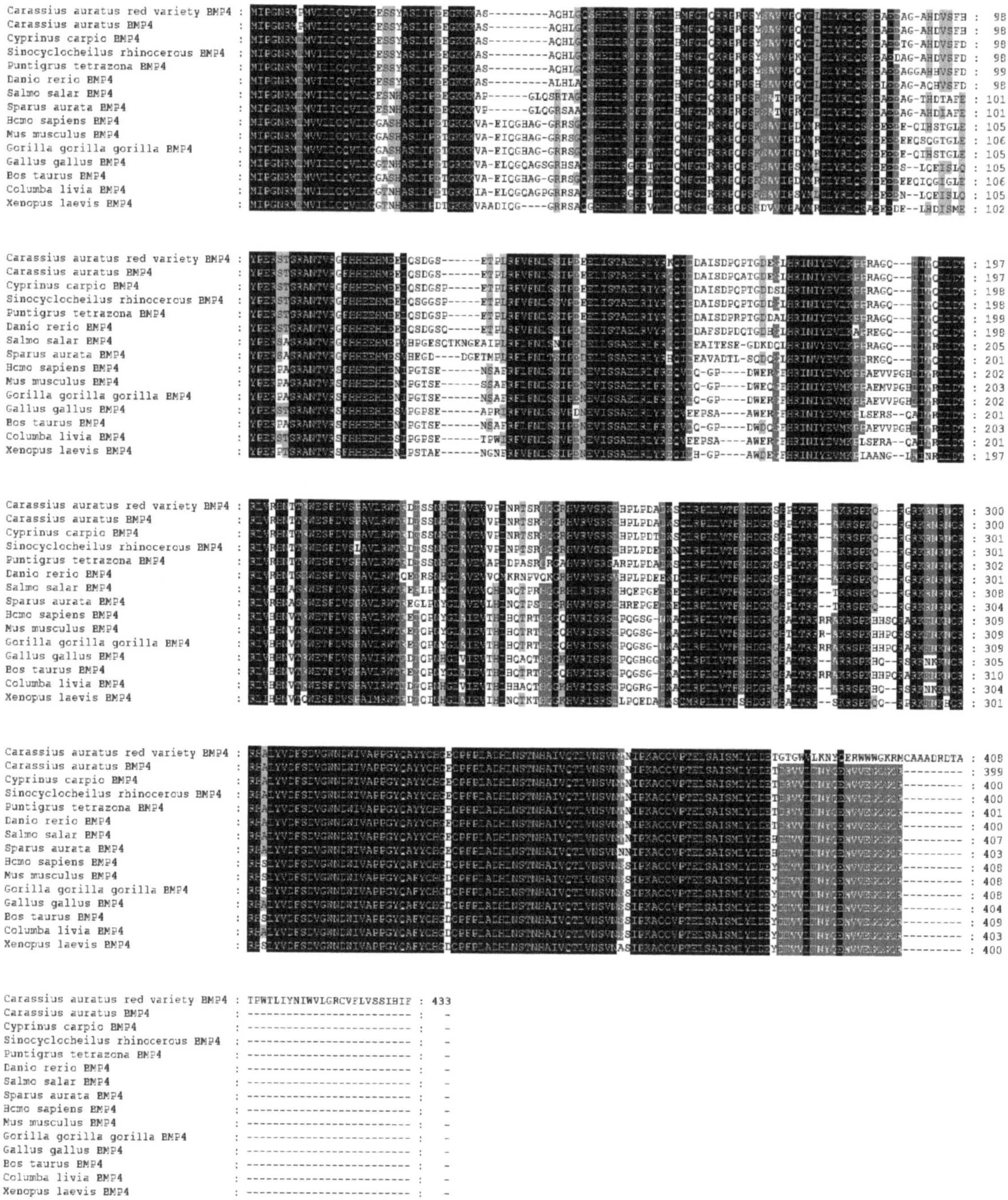

Figure 4. Multiple amino acid sequence alignment of BMP4 from *C. auratus* red var. and other species. The same amino acids are marked with black, and amino acids with more than 75% and 50% similarity are, respectively, shown in dark gray and light gray.

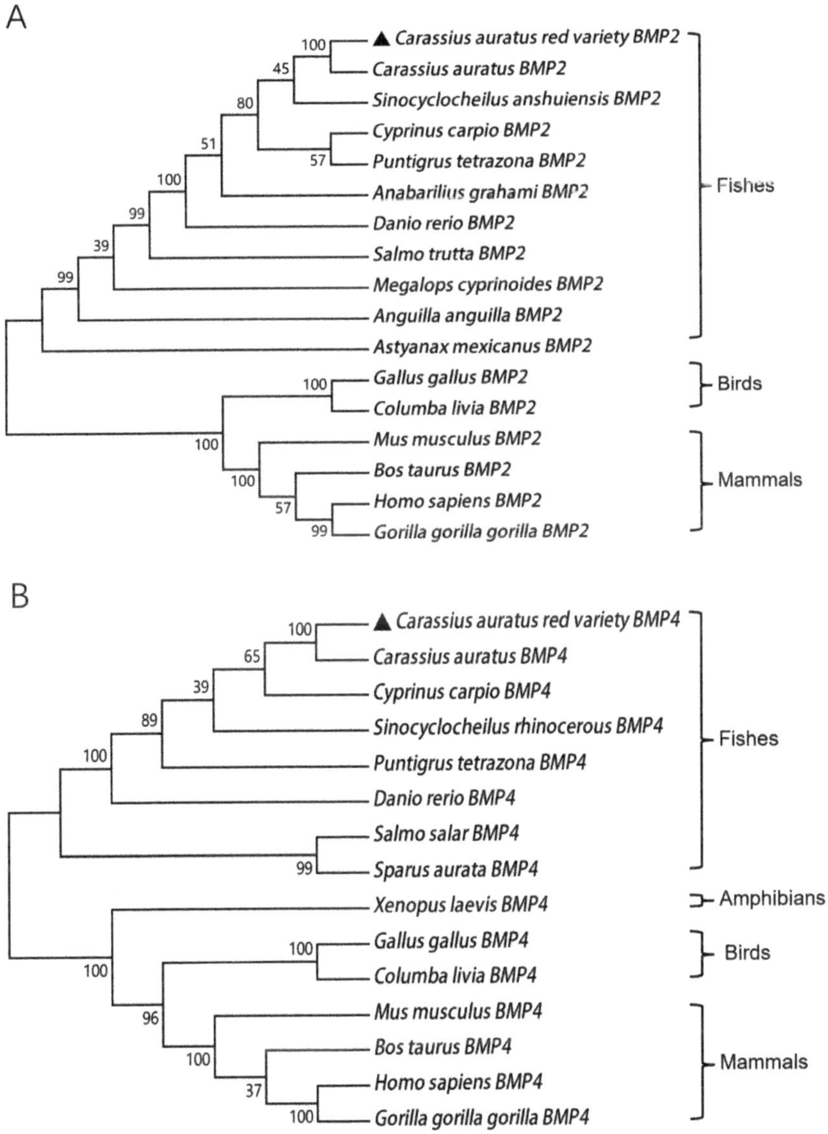

Figure 5. Phylogenetic tree of BMP2 (**A**) and BMP4 (**B**) sequences in different vertebrates (*C. auratus* red var. BMP2 and BMP4 are marked with black triangles). The phylogenetic tree was constructed using the neighbor-joining (NJ) algorithm. Node values represent the bootstrap values obtained after 1000 replications. The GenBank accession numbers of BMP2 and BMP4 are listed in Table 2.

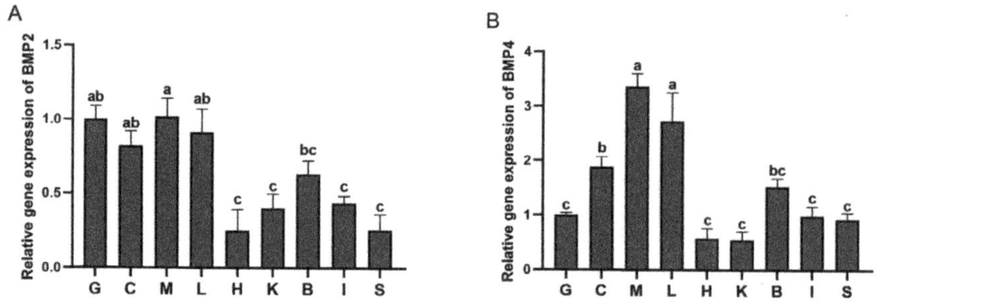

Figure 6. Relative expression of *C. auratus* red var. *BMP2* and *BMP4* in the gill (G), caudal fin (C), heart (H), intestine (I), kidney (K), liver (L), muscle (M), brain (B), and spleen (S). Transcriptional fold-changes of *BMP2* and *BMP4* in different tissues were calculated compared to gill tissue, and β-actin was employed as the internal reference. Different lowercase letters represent significant differences ($p < 0.05$, $n = 3$).

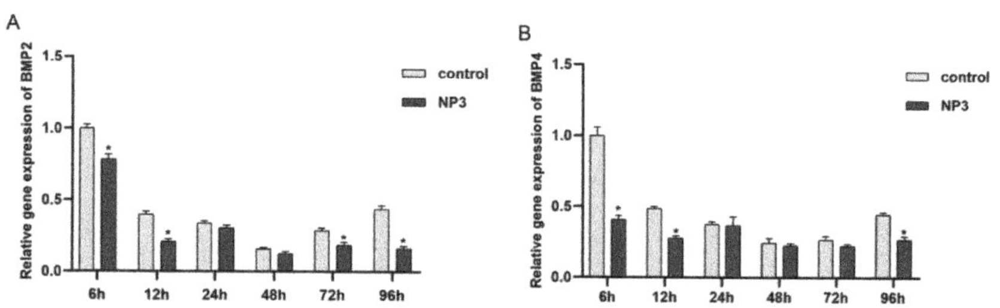

Figure 7. Expression levels of *BMP2* (**A**) and *BMP4* (**B**) in the treatment and control groups at six developmental stages of *C. auratus* red var. embryos (control: 0 μmol/L NP-treated embryos; NP3: 3 μmol/L NP-treated embryos). β-actin was used as the internal reference. Transcriptional fold-changes in the target gene at different time points were calculated compared with the control (6 h). Significant differences at different time points after NP exposure compared with corresponding control groups are indicated by asterisks (*: $p < 0.05$; $n = 3$).

3.5. Impact of NP and LDN193189 on BMP-Smad Pathway Gene Expression

To assess the impact of NP exposure on *BMP2*, *BMP4*, and other BMP-Smad pathway-related genes, the expression levels of *BMPRI*, *BMP2*, *BMP4*, *Runx2*, and *Osterix* were quantified in the caudal fin of *C. auratus* red var. at 12, 24, 48, 72, and 96 h post-exposure (Figure 8). At 12 h, a significant upregulation of *BMPRI*, *BMP4*, and *Runx2* was observed under a low NP concentration ($p < 0.05$), whereas *BMP2* expression was downregulated ($p < 0.05$). At 24 h, under a low NP concentration, *BMPRI*, *BMP2*, *Runx2*, and *Osterix* mRNA levels were significantly increased ($p < 0.05$). At 12 h under high NP exposure, *BMPRI* and *Runx2* mRNA expression increased significantly ($p < 0.05$), whereas *BMP2* expression decreased ($p < 0.05$). At 24 h, the expression of *BMPRI*, *BMP2*, and *Osterix* increased significantly under high NP exposure ($p < 0.05$), while *BMP4* expression was significantly reduced ($p < 0.05$). Expression of all genes analyzed progressively decreased with extended NP exposure time until 96 h, with the most significant differences observed at 48 h, where *BMPRI*, *BMP2*, *BMP4*, *Runx2*, and *Osterix* expression levels were significantly downregulated ($p < 0.05$). These findings indicated that the 251.3 μg/L NP exposure group exerted a more pronounced effect on the expression of the *BMPRI*, *BMP2*, *BMP4*, *Runx2*, and *Osterix* genes compared with the 753.9 μg/L NP exposure group.

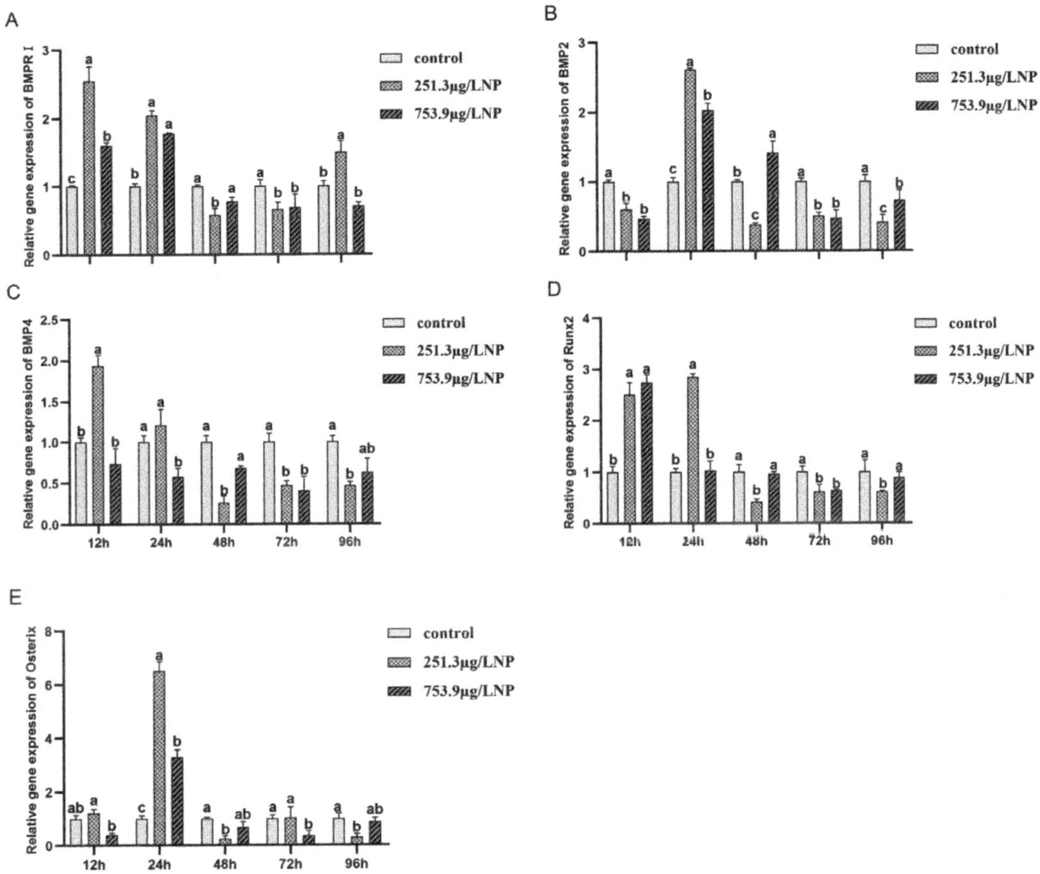

Figure 8. Effect of different concentrations of NP stress on the expression levels of *BMPRI* (**A**), *BMP2* (**B**), *BMP4* (**C**), *Runx2* (**D**), and *Osterix* (**E**) mRNA in caudal fin (0.01% ethanol-treated control group; 251.3 μg/L NP-treated group; 753.9 μg/L NP-treated group). β-actin was used as the internal reference. The transcriptional fold-changes in the target genes were calculated compared with the corresponding control group. Statistically significant differences ($p < 0.05$) are indicated by different lowercase letters.

To further elucidate the effects of NP stress on BMP-Smad pathway-related gene expression, adult fish were treated with the specific *BMPRI* inhibitor LDN193189 followed by combined NP exposure. After a 12 h injection with LDN193189, fish were then exposed to 251.3 μg/L of NP for an additional 48 h. The results revealed that LDN193189 significantly decreased the activity of *BMPRI, BMP2, BMP4, Runx2,* and *Osterix* ($p < 0.05$). In comparison with the control group, the combined treatment with LDN193189 and NP led to a greater reduction in the expression levels of *BMPRI, BMP2, BMP4, Runx2,* and *Osterix* than either treatment alone (Figure 9).

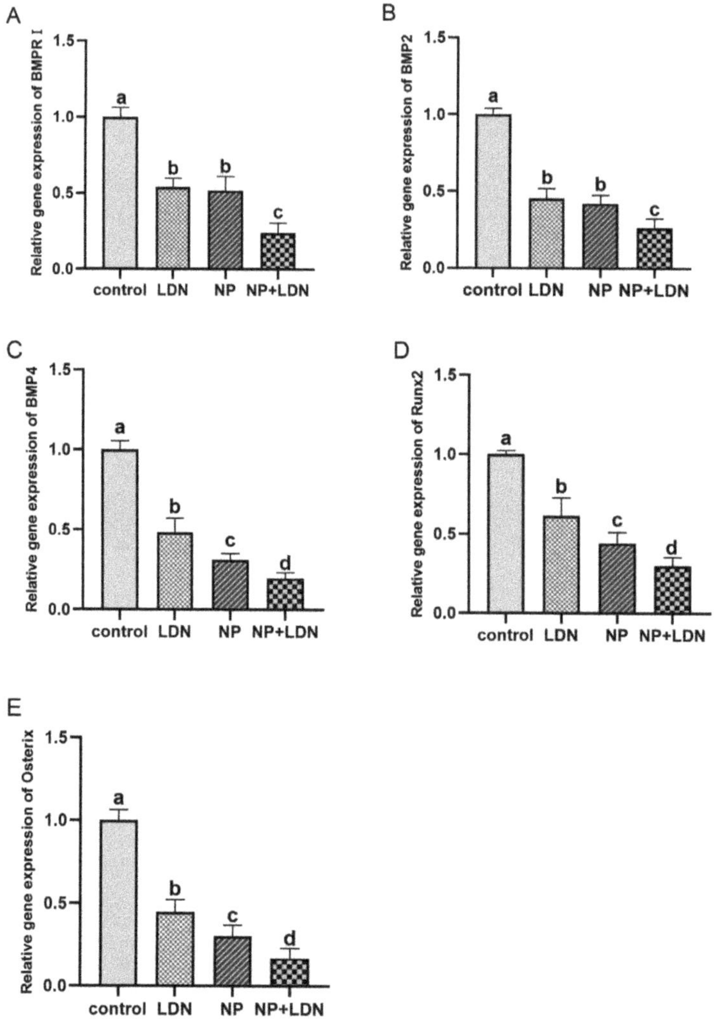

Figure 9. Effects of NP exposure and LDN193189 injection on the relative expression of *BMPRI* (**A**), *BMP2* (**B**), *BMP4* (**C**), *Runx2* (**D**), and *Osterix* (**E**) mRNA in caudal fin of *C. auratus* red var. Control: exposed to 0.01% ethanol and injected with 0.01% DMSO; LDN: injected with 0.625 mg/kg LDN193189 and exposed to 0.01% ethanol; NP: injected with 0.01% DMSO and exposed to 251.3 μg/L NP; NP+LDN: injected with 0.625 mg/kg LDN193189 and exposed to 251.3 μg/L NP. β-actin was used as the internal reference. Transcriptional fold-changes in target gene were calculated compared with the control. The different lowercase letters indicate statistically significant differences ($p < 0.05$).

4. Discussion

The bone morphogenetic protein (BMP) family plays a crucial role in regulating cellular activities and is involved in almost all tissue development processes [33,34]. BMPs are essential for bone formation during mammalian development and exhibit diverse functions within the body [35]. Disruptions of TGF-β/BMP signaling have been implicated in multiple bone diseases [36]. In this study, the BMP2 and BMP4 of *C. auratus* red var. contained seven conserved cysteine residues, which are typical features of the BMP family [37]. Protein domain analysis revealed both the BMP2 and BMP4 proteins of *C. auratus* red var. possessed two TGF-β domains, which is a mature peptide that controls proliferation, differentiation, and other functions in many cell types [38]. Multiple alignments of the vertebrate BMP2 and BMP4 protein sequences available revealed the high conservation of amino acid residues. Phylogenetic analysis showed that the amino acid sequences of BMP2 and BMP4 were conserved among the analyzed fish species, and *C. auratus* red var. sequences were closely related to those of *C. auratus*. The structure and sequence similarity of BMP2 and BMP4 between *C. auratus* red var. and other fish suggests that cloned *C. auratus* red var. BMPs belong to the teleost BMP family.

Research has indicated that the *BMP2* and *BMP4* genes are broadly expressed across various tissues in fish, with expression patterns exhibiting species-specific variability. For instance, in *C. carpio*, both the *BMP2* and *BMP4* genes are expressed at high levels in several healthy tissues, including the gill, intestine, liver, spleen, brain, and blood [39]. In *C. carpio* var. Jian, the expression of *BMP2* is most highly expressed in muscle, with subsequent levels observed in the liver; however, lower expression levels have been detected in both the heart and brain [26]. In barbel steed (*Hemibarbus labeo*), *BMP2* shows preferential expression in the gill, with significant levels also detected in muscle and the liver, while expression in the heart is lower [27]. In the current study, qRT-PCR was employed to assess the expression of *BMP2* and *BMP4* in various tissues of *C. auratus* red var. Expression was detected in all sampled tissues, with the muscle and liver exhibiting higher levels, followed by the caudal fin and lower levels in the heart. This widespread distribution suggests diverse physiological roles for *BMP2* and *BMP4* in *C. auratus* red var. For example, higher expression in muscle suggests that *BMP2* and *BMP4* might be related to the occurrence of intramuscular spines [26]. In the liver, their higher expression indicates the importance of the BMP pathway in liver development, where it is closely related to iron balance in the liver [27,40]. High expression in the caudal fin suggests that *BMP2* and *BMP4* are involved in the process of bone formation [41]. Comparatively, the tissue-specific expression pattern in *C. auratus* red var. closely aligns with that observed in *Cyprinus carpio var.* Jian, another cyprinid species. This conservation of expression patterns within cyprinids suggests that *BMP2* and *BMP4* may play evolutionarily conserved roles in these fishes, with potential implications for understanding their physiological functions across different tissues.

Nonylphenol (NP) has been examined for its toxicological effects across multiple tissues, with particular emphasis on the skeletal system [4,6]. It has been demonstrated that NP plays a critical role in bone, which can inhibit osteoclast formation and lead to skeletal disorders [42]. Furthermore, NP may modulate the molecular mediators involved in osteoblast differentiation through its interaction with estrogen receptors [43]. The BMP signaling pathways are known to be important in osteoblast differentiation [17]. This suggests that NP may influence BMP pathway-related genes through its interaction with estrogen receptors. As prominent ligands of this pathway, BMP2 and BMP4 activate it by binding to BMP type I (BMPRI) and type II (BMPRII) receptors, triggering the phosphorylation of downstream proteins such as Smad1/5/8 [44]. Subsequently, the phosphorylated Smad complex with Smad4 translocates to the nucleus and acts as a transcription factor, regulating genes such as *Runx2* and *Osterix*, which are crucial for bone formation and maintenance [45]. In our study, we found that NP exposure can significantly depress the expression of *BMP2* and *BMP4* in *C. auratus* red var. embryos and adults. BMP plays a crucial role in the formation of the body axis, particularly in the development of the ventral side during early life stages [46]. NP's effects on embryos can lead to the downregulation of *BMP2* and *BMP4* expression, resulting in morphological changes such as the abnormal development of tail bones during the early life stages [30]. In addition, in the cotreatment group, an intraperitoneal injection of the *BMPR*I-specific inhibitor LDN193189 was administered to disrupt gene expression, along with exposure to 251.3 μg/L of NP; the expression levels of *BMPRI*, *BMP2*, *BMP4*, *Runx2*, and *Osterix* were downregulated more than in the other treatment groups. To date, there has been limited research exploring the effect of NP on the BMP-Smad signaling pathway in fish. Our results align with those of previous research, which has demonstrated that 2.5 μM of NP reduces the expression of BMPs during the osteogenic differentiation of mesenchymal stem cells into osteoblasts in rats [47]. The results suggest that NP may inhibit osteoclast formation by interfering with the expression of BMP pathway-related genes, which will provide cues for further research on the molecular mechanism of the NP effect on bone development. Studies of gobiid fish (*Odontobutis potamophila*) have shown that long-term exposure to NP can affect the tissue structure of the testes, leading to cell necrosis and fibrosis [48]. In this study, an acute exposure experiment was conducted that does not provide a thorough understanding of the effects of long-term NP exposure on fish.

Recent studies have increasingly uncovered the potential hazards of endocrine-disrupting chemicals (EDCs) on the skeletal development of aquatic organisms. Evidence suggests that EDCs can inhibit the expression of key skeletal development genes, such as *BMP2*, *BMP4*, *Sox9*, and *Runx2*. This study further validates these findings by experimentally observing that NP, a ubiquitous EDC, significantly reduces the expression levels of the *BMP2* and *BMP4* genes in *Carassius auratus* red var. This discovery is consistent with the inhibitory effects observed previously in other fish exposed to EDCs, such as bisphenol A (BPA) [49] and benzyl butyl phthalate (BBP) [50]. These findings underscore the crucial role of the BMP-Smad signaling pathway in the adaptive mechanisms of teleost fish in response to environmental toxicants. Concurrently, they highlight the significance of safeguarding aquatic ecosystems against EDC pollution. Hence, this research establishes a scientific framework for the formulation of environmental management strategies that specifically target EDCs, aiming to alleviate their potential detrimental effects on aquatic organisms and ecological systems.

5. Conclusions

In the present study, full-length cDNAs of BMP2 and BMP4 were successfully isolated and characterized from *Carassius auratus* red var. Homology and phylogenetic analyses indicated that BMP2 and BMP4 were most closely related to BMP2 and BMP4 from *C. auratus* and were relatively conservative between different species. Furthermore, tissue-specific distributions showed that these two proteins were expressed in all tissues examined and highly expressed in muscle, the liver, and the caudal fin. The levels of BMP2 and BMP4 mRNA were downregulated at different developmental stages of embryos after NP exposure. Additionally, the different concentrations of NP stress on the caudal fins of adult fish affected the expression of BMP2, BMP4, and other BMP-Smad pathway-related genes. Importantly, the results of the independent and combined treatments with NP and the BMPRI specific inhibitor suggested that NP reduced the expression of BMPRI, BMP2, BMP4, Runx2, and Osterix. Our findings demonstrate that BMP2 and BMP4 may play a regulatory role via the BMP-Smad pathway under the exposure of NP and advance our understanding of how NP stress affects BMP2 and BMP4.

Supplementary Materials: The following supporting information can be downloaded at: https://www.mdpi.com/article/10.3390/fishes9050159/s1, Figure S1: Morphological changes of the *Carassius auratus* red var. embryos at different developmental stages exposed to NP exposure; Table S1: Embryo sampling details for NP exposure experiments.

Author Contributions: Y.S. and X.C. conceived and designed the study. D.L., S.C., J.X. and Y.L. performed the experiments. D.L. and Q.Z. analyzed the data and discussed the main findings. D.L., Q.Z. and Y.S. wrote and revised the manuscript. All authors have read and agreed to the published version of the manuscript.

Funding: This research was funded by the Scientific Research Fund of the Hunan Provincial Education Department (Grant No. 21B0923) and the Postgraduate Scientific Research Innovation Project of Hunan Province (Grant No. CX20221061).

Institutional Review Board Statement: The study was approved by the Animal Ethical Review Committee (AERC) of Hunan Normal University (Approval code: 20221013009; Approval date: 20 October 2022) and followed the guidelines statement of the Administration of Affairs Concerning Animal Experimentation of China. This manuscript does not involve the use of any human data or tissue. The animals used in this study came from Hunan Normal University, and we obtained written consent from Hunan Normal University to use these animals in our research.

Informed Consent Statement: Not applicable.

Data Availability Statement: The datasets used and/or analyzed during the current study are available from the corresponding author on reasonable request.

Acknowledgments: We thank the laboratory members for their technical assistance and constructive comments.

Conflicts of Interest: The authors declare no conflicts of interest.

References

1. Nishie, T.; Komaru, A.; Shiroguchi, S.; Yamaizumi, T.; Ono, Y.; Motomochi, A.; Tooyama, I.; Fujioka, Y.; Sakai, N.; Higaki, S.; et al. Nonylphenol reduced the number of haploids in in vitro spermatogenesis of the endangered cyprinid *Gnathopogon caerulescens*. *Toxicol. Vitro.* **2023**, *89*, 105565. [CrossRef] [PubMed]
2. Hart, C.E.; Lauth, M.J.; Hunter, C.S.; Krasny, B.R.; Hardy, K.M. Effect of 4-nonylphenol on the immune response of the pacific oyster *Crassostrea gigas* following bacterial infection with *Vibrio campbellii*. *Fish. Shellfish. Immunol.* **2016**, *58*, 449–461. [CrossRef] [PubMed]
3. Xia, Y.Y.; Zhan, P.; Wang, Y. Effects of nonylphenol on brain gene expression profiles in F1 generation rats. *Biomed. Environ. Sci.* **2008**, *21*, 1–6. [CrossRef] [PubMed]
4. Ni, C.; Pan, K.; Xu, J.; Long, X.; Lin, F.; Nie, Y.; Yang, Y.; Yu, J. Effects and mechanism of perinatal nonylphenol exposure on cardiac function and myocardial mitochondria in neonatal rats. *Ecotoxicol. Environ. Saf.* **2023**, *258*, 114977. [CrossRef] [PubMed]
5. Suna, P.A.; Cengiz, O.; Ceyhan, A.; Atay, E.; Ertekin, T.; Nisari, M.; Yay, A. The protective role of curcumin against toxic effect of nonylphenol on bone development. *Hum. Exp. Toxicol.* **2021**, *40*, S63–S76. [CrossRef] [PubMed]
6. Agas, D.; Sabbieti, M.G.; Marchetti, L. Endocrine disruptors and bone metabolism. *Arch. Toxicol.* **2013**, *87*, 735–751. [CrossRef] [PubMed]
7. Miyawaki, J.; Kamei, S.; Sakayama, K.; Yamamoto, H.; Masuno, H. 4-tert-octylphenol regulates the differentiation of C3H10T1/2 cells into osteoblast and adipocyte lineages. *Toxicol. Sci.* **2008**, *102*, 82–88. [CrossRef] [PubMed]
8. Zhong, Z.; Ethen, N.J.; Williams, B.O. Wnt signaling in bone development and homeostasis. *Wiley Interdiscip. Rev. Dev. Biol.* **2014**, *3*, 489–500. [CrossRef] [PubMed]
9. Mukherjee, U.; Samanta, A.; Biswas, S.; Ghosh, S.; Das, S.; Banerjee, S.; Maitra, S. Chronic exposure to nonylphenol induces oxidative stress and liver damage in male zebrafish (*Danio rerio*): Mechanistic insight into cellular energy sensors, lipid accumulation and immune modulation. *Chem. Biol. Interact.* **2022**, *351*, 109762. [CrossRef]
10. Desai, J.K.; Trangadia, B.J.; Patel, U.D.; Patel, H.B.; Kalaria, V.A.; Kathiriya, J.B. Neurotoxicity of 4-nonylphenol in adult zebrafish: Evaluation of behaviour, oxidative stress parameters and histopathology of brain. *Environ. Pollut.* **2023**, *334*, 122206. [CrossRef]
11. Horie, Y.; Kanazawa, N.; Takahashi, C.; Tatarazako, N.; Iguchi, T. Gonadal soma-derived factor expression is a potential biomarker for predicting the effects of endocrine-disrupting chemicals on gonadal differentiation in Japanese medaka (*Oryzias latipes*). *Environ. Toxicol. Chem.* **2022**, *41*, 1875–1884. [CrossRef] [PubMed]
12. Chaube, R.; Gautam, G.J.; Joy, K.P. Teratogenic effects of 4-nonylphenol on early embryonic and larval development of the catfish *Heteropneustes fossilis*. *Arch. Environ. Contam. Toxicol.* **2013**, *64*, 554–561. [CrossRef] [PubMed]
13. Zhang, H.; Jiang, J.L.; Shan, Z.J.; Bu, Y.Q.; Tian, F. Toxic effects of 4-nonylphenol on embryo/larva of zebrafish (*Danio rerio*). *J. Ecol. Rural. Environ.* **2017**, *33*, 737–742. [CrossRef]
14. Zhang, Q.Y.; Sun, Y.D.; Wang, Z.J.; Hu, X.J.; Kui, X. Preliminary study on toxicity of nonylphenol to embryo development of goldfish (*Carassius auratus*). *Prog. Mod. Biomed.* **2016**, *16*, 3040–3043. [CrossRef]
15. Shu, B.; Zhang, M.; Xie, R.; Wang, M.; Jin, H.; Hou, W.; Tang, D.; Harris, S.E.; Mishina, Y.; O'Keefe, R.J.; et al. BMP2, but not BMP4, is crucial for chondrocyte proliferation and maturation during endochondral bone development. *J. Cell Sci.* **2011**, *124*, 3428–3440. [CrossRef] [PubMed]
16. Yoon, B.S.; Lyons, K.M. Multiple functions of BMPs in chondrogenesis. *J. Cell Biochem.* **2004**, *93*, 93–103. [CrossRef] [PubMed]
17. Wan, M.; Cao, X. BMP signaling in skeletal development. *Biochem. Biophys. Res. Commun.* **2005**, *328*, 651–657. [CrossRef] [PubMed]
18. Salazar, V.S.; Gamer, L.W.; Rosen, V. BMP signalling in skeletal development, disease and repair. *Nat. Rev. Endocrinol.* **2016**, *12*, 203–221. [CrossRef] [PubMed]
19. Hwang, J.K.; Min, K.H.; Choi, K.H.; Hwang, Y.C.; Jeong, I.K.; Ahn, K.J.; Chung, H.Y.; Chang, J.S. Bisphenol A reduces differentiation and stimulates apoptosis of osteoclasts and osteoblasts. *Life Sci.* **2013**, *93*, 367–372. [CrossRef]
20. Ju, L.; Tang, K.; Guo, X.R.; Yang, Y.; Zhu, G.Z.; Lou, Y. Effects of embryonic exposure to polychlorinated biphenyls on zebrafish skeletal development. *Mol. Med. Rep.* **2012**, *5*, 1227–1231. [CrossRef]
21. Bandyopadhyay, A.; Tsuji, K.; Cox, K.; Harfe, B.D.; Rosen, V.; Tabin, C.J. Genetic analysis of the roles of BMP2, BMP4, and BMP7 in limb patterning and skeletogenesis. *PLoS Genet.* **2006**, *2*, e216. [CrossRef] [PubMed]
22. Li, X.K.; Huang, Q.K.; Feng, L.L.; Guo, Y.F.; Liang, J.; Lan, G.Q. Sequence and expression differences of BMP2 and FGFR3 genes in Guangxi bama mini pig and landrace pig. *J. South. Agric.* **2021**, *52*, 1709–1718. [CrossRef]
23. Li, M.; Liu, X.; Liu, X.; Ge, B. Calcium phosphate cement with BMP-2-loaded gelatin microspheres enhances bone healing in osteoporosis: A pilot study. *Clin. Orthop. Relat. Res.* **2010**, *468*, 1978–1985. [CrossRef] [PubMed]

24. Martínez-Barberá, J.P.; Toresson, H.; Da Rocha, S.; Krauss, S. Cloning and expression of three members of the zebrafish BMP family: BMP2A, BMP2B and BMP4. *Gene* **1997**, *198*, 53–59. [CrossRef] [PubMed]
25. Zeng, X.; Shi, Z.Y.; Chen, X.W.; Cheng, Q.Q. Skeleton development of Japanese flounder (*Paralichthy solivaceus*) and expression analysis of related genes (SOX9, BMP4 and BMP2) during metamorphosis. *Mar. Fish.* **2009**, *31*, 337–346. [CrossRef]
26. Ma, L.X.; Dong, Z.J.; Su, S.Y.; Zhang, J.Q.; Liu, W.; Li, L.L.; Yuan, X.H. Cloning of fragments of bone morphogenetic protein gene (BMP2B) of *Cyprinus carpio* var. Jian and its expression analysis. *Jiangsu J. Agr. Sci.* **2013**, *29*, 370–378. [CrossRef]
27. Chen, J.; Lü, Y.P.; Dai, Q.M.; Zhang, L.; Xu, C.B.; Lu, J. Molecular characterization of a BMP2A homologue in barbel steed (*Hemibarbus labeo*) and its involvement in intermuscular bone development. *Acta Hydrobiol. Sin.* **2021**, *45*, 8–13. [CrossRef]
28. Cao, X.Y.; Ma, C.X.; Xia, X.; Zhang, R.Q.; Chen, X.W.; Zhao, J.L. Cloning and expression analysis of jaw remodeling developmental gene BMP4 in *Siniperca chuatsi*. *Genom. Appl. Biol.* **2020**, *39*, 5075–5083. [CrossRef]
29. Huang, L.; Deng, X.; Yang, X.; Tang, Z.; Fan, S.; Zhou, Z.; Tao, M.; Liu, S. Cloning, distribution, and effects of growth regulation of MC3R and MC4R in red crucian carp (*Carassius auratus* red var.). *Front. Endocrinol.* **2024**, *14*, 1310000. [CrossRef]
30. Tian, Y.S.; Sun, Y.D.; Ou, M.; Liu, Y.F.; Cui, X.J.; Zhou, D.G.; Che, W.A.; Chen, K.C. Preliminary studies on the mechanism of nonylphenol-induced malformation of *Carassius auratus* red var. *J. Fish. China* **2020**, *44*, 1619–1636. [CrossRef]
31. Lu, X.H.; Gu, Y.; Song, Y. Toxicity and tissue accumulation of nonylphenol in *Carassius auratus* red variety, *Grass Carp* and *Sliver Carp*. *J. Hyg. Res.* **2012**, *41*, 785–789. [CrossRef]
32. Rajaram, S.; Patel, S.; Uggini, G.K.; Desai, I.; Balakrishnan, S. BMP signaling regulates the skeletal and connective tissue differentiation during caudal fin regeneration in sailfin molly (*Poecilia latipinna*). *Dev. Growth Differ.* **2017**, *59*, 629–638. [CrossRef] [PubMed]
33. Schlange, T.; Andrée, B.; Arnold, H.H.; Brand, T. BMP2 is required for early heart development during a distinct time period. *Mech. Dev.* **2000**, *91*, 259–270. [CrossRef] [PubMed]
34. Du, Y.; Xiao, Q.; Yip, H.K. Regulation of retinal progenitor cell differentiation by bone morphogenetic protein 4 is mediated by the smad/id cascade. *Invest. Ophthalmol. Vis. Sci.* **2010**, *51*, 3764–3773. [CrossRef] [PubMed]
35. Katagiri, T.; Takahashi, N. Regulatory mechanisms of osteoblast and osteoclast differentiation. *Oral. Dis.* **2002**, *8*, 147–159. [CrossRef]
36. Miyazono, K.; Maeda, S.; Imamura, T. BMP receptor signaling: Transcriptional targets, regulation of signals, and signaling cross-talk. *Cytokine Growth Factor. Rev.* **2005**, *16*, 251–263. [CrossRef] [PubMed]
37. Bragdon, B.; Moseychuk, O.; Saldanha, S.; King, D.; Julian, J.; Nohe, A. Bone morphogenetic proteins: A critical review. *Cell Signal.* **2011**, *23*, 609–620. [CrossRef] [PubMed]
38. Hinck, A.P.; Archer, S.J.; Qian, S.W.; Roberts, A.B.; Sporn, M.B.; Weatherbee, J.A.; Tsang, M.L.; Lucas, R.; Zhang, B.L.; Wenker, J.; et al. Transforming growth factor beta 1: Three-dimensional structure in solution and comparison with the X-ray structure of transforming growth factor beta 2. *Biochemistry* **1996**, *35*, 8517–8534. [CrossRef] [PubMed]
39. Chen, L.; Dong, C.; Kong, S.; Zhang, J.; Li, X.; Xu, P. Genome wide identification, phylogeny, and expression of bone morphogenetic protein genes in tetraploidized common carp (*Cyprinus carpio*). *Gene* **2017**, *627*, 157–163. [CrossRef]
40. Shin, D.; Shin, C.H.; Tucker, J.; Ober, E.A.; Rentzsch, F.; Poss, K.D.; Hammerschmidt, M.; Mullins, M.C.; Stainier, D.Y. Bmp and Fgf signaling are essential for liver specification in zebrafish. *Development* **2007**, *134*, 2041–2050. [CrossRef]
41. Rafael, M.S.; Laizé, V.; Cancela, M.L. Identification of *Sparus aurata* bone morphogenetic protein 2: Molecular cloning, gene expression and in silico analysis of protein conserved features in vertebrates. *Bone* **2006**, *39*, 1373–1381. [CrossRef]
42. Hagiwara, H.; Sugizaki, T.; Tsukamoto, Y.; Senoh, E.; Goto, T.; Ishihara, Y. Effects of alkylphenols on bone metabolism in vivo and in vitro. *Toxicol. Lett.* **2008**, *181*, 13–18. [CrossRef] [PubMed]
43. Sabbieti, M.G.; Agas, D.; Palermo, F.; Mosconi, G.; Santoni, G.; Amantini, C.; Farfariello, V.; Marchetti, L. 4-nonylphenol triggers apoptosis and affects 17-β-estradiol receptors in calvarial osteoblasts. *Toxicology* **2011**, *290*, 334–341. [CrossRef] [PubMed]
44. Nohe, A.; Hassel, S.; Ehrlich, M.; Neubauer, F.; Sebald, W.; Henis, Y.I.; Knaus, P. The mode of bone morphogenetic protein (BMP) receptor oligomerization determines different BMP-2 signaling pathways. *J. Biol. Chem.* **2002**, *277*, 5330–5338. [CrossRef]
45. Bruderer, M.; Richards, R.G.; Alini, M.; Stoddart, M.J. Role and regulation of RUNX2 in osteogenesis. *Eur. Cell Mater.* **2014**, *28*, 269–286. [CrossRef]
46. Wu, Y.; Sun, A.; Nie, C.; Gao, Z.X.; Wan, S.M. Functional differentiation of BMP2A and BMP2B genes in zebrafish. *Gene Expr. Patterns* **2022**, *46*, 119288. [CrossRef]
47. Hussein, A.M.; Sina, M. p-Nonylphenol impairment of osteogenic differentiation of mesenchymal stem cells was found to be due to oxidative stress and down-regulation of RUNX2 and BMP. *Endocr. Metab. Immune Disord. Drug Targets.* **2020**, *20*, 1336–1346. [CrossRef] [PubMed]
48. Li, X.J.; Zhou, Z.L.; Gu, J.H. Effect of nonylphenol on the gonadal differentiation and development in gobbid fish (*Odontobutis potamophila*). *Fish. Sci.* **2009**, *28*, 15–19. [CrossRef]
49. Fan, X.; Wu, L.; Hou, T.; He, J.; Wang, C.; Liu, Y.; Wang, Z. Maternal bisphenol A exposure impaired endochondral ossification in craniofacial cartilage of rare minnow (*Gobiocypris rarus*) offspring. *Ecotoxicol. Environ. Saf.* **2018**, *163*, 514–520. [CrossRef]
50. Pu, S.Y.; Hamid, N.; Ren, Y.W.; Pei, D.S. Effects of phthalate acid esters on zebrafish larvae: Development and skeletal morphogenesis. *Chemosphere* **2020**, *246*, 125808. [CrossRef]

MDPI AG
Grosspeteranlage 5
4052 Basel
Switzerland
Tel.: +41 61 683 77 34

Fishes Editorial Office
E-mail: fishes@mdpi.com
www.mdpi.com/journal/fishes

www.ingramcontent.com/pod-product-compliance
Lightning Source LLC
Chambersburg PA
CBHW061601210926
51850CB00009B/2176

* 9 7 8 3 7 2 5 8 6 0 1 9 7 *